RHEOMETRY OF PASTES, SUSPENSIONS, AND GRANULAR MATERIALS

RHEOMETRY OF PASTES, SUSPENSIONS, AND GRANULAR MATERIALS

Applications in Industry and Environment

Philippe Coussot

A JOHN WILEY & SONS, INC., PUBLICATION

Published by John Wiley & Sons, Inc., Hoboken, New Jersey.
Published simultaneously in Canada.

For general information on our other products and services please contact our Customer Care Department within the U.S. at 877-762-2974, outside the U.S. at 317-572-3993 or fax 317-572-4002.

Wiley also publishes its books in a variety of electronic formats. Some content that appears in print, however, may not be available in electronic format.

Library of Congress Cataloging-in-Publication Data:

Coussot, Philippe.
    Rheometry of pastes, suspensions, and granular materials : applications in industry and environment / Philippe Coussot.
        p. cm.
    Includes index.
    ISBN 13 978-0-471-65369-1

1. Amorphous substances. 2.   Rheology. I. Title.
    TA418.9.A58C68 2005
    620.1′1—dc22   2004021908

CONTENTS

PREFACE

Although the behavior of clays and paints initially prompted Bingham [1] to introduce rheology as a new scientific field in the early 1920s, the main developments in that field since then have concerned polymers. The internal structure of polymers can now be easily characterized using various techniques and shows obvious physical similarities between materials. As a consequence, a general descriptive framework for these materials is available now, allowing us to determine fairly accurately the physicochemical origins of the mechanical behavior of polymers [2,3].

Besides polymers, industry and nature provide us with a vast range of materials composed of complex polymeric elements in suspension in a liquid: emulsions, foams, and suspensions of solid particles. When the concentration of suspended elements is weak, the behavior of such materials is qualitatively similar to that of the interstitial liquid, which is generally Newtonian. At sufficiently high concentrations, the suspended elements develop specific mutual interactions and hence we usually deal with a "pasty" material, incapable of flow when the force exerted onto it is below a critical value. We currently encounter this type of material in our daily lives in products such as shaving foam, Chantilly, solar cream, cosmetic cream, mayonnaise, puree, paint, modeling paste, peanut butter, marmalade, and hair gel, or in natural phenomena such as mining slurries, mudflows, debris flows, lavas, snow, and lahars (water–ash mixtures on volcanic hill slopes). Civil engineering, food, and cosmetic industries also use pasty or granular materials for transporting or storing solid matter or for product forming, with agents such as drilling fluids, cement paste, concrete, mortar glues, ceramic slip, foodstuff pastes, sewage sludges, sand, grains, and powders.

In contrast to polymers, all these materials have extremely different internal structures, ranging from the packing of submicrometric, platelet, clay particles for muds, to the crowding of soft bubbles (with diameters in the high micrometer range) for foams. More critical is the fact that most industrial materials contain a wide range of elements of various sizes and interaction patterns (polymers in liquid suspensions or adsorbed by solid particles, colloidal particles interacting at a distance via electrostatic forces, formation of viscoelastic droplets or bubbles, grains interacting via friction or collision through the liquid, etc.). In this context one often focuses on a specific material and notes that it exhibits unique characteristics, which certainly is true from a physicochemical perspective. Such individual approaches unfortunately do not promote the development of a unifying approach to define the relationship between internal structure and mechanical properties for this wide class of materials. A new concept has emerged in physics, known as "jamming" [4], in which all these materials are perceived as "jammed" systems. Although this concept is not yet precisely defined, it concerns the "pasty" materials as defined above since this internal jamming implies that these materials cannot flow unless some outside force is applied to "unjam" this structure. In this field physicists essentially seek generic thermodynamic properties at the origin of the link between this jamming and the evolutions of the internal structure of these materials.

As I was studying various materials, I increasingly concluded that, considering the variety of internal structures of jammed materials, generic laws, if they exist, should be sought in the mechanical behavior of these materials. Indeed, in mechanics the specificities of the internal structure of each material may be put aside in favor of the qualitative characteristics of this internal structure, which induces different mechanical trends. The primary objective of this book is to propose a synthetic and general approach to define the mechanical behavior of pastes and granular materials.

In practice, it is critical to apply reliable rheometrical techniques for relevant characterization of materials, specifically, to measure some physical parameters related to the effective rheological behavior of the material. For pasty and granular materials this approach still constitutes a challenge for several reasons; for instance, these materials may exhibit a strongly nonlinear behavior [they behave either as solids or liquids depending on flow conditions], several experimental problems may occur (wall slip, fracture, drying, etc.) with these materials, and for some of these substances (granular materials) there may exist no constitutive equation intrinsic to the material (i.e., independent of boundary conditions). Finally, probably because of these difficulties, a variety of practical techniques for characterization of these materials in a rapid and robust way (slump test, Marsh cone, penetrometric measurement, etc.) involve "nonviscometric flows," specifically, flows that are more complex and controlled than the so-called viscometric flows used in rheometry. Often, each industrial field developed its own techniques that in fact involved the same basic flow types found in other fields. In this context the second major objective of this book was to review the experimental problems encountered with such materials and to examine the techniques used in different fields through the basic nonviscometric flows they involved, in order to propose theoretical analysis that would enable one to extract relevant rheological parameters from such tests.

This book is merely a sketch of what might be written on the topic of rheology. In the near future the development of new techniques for internal exploration of materials will probably give us a deeper insight into the behavior of pastes or granular materials, which will make rheometrical analysis even more accurate. However, the tools described here may provide an at least initially comprehensive approach to this field.

This work results from the research I have carried out since 1989. At first my motivation for the study of pasty materials was to provide protection against mudflows in mountain streams, initiated by M. Meunier in Cemagref and supervised by J. M. Piau in the Laboratory of Rheology in Grenoble. I undoubtedly enjoyed a perfect launching site and hopefully I was worthy of it since these individuals provided me with much of my scientific formation in different ways. Another key stage was my meeting with O. Coussy, the then head of LMSGC. Not content with teaching me the details of research management, he also boosted rheology research by creating a mainline around rheophysics in our laboratory.

During these years I had the opportunity to work with researchers whose ideas are reflected at various points in this book: C. Ancey (EPFL, Lausanne), who, among other things, introduced me to the specificities of granular materials; D. Bonn (LPS, ENS Paris), with whom we developed the very important concept of viscosity bifurcation; J. C. Baudez (Cemagref), who had a remarkable approach to the reconstruction of a velocity profile; and X. Chateau (LMSGC), with whom I am still daily discovering the subtleties of mechanics. Several individuals also kindly agreed to review parts of this book, expressed comments, and provided enlightening advice; I thank them warmly here, and hope that this work will be of some personal benefit to them as well: N. Alderman (Aspen Technology), G. Ovarlez (LMSGC), X. Chateau (LMSGC), B. Herzhaft (IFP), F. Chevoir (LMSGC), and S. Rodts (LMSGC). I would also like to thank John Wiley & Sons, Inc. for trusting me. Eventually, last but not least, many thanks to my wife, who managed once again to put up with seeing me somewhat tense, focused on a new objective for about 2 years.

REFERENCES

1. E. C. Bingham, *Fluidity and Plasticity*, McGraw-Hill, New York, 1922.
2. J. D. Ferry, *Viscoelastic Properties of Polymers*, Wiley, New York, 1970.
3. P.-G. de Gennes, *Scaling Concepts in Polymer Physics*, Cornell Univ. Press, Ithaca, NY, 1979.
4. A. J. Liu and S. R. Nagel, *Jamming and Rheology*, Taylor & Francis, New York, 2001.

NOTATION

Roman Symbols

a	acceleration vector
b	separation distance between the centers of two elements; half-height of a cylinder of material
b	volume force
Ba	Bagnold number (see Section 2.3)
B	magnetic field; buoyancy force
d	strain rate in elongational flow; characteristic element length (equivalent average diameter)
ds	surface element
dv	volume element
D	fluid width
D_I, D_{II}, D_{III}	invariants of the strain rate tensor
D	strain rate tensor
E	elastic modulus of a solid particle
f	friction coefficient between two solid surfaces
F_c	critical force for incipient motion or stoppage
F_0	imposed force in squeeze flow
Fr	Froude number [equation (5.57)]
$\mathbf{F}_D$	drag force
$\mathbf{F}_t$	relative configuration gradient [equation (1.2)]
g	gravity
G	elastic modulus for Maxwell model
G', G''	elastic and viscous moduli
G	magnetic field gradient

h	thickness of fluid layer; separation distance between two solid surfaces
h_c	critical fluid thickness for incipient motion or stoppage
h_{stop}	critical granular material thickness at stoppage
h_0	uniform flow thickness for a granular material
H	gap between plates or disks; dimensionless fluid thickness; height
$H_{\vartheta<t}$	functional of a variable history (up to t)
i	slope angle of an inclined plane
i_c	critical slope for incipient motion or stoppage
$\mathbf{I}$	identity tensor
J_e	rate of evaporation
$k = \tan\varphi$	coefficient of friction of a granular material; permeability of a porous material; drag coefficient
k_c	critical drag coefficient
L	flow or object length
Le	Leighton number [equation (2.27)]
$\mathbf{L}$	velocity gradient tensor
M	torque
$\mathbf{M}$	magnetic moment (spin)
n	vapor density
$\mathbf{n}$	normal vector
N	normal force
p	pressure
p_f	pressure within the interstitial fluid of a granular paste
p_0	ambient gas pressure
P	energy stored or dissipated by mechanical action
q	flow rate by unit surface
$\mathbf{q}$	heat density flux
Q	flow rate (fluid discharge)
$\mathbf{Q}$	rotation (orthogonal) tensor
r_c	critical radius
r_0	radius of a capillary
r_1, r_2	inner and outer cylinder radii of coaxial cylinders
R	radius of a cylindrical object or material; dimensionless radial distance
Re	Reynolds number [equation (3.80)]
S	surface
St	Stokes number [equation (2.26)]
t	time
$\mathbf{t}$	stress vector
T	tangential force; temperature
$\mathbf{T}$	deviatoric stress tensor [equation (1.34)]
u	velocity in a simple shear between parallel plates
U	mean fluid velocity; maximum fluid velocity
U_s	slip velocity

v_x, v_y, v_z	velocity components in a Cartesian frame
v_r, v_θ, v_z	velocity components in a cylindrical frame
v_φ, v_θ, v_z	velocity components in a spherical frame
$\mathbf{v}$	velocity vector
V	dimensionless velocity; velocity of a solid object through a fluid; volume
W	work
$\mathbf{x}$	position at time t
$\mathbf{x}^*$	position in another frame of observation
y_c	critical height separating sheared and unsheared regions
x, y, z	coordinates in a Cartesian frame
r, θ, z	coordinates in a cylindrical frame
φ, θ, z	coordinates in a spherical frame
∇	gradient operator

Greek Symbols

η	apparent viscosity of a material
ς	shear stress function
β	yielding criterion coefficient
γ	simple shear deformation (or strain); gyromagnetic ratio
γ_c	critical deformation
γ_0	strain amplitude
γ_{LG}	interfacial tension between a gas and a liquid
$\dot{\gamma}$	shear rate
$\dot{\gamma}_{app}$	apparent (macroscopic) shear rate
$\dot{\gamma}_c$	critical shear rate
$\dot{\gamma}_{eff}$	effective (local) shear rate
$\dot{\gamma}_R$	shear rate at the periphery of a cylindrical sample
δ	slip layer thickness
Δp	pressure difference in a capillary flow
ϕ	volume fraction
ϕ_m	maximum packing fraction
φ	angle of friction of a granular material; phase shift
Φ	potential energy of interaction
λ	structure parameter
λ_m	maximum growth rate of instability
μ	viscosity of a Newtonian fluid; generalized coefficient of friction for granular flows
μ_0	viscosity of a Newtonian interstitial fluid
ω	rotation velocity of a fluid element; oscillation frequency; Larmor frequency
Ω	rotation velocity of a cylinder, a cone or a disk; fluid volume; particle volume
ρ	fluid density

ρ_p	solid particle density
ρ_0	interstitial fluid density
σ	normal stress
$\sigma_{xx}, \sigma_{xy}, \sigma_{xz}$, (etc.)	components of stress tensor
$\Sigma_I, \Sigma_{II}, \Sigma_{III}$	invariants of stress tensor
$\boldsymbol{\Sigma}$	stress tensor
$\boldsymbol{\Sigma}_f$	stress tensor component of interstitial fluid
$\boldsymbol{\Sigma}_g$	stress tensor component of granular phase
$\boldsymbol{\Sigma}_0$	stress tensor at initial time
τ	shear stress [equation (1.43)]
τ_c	critical or yield stress
τ_p	shear stress at wall
τ_0	shear stress amplitude
ϑ	time
χ	shear rate function
ξ	position vector at time ϑ

INTRODUCTION

The objective of this book is to provide useful tools for characterizing environmental or industrial pastes, slurries, and granular materials from a mechanical perspective. In particular, we aim at going beyond the basic rheometrical techniques with the help of sophisticated analysis of the rheological behavior of pastes and experimental artefacts and beyond the various existing qualitative approaches or comparative techniques by providing, as far as possible, relevant rheological interpretation of these practical tests.

Any rheological approach relies on a description of the motion within the frame[1] of continuum mechanics, and on the assumption of some mechanical behavior intrinsic to the material under study. In practice, very simple flows, specifically, viscometric flows, must be used because they provide straightforward relations between stress components and flow history. Although the validity of the underlying assumptions for such an approach (continuum medium, simple fluid) is clear for simple liquids, such as water, alcohol, and oil, it is not obvious for more complex materials such as pastes or granular materials. In Chapter 1 we review the tools used in continuum mechanics and the principles for expressing the constitutive equation of any material for the more general case. This, in particular, places pastes, slurries, and granular materials in an intermediate class, which may be referred to as "jammed systems," between solids and fluids. The

[1] The term "frame" is used throughout the text to denote *frame of reference, perspective, framework, context,* or similar.

Rheometry of Pastes, Suspensions, and Granular Materials: Applications in Industry and Environment
By Philippe Coussot   Copyright © 2005 John Wiley & Sons, Inc.

peculiarity of jammed systems is that their constitutive equations can be of either the solid or fluid type depending on flow history. The usual viscometric flows and the corresponding methods of analysis are also reviewed taking into account as far as possible the specific behavior patterns of these materials.

Before studying the rheometrical techniques appropriate for jammed systems, it is necessary to have a clear idea about the rheological trends that must be accounted for. In Chapter 2 we propose a classification of the materials as a function of the qualitative rheological trends that they exhibit. We start by reviewing the main interactions between the different possible elementary components of these systems, and show how they qualitatively affect the rheological behavior of the materials. Finally, three main classes emerge:

1. *Pastes* (soft jammed systems), in which mainly soft interactions predominate and that are basically thixotropic, yield stress fluids
2. *Granular materials*, in which direct (hard) interactions between grains predominate, and whose "frictional regime" behavior is somewhat analogous with the behavior of two solids in contact and in relative motion
3. *Granular pastes*, in which either soft or direct interactions may predominate depending on the flow regime, and which thus can exhibit either a pasty or granular behavior depending on flow conditions

Setting up the constitutive equation of granular pastes (class 3 above) is still a tricky task. In this book the main developments concern classes 1 and 2 (i.e., pastes and granular materials), for which a significant set of knowledge appears to be available, bearing in mind that the behavior of granular pastes can be extrapolated from the behavior of the two extreme cases described above (classes 1 and 2), once their flow regimes have been determined. Moreover, only pasty materials exhibit an intrinsic behavior, independent of boundary conditions. Consequently, in Chapters 3–5 we focus on different aspects of paste rheometry. The granular materials in a frictional regime are considered in Chapter 6. Practical rheometrical tests for both material types are reviewed in Chapter 7.

Rheometry with pastes appears to be a difficult task because of their complex rheological behavior, which may include elasticity, solid–liquid transition, and thixotropy. Chapter 3 focuses on the rheometrical procedures for best-guess determination of these properties, and on the various measurement problems that may occur, such as wall slip, shear localization, surface tension, drying, phase separation, crack, temperature effects, and inertia effects, and may preclude a relevant interpretation of the data in terms of the effective, macroscopic behavior of the homogeneous material.

Soft jammed systems (and also granular materials) exhibit strong nonlinear rheological trends that may lead to severe heterogeneities or time variations of the flow field in viscometric flows. In general these phenomena cannot be properly appreciated from usual macroscopic measurements. A more sophisticated rheometrical approach, which we refer to as "local rheometry," consists in measuring and interpreting the flow field within the rheometer. In Chapter 4 we review the principles of the most versatile technique that may be used for this purpose: magnetic resonance imaging (MRI) velocimetry. Methods for interpreting measured velocity profiles within Couette geometry in terms of the constitutive equation of the material are then discussed. Finally, an alternative technique, "velocity profile reconstruction from rheometry," which enables us to directly view the velocity profiles over short timescales from the usual rheometrical tests, is presented.

In practice, it is seldom easy to carry out rheometrical tests under ideal conditions as required by theoretical viscometry because a sample must be taken from the fabrication process and placed in a laboratory rheometer of appropriate size (for the continuum assumption to be valid). This has led to the development of various practical techniques often involving flows that are nonviscometric [i.e., not characterized by relative motion (gliding) of parallel or concentric fluid layers over one another] but that nevertheless have some sufficiently simple properties for the extrapolation of paste rheological behavior under specific conditions. This is the case for the displacement of an object relative to the fluid, for the "squeeze" flow, or for the coating or spreading flows. In Chapter 5 we review the characteristics of these flows in the case of pastes and focus on how their macroscopic characteristics may be related to yield stress.

The rheology of granular materials cannot be determined from the usual rheometrical techniques, since they do not exhibit an intrinsic constitutive equation, independent of boundary conditions; hence granular flows can be described only under certain flow conditions. In Chapter 6, we review the characteristics of granular flows for which a frictional regime may take place: viscometric flows, free surface flows, conduit flows, compression flows, and free surface flows for granular pastes. In most cases, simple relationships describing flow characteristics in terms of the friction coefficient can be established. These flows can then be used as rheometrical tests for granular materials.

Many practical tests have been developed in different industries for characterizing the rheological behavior of pastes or granular materials. Most of them provide one single parameter that may be used for comparison between materials but does not find a clear rheological interpretation. In Chapter 7 we review the main different practical tests and show that for most of these tests it is possible to find a relevant rheological interpretation of the measured parameter if an appropriate experimental procedure is respected. Afterward we review the main types of industrial and environmental materials, discuss their classification (within the three main categories of materials: pastes, granular pastes, or granular materials), and suggest the most appropriate practical rheometrical test to be used with each type.

CHAPTER 1

MATERIAL MECHANICS

1.1 INTRODUCTION

Any rheological approach relies on a description of the motion within the frame of continuum mechanics, and on the assumption of some mechanical behavior intrinsic to the material under study. In practice, in order to determine this behavior, one uses viscometric flows that, because of their simplicity, provide straightforward relations between stress components and flow history under certain conditions. The theory of viscometric flows, which provides a complete theoretical frame for analyzing such data in rheological terms, has nevertheless been developed within the frame of a specific class of materials, *simple fluids*, which have a vanishing memory.

Although the validity of these assumptions (continuum medium, simple fluid) is clear for simple liquids, such as water, alcohol, and oil, it is not obvious for more complex materials such as pastes or granular materials. Indeed, we often observe these materials to be strongly heterogeneous at our scale of observation, and thus may question the validity of the continuum assumption. Moreover these materials may have a dual behavior, behaving like solids under some conditions and liquids under other conditions. For example, powders or foams remain stationary like a solid under the action of gravity alone, but begin to flow like a liquid under vibration (powders) or squeezing (foams). Various pasty materials also exhibit time-dependent properties; their viscosity may continuously increase when left at rest but, during flow, may decrease in time. Thus, pastes and granular materials cannot be considered a priori as simple fluids with vanishing memory.

Rheometry of Pastes, Suspensions, and Granular Materials: Applications in Industry and Environment
By Philippe Coussot   Copyright © 2005 John Wiley & Sons, Inc.

There is thus a need to review the basic tools of (continuum) fluid mechanics and rheometry in a more general frame including the specificities of pastes and granular materials. In this context we first review the definitions and principles of continuum mechanics and their physical origin, focusing on the validity of the continuum assumption for pastes and granular materials (Section 1.2). Then we examine the role and structure of the constitutive equation, and show that pastes and granular materials belong to a class intermediate between solids and fluids (Section 1.3). Usual viscometric flows and the corresponding methods of analysis are reviewed in Section 1.4 taking into account as far as possible the specific behavior pattern of these materials.

1.2 CONTINUUM MECHANICS

1.2.1 Definition of a Material

A *material* can be defined as the ensemble of elements of matter contained in a connex volume. The fact that we identify this ensemble as a single material means that in any two parts of this volume we would find similar ensembles of elements of matter. Nevertheless, for any given material, this definition would probably fail if viewed on the scale of some basic components of this matter (atoms, molecules, bubbles, solid particles, etc.), since on this scale the material clearly varies from one (very small) part to another. This implies that our evaluation of the material in a given part must account for a certain number of basic components over which we can proceed to some average of a physical property q (see Figure 1.1). It follows that the minimum scale (l) on which one can reasonably "observe" the system and effectively consider it as a material

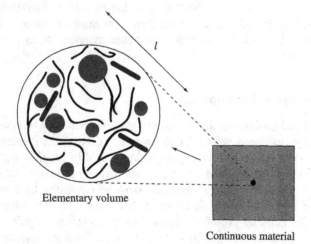

Elementary volume

Continuous material

Figure 1.1 Aspect of an elementary volume (left) of a material, continuous on our scale of observation (right), but including different types of matter.

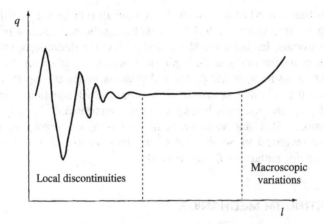

Figure 1.2   Variation of a physical property (q) averaged over a given volume of material as a function of the extent of this volume (l).

(i.e., the minimum volume of a given part of the system) corresponds to the point beyond which the result of this (q) average no longer varies when this scale further increases (see Figure 1.2). These volumes will be referred to as *elementary parts* (or *volumes*) of the material, while the basic components identified above will be referred to as the *elements*. However, it is worth noting that, when further increasing the scale of observation, macroscopic variations resulting from thermal and/or mechanical constraints imposed to the material begin to play a role (see Figure 1.2). These variations are precisely the phenomena that continuum mechanics or thermodynamics intends to describe with the help of mathematical concepts. The appropriate scale of elementary parts is thus situated between the range of rapid variations of q due to matter discontinuity and the range of macroscopic variations. When these two ranges recover, it is not possible to consider the system under the continuum assumption, as the flow is in a "discrete" regime.

1.2.2   Continuum Assumption

The material can thus be considered as *continuous* when the averages of physical properties (density, force, velocity, temperature, etc.) over the elementary parts of the material slowly vary from one such part to one of its neighboring parts. More precisely, the continuum assumption implies that the physical variables, which assume the values of property averages, may be described by continuous and continuously derivable functions of space and time. This will make it possible to describe the material evolutions from a set of equations relating an ensemble of variables and their changes in time and space. Such an approach opens the way to a quantitative description of the material's properties and, in particular, its mechanical behavior.

Now let us examine the conditions under which the continuum assumption is valid. In practice, anticipating the definitions of some variables in Section 1.2.3, it appears that the continuities of density and velocity are the two main factors to consider. The density continuity is the easiest to predict and control because it relies on geometric considerations; we can define a minimum scale beyond which the continuum assumption is valid for the density from the direct observation of the distribution of material components in space. A practical means consists in computing the ratio of some characteristic flow length to the dimensions of the largest elements. The description of material properties with the continuum assumption becomes more precise as this ratio increases. It is difficult to propose a general rule, but in practice, as a general rule of thumb, a ratio of the order of 30 should ensure a good approximation of reality using the continuum approach.[1]

The problem is less simple for the velocity continuity, because the velocity field results from an interplay between the material behavior, the boundary conditions, and the thermomechanical constraints. Flow instabilities or discontinuities may result from this coupling, the effects of which can seldom be predicted a priori. Typically, slow flows of pastes or granular materials give rise to shear localization or wall slip (see Chapter 3). In that case the flow is composed of at least two regions, one in which the material flows and the other in which it remains rigid. At a sufficiently large scale of observation (L_1), the apparent thickness of the region of flow is so small that the velocity field may seem to exhibit a discontinuity (see Figure 1.3). Generally, this velocity field in fact appears as continuous on a sufficiently small scale, but the continuum assumption is not valid at this scale if the geometric continuity (see above) fails when using the thickness of the flowing region as the characteristic flow length, that is, when the ratio of this thickness to the typical particle size (L) remains sufficiently large. Otherwise, the sheared region does not correspond to a flow of the homogeneous material and may be regarded as an interface through which the velocity is discontinuous.

Finally, when localization occurs, because the thickness of the sheared region and thus the effective continuity of the variables depend on several material and flow characteristics in possible interplay, there is no general criterion that may

---

[1] This result may be obtained from the following rough approach. Let us consider a suspension of solid particles (of elementary volume V) in a gas (solid volume fraction ϕ); measuring the density of the material over a volume of material Ω takes into account $\phi\Omega/V$ particles, a number that may typically fluctuate from one volume to another in the suspension. The error with respect to the density is thus of the order of 10% for $\phi\Omega/V \approx 10$; for $\phi = 35\%$, we get the ratio for the dimensions of an elementary volume of material to the dimensions of the particle $b/a \approx (\Omega/V) \approx 3$; now, for a relevant mathematical description of the flow characteristics, there must exists at least ~ 10 such elementary volumes along a typical length, which leads to a ratio of the characteristic flow length to the dimensions of the largest elements of 30.

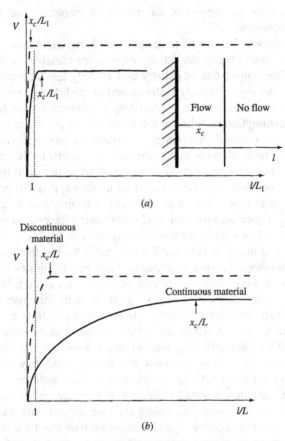

(a)

(b)

Figure 1.3 Possible aspects of the velocity field at a sufficiently large scale of observation (L_1) (a) and on an observation scale covering only the basic components (L) (b) in a material for which shear localization occurs, that is, where flow occurs only in a region of thickness x_c from the solid–fluid interface. Both velocity fields appear to be discontinuous at the scale L_1, but the first one (unbroken line) is continuous with respect to density, while the second one (dashed line) is discontinuous on the smaller scale L.

ensure a priori the validity of the continuum assumption for a given material. In practice, the validity of the continuum assumption must be checked a posteriori. In this book we will assume that the continuum assumption is satisfied and, in specific cases, discuss the origin of possible discontinuities in terms of flow or material characteristics.

1.2.3 Main Variables

Density and Concentration

The first physical property that can be identified directly derives from the definition of the material; this is the spatial distribution of matter within the volume.

In order to neglect the local specificities of matter, it is logical to use a generic property, specifically, the *mass*, which we will compute for a given volume of material by adding the mass values for all the individual elements of matter contained in that material. Under the continuum assumption, that is, on a scale sufficiently small to avoid macroscopic variations but sufficiently larger than the local discontinuities, this mass is simply proportional to the volume considered. We can thus define a locally constant quantity, namely, the *density*, as the ratio of this mass to the corresponding volume, and we will denote it as ρ. Note that the different elements of matter can obviously be in motion; in that case the preceding definition must be seen as describing the instantaneous, spatial distribution of the material mass.

A related quantity, which will play an important role when studying materials such as suspensions, is the *volume fraction* (ϕ) of a species of elements (bubbles, particles, etc.) in the remaining material. This is defined as the ratio of the volume occupied by this species to the corresponding volume of material. In some cases variables related to the mass fraction are used instead. However, in general, the volume fraction has a more straightforward rheological meaning than does mass concentration, since it provides a direct indication of the fraction of volume of material that can flow around the species in question, which is at the origin of the apparent viscosity of suspensions (see Chapter 2). However, it is worth noting that since gravity effects often play a major role in the flow properties of granular materials, mass concentration can also be a relevant parameter in view of rheological considerations for such materials. In particular, slight changes in density may induce significant changes in their rheological behavior; dry granular systems appear to suddenly liquefy at a critical density value. This effect has some similarities with the structure breakdown of pasty materials, where this is basically the local configuration of elements and not the density, which changes; since these materials are generally composed of various elements immersed in a liquid, significant variations of density rarely occur because this would imply some variations of the liquid density, which are generally negligible. Macroscopic density variations may nevertheless occur in such materials as a result of some collective migration of elements through the liquid (phase separation; see Section 3.6). Such an effect can be accounted for in the mechanical description with respect to its implications on the local rheological behavior of the material. In the following continuum mechanics treatment phase separation phenomena are assumed negligible, and if some of them occur, they will be considered as perturbating effects (see Section 3.6).

Temperature and Local Displacement

Now we can consider the possible displacements of the different elements of a material. When directly observing the small elements in the material, we can observe various motions in different directions resulting from the various interactions on different scales between the elements, from thermal agitation of the liquid or gas molecules to the motion fluctuations of the colloidal particles or grains; similar to the density, on a sufficiently small scale, the velocity fluctuates in space and time. Once again, if we increase the spatiotemporal scales of observation (i.e., the dimensions of the elementary volume and the time over which one measures the quantity), the average motion in this volume should in principle reach a nonfluctuating value, varying only slowly in space and time because of thermomechanical constraints. The displacement of an elementary volume on this scale will be considered as the *local displacement* at a given point in the continuous material. The more local and rapid fluctuations of velocity within the volume of observation reflect the *temperature* (T) of the sample, but a more precise definition of this notion is beyond the scope of this book. In addition, some fluctuations significant on any scale of observation intermediate between that of the elementary parts and the macroscopic scale may occur under some conditions as a result of inertia effects; this is termed *turbulence* (see Section 1.2.4).

Deformation and Velocity

Here, we consider the deformation in time of a material. This deformation is the result of the relative displacements of the different elementary volumes of the material. These motions induce deformations and relative displacements of the elementary volumes that contain various such elements. In order to quantify the flow characteristics of the material, it is necessary to give a mathematical description of these motions in both space and time. Here we will follow the evolutions in time of the distribution of elementary volumes in space; this is the *material configuration*. The mathematical approach adopted is that of Coleman et al. [1].

Let us first consider the displacement of a given elementary volume of material situated in **x** at time t. We will designate ξ as the position of this elementary part at time $\vartheta < t$. At each time t, we then can define a function χ_t that gives the position ξ at ϑ of the elementary part, which is in the position **x** at t (see Figure 1.4):

$$\xi = \chi_t(\mathbf{x}, \vartheta) \tag{1.1}$$

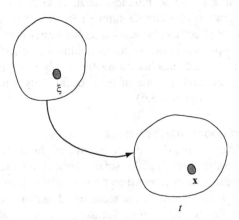

Figure 1.4   Configuration at time ϑ as a function of the configuration at the actual time t.

Now we can quantify the changes in the relative positions of the elementary parts of the material. A global description of the changes in their spatial configurations in time would be difficult. It is more appropriate and straightforward to describe these relative motions on a local scale, only around the elementary part considered, through the *relative configuration gradient*, defined as

$$\mathbf{F}_t(\mathbf{x}, \vartheta) = \nabla \chi_t(\mathbf{x}, \vartheta) \tag{1.2}$$

in which the gradient is taken relative to the actual configuration $\mathbf{x}$. It is usual to drop $\mathbf{x}$ and write $\mathbf{F}_t(\mathbf{x}, \vartheta)$ as $\mathbf{F}_t(\vartheta)$. This tensor, for example, makes it possible to determine the relative configuration change, since the time ϑ, of a small material volume represented by the vector $d\mathbf{x}$ around $\mathbf{x}$ at the time t: it is equal to $\mathbf{F}_t(\mathbf{x}, \vartheta) \, d\mathbf{x}$. Note that obviously if $\vartheta = t$, or if the material volumes do not move relative to each other around $\mathbf{x}$ between ϑ and t (i.e., $\mathbf{x} = \boldsymbol{\xi}$), we have $\mathbf{x} = \chi_t(\mathbf{x}, \vartheta)$, so that $\mathbf{F}_\vartheta(\vartheta) = \mathbf{I}$.

The instantaneous *velocity* $\mathbf{v}$ of the elementary volume situated in $\mathbf{x}$ at t can also be defined from the function χ_t:

$$\mathbf{v}(\mathbf{x}, t) = \frac{\partial}{\partial \vartheta}(\chi_t(\mathbf{x}, \vartheta))_{\vartheta = t} \tag{1.3}$$

It is worth noting that the function $\boldsymbol{\xi}(\vartheta) = \chi_t(\mathbf{x}, \vartheta)$ can be determined from knowledge of the velocity field $v(\mathbf{x}, \vartheta)$ at any time $\vartheta < t$. Indeed, by definition we have $\boldsymbol{\xi} = \chi_\vartheta(\boldsymbol{\xi}, \vartheta)$ for each ϑ, so that writing equation (1.3) for $\mathbf{x} = \boldsymbol{\xi}$ and $t = \vartheta$ gives the differential equation

$$\dot{\boldsymbol{\xi}}(\vartheta) = \mathbf{v}(\boldsymbol{\xi}(\vartheta), \vartheta) \tag{1.4}$$

from which $\boldsymbol{\xi}(\vartheta)$ is the solution with the initial condition $\boldsymbol{\xi}(t) = \mathbf{x}$.

The instantaneous *acceleration* vector $\mathbf{a}$ of this elementary volume is obtained in a similar way:

$$\mathbf{a}(\mathbf{x}, t) = \frac{\partial^2}{\partial \vartheta^2}(\chi_t(\mathbf{x}, \vartheta))_{\vartheta = t} \tag{1.5}$$

Using (1.3) and (1.4) in (1.5), we find the expression for the acceleration as a function of velocity:

$$\mathbf{a} = \frac{\partial \mathbf{v}}{\partial \vartheta} + \nabla \mathbf{v} \cdot \mathbf{v} \tag{1.6}$$

It is worth noting that these definitions for the velocity and the acceleration correspond to the so-called Eulerian description, in which we follow at each time the physical properties of elementary fluid volumes as a function of their different spatial positions.

We can also define the *velocity gradient tensor* as

$$\mathbf{L}(t) = \nabla \mathbf{v}(\mathbf{x}, t) \tag{1.7}$$

which expresses the spatial variations of local velocity. Note that, as expected, since we described spatial and temporal variations of the local positions, in both cases, this velocity gradient tensor is simply equal to the time derivative of the configuration gradient history at the current instant:

$$\dot{\mathbf{F}}_{\vartheta < t}(t) = \frac{\partial}{\partial \vartheta}\mathbf{F}_t(\vartheta)_{\vartheta = t} = \nabla\left(\frac{d}{d\tau}\chi_t(\mathbf{x}, \tau)_{\tau = t}\right) = \mathbf{L} \qquad (1.8)$$

Thus $\mathbf{L}$ expresses the rate of configuration change of the material around $\mathbf{x}$ at t. It may be expressed as the sum of an antisymmetric tensor ($\mathbf{\Omega} = \frac{1}{2}(\mathbf{L} - \mathbf{L}^T)$), which expresses the rotations of the configuration, and a symmetric tensor:

$$\mathbf{D} = \frac{1}{2}(\mathbf{L} + \mathbf{L}^T) \qquad (1.9)$$

This is the *strain rate tensor*, which expresses solely the deformation of the fluid around the point under consideration.

Forces

Obviously a material is not left to evolve by itself; there are various external and internal actions that condition its transformations. Each of these actions is a *force*, which is loosely defined as an action that would tend to increase the velocity of an elementary part of material if it were acting solely on that part. Different types of forces are exerted on the elements. In general, the external, long-range actions due to gravity or electromagnetic effects do not significantly vary from one position to another close position in the material, so the resulting forces over close, similar elements are similar. It follows that the resulting forces over two neighboring elementary volumes containing a similar set of various elements are similar. The total force resulting from such actions is simply proportional to the small volume of material (dV) under consideration, which we can express as

$$\mathbf{b}\, dV \qquad (1.10)$$

where $\mathbf{b}$ is the force density. Note that in the presence of gravity only, the force density is $\rho\mathbf{g}$.

In parallel there exist short-range forces between the elements, due to intermolecular or interparticle (electrochemical or frictional) forces. Existing theoretical expressions of these forces [2] generally concern simple situations (spherical or platelet particles, uniform ionic strength, etc.) but give an idea of typical variations of the force intensity with distance. These forces decrease rapidly with distance; thus, as a first approximation, their effects can be considered as restricted to actions between neighboring elements. The net force exerted between the elements from one side of a surface and those on the other side is as a first approximation simply equal to the sum of the combined forces between neighboring elements on both sides of this surface. It follows that the net force between two small ensembles of elementary volumes of material through some surface

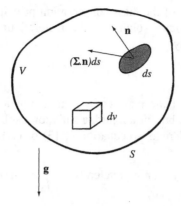

Figure 1.5 Volume and surface elements in a material.

is proportional to the extent of this surface (ds). It is then possible to show (see, e.g., Ref. 3) that, in the absence of internal torque acting on the elements, this force may be expressed as $(\mathbf{\Sigma} \cdot \mathbf{n})\,ds$, where $\mathbf{\Sigma}$ is the stress tensor, which is symmetric ($\mathbf{\Sigma} = \mathbf{\Sigma}^T$), and $\mathbf{n}$ is the normal vector to the elementary surface under consideration. The stress vector $\mathbf{t}$ on this surface element (Figure 1.5) is thus defined as follows:

$$\mathbf{t} = \mathbf{\Sigma} \cdot \mathbf{n} \tag{1.11}$$

1.2.4 Conservation Laws

The three variables (density, velocity, temperature) provide a spatiotemporal description of the distribution of the elements of material in statistical terms (due to the averaging over the elementary volumes). The stress tensor and the density of external forces provide information concerning the different actions exerted on these elements. The objective of thermodynamics is to predict the evolution of the abovementioned variables in time and space under the action of these forces. In this frame we can set up three principles of conservation of physical quantities: mass, momentum, and energy. Here, only the basic formulas useful for the developments in this book are directly presented; the reader being referred to the text by Bird et al. [4] for a much more complete approach.

Mass Conservation
The *mass conservation* principle expresses the fact that in the absence of any source of matter, the amount of matter remains constant during any displacement, possibly with some density change. More precisely, let us consider a fixed volume V with an external surface S; a net flux of matter through this volume during a short time dt must be balanced by a variation in the local density within the volume V, given by

$$\frac{d}{dt} \left(\int_V \rho \, dv \right) = - \int_S (\rho \mathbf{v} \cdot \mathbf{n}) \, ds = - \int_V \nabla \cdot \rho \mathbf{v} \, dv \tag{1.12}$$

This is the integral form of the mass conservation principle. The validity of this equation in each elementary volume of the material implies that we have the following formula locally:

$$\frac{\partial \rho}{\partial t} + \nabla \cdot \rho \mathbf{v} = 0 \tag{1.13}$$

This is the local form of the mass conservation. For a gas, density variations under flow may be wide, but for a liquid, a solid, or a solid–liquid mixture, these variations are often negligible, so equation (1.13) or (1.12) generally reduces to

$$\nabla \cdot \mathbf{v} = 0 \quad \text{or, equivalently,} \quad \int_S (\mathbf{v} \cdot \mathbf{n}) \, ds = 0 \tag{1.14}$$

Momentum Equation

The *momentum equation* expresses the fact that the fundamental principle of dynamics acts on each material element, that is, that the acceleration of a material element is proportional to the sum of forces acting on it. Let us apply this principle to a volume of material V of external surface S. The sum of forces exerted on its different components is equal to the sum of external forces acting on all the elements composing it and the sum of all surface forces acting on all the elementary surfaces in this volume. For each elementary surface surrounded by material (i.e., not situated along the boundary), the sum of surface forces acting from one side to the other (and reciprocally) cancels, so that only the surface forces acting along the external surface of the volume remain. Finally the momentum equation in integral form is expressed as follows:

$$\int_V \rho \mathbf{a}(\mathbf{x}, t) \, dv = \int_V \left(\rho \frac{\partial \mathbf{v}}{\partial t} + \rho \nabla \mathbf{v} \cdot \mathbf{v} \right) dv = \int_V \rho \mathbf{b} \, dv + \int_S (\mathbf{\Sigma} \cdot \mathbf{n}) \, ds \tag{1.15}$$

The validity of this equation for each material volume leads to the local expression of the momentum equation:

$$\rho \frac{\partial \mathbf{v}}{\partial t} + \rho \nabla \mathbf{v} \cdot \mathbf{v} = \rho \mathbf{b} + \mathrm{div} \mathbf{\Sigma} \tag{1.16}$$

Note that in the absence of inertia effects (see Section 3.9) the left-hand side of (1.15) or (1.16) can be neglected.

In rheometry the material is often submitted to a torque and its flow characteristics are symmetric by rotation around an axis (see Section 1.4). Here, in the absence of inertia effects, and if the volume force can be taken to be homogeneous ($\mathbf{b} = $ constant), equation (1.16) may be rewritten

$$\int_S \mathbf{r} \times (\mathbf{\Sigma} \cdot \mathbf{n}) \, ds = 0 \tag{1.17}$$

which provides a simplified form of the kinetic momentum theorem.

Energy Conservation

There exist some solicitations, such as heat fluxes or chemical reactions, which generally influence the material temperature. These processes are taken into account in the first principle of thermodynamics, which expresses the balance between the time variation of the total energy (e) of the system, which includes all types of element motion (fluctuations and average displacements), and the heat and work imparted to the material by external actions:

$$\frac{de}{dt} = \mathrm{Tr}(\mathbf{\Sigma} \cdot \mathbf{D}) + r - \nabla \cdot \mathbf{q} \qquad (1.18)$$

in which r is the volume density of internal heat (as a result, e.g., of chemical reactions) and $\mathbf{q}$ is the heat density flux. The first term of the right-hand side [$\mathrm{Tr}(\mathbf{\Sigma} \cdot \mathbf{D})$] results solely from relative motions between material elements so that it corresponds to the energy stored or dissipated by mechanical actions per unit time within the material. In the absence of elastic effects, it is generally referred to as *energy dissipation* (per unit time). The corresponding total energy dissipation in a volume V per unit time gives

$$P = \int_V \mathrm{Tr}(\mathbf{\Sigma} \cdot \mathbf{D}) \, ds \qquad (1.19)$$

We should add the second law of thermodynamics to this equation. However, in rheometry, we are concerned primarily with the determination of mechanical behavior under isothermal conditions. Thermal processes may play a role as an artifact or simply because the temperature constitutes one parameter of the constitutive equation (see Section 3.8) but, in practice, thermodynamic laws are rarely used for determining the constitutive equation.

Boundary Conditions

The boundary conditions provide the values of the flow characteristics at the frontiers of the sample. They are absolutely necessary for solving a flow problem since the conservation equations retain a given form only over a volume in which the material characteristics do not change. In order to solve a given set of differential equations, it is thus necessary to know some specific values assumed by the variables at some place and time. This corresponds to the boundary conditions. In practice we need to know three independent components of $\mathbf{u}$ and $\sigma \cdot \mathbf{n}$ along the boundary of the material (surface Σ).

Typical boundary conditions are (1) the stress tensor at the interface between a gas and a more complex fluid is known, whereas the velocity is unknown (this follows from conservation equations); for example, when gas motions are negligible, we have

$$\mathbf{\Sigma} \cdot \mathbf{n} = -p_0 \mathbf{n} \qquad (1.20)$$

along the boundary (in which p_0 is the gas pressure); and (2) at the interface between a solid and a fluid, the stress is unknown but the velocity is continuous

($\mathbf{v}_F = \mathbf{v}_S$, in which $\mathbf{v}_S$ is given along the boundary). Note that in case (1) there may be some difference between the stress components at the interface as a result of surface tension effects (see Section 3.4). Similarly, in case (2), there may be some velocity variations at the interface if wall slip occurs (see Section 3.2).

Constitutive Equation

If we disregard possible thermodynamic changes, the flow characteristics should be determined from the set of equations including the mass conservation and the momentum equation along with boundary and initial conditions. However, although the external forces $\mathbf{b}$ generally are known, this is not the case for the stress tensor. This means that the number of unknown variables in this set of equations is equal to 1 (density) + 3 (velocity) + 6 (stress tensor) = 10, which is much larger than the number of equations (four). Knowledge of some additional relation between the stress tensor and the velocity, or more generally the history of deformation, thus appears to be necessary in order to solve the problem. This is the main objective of rheometry to determine at best this relation, in fact the *constitutive equation* of the material, and we discuss its origin and expected structure in detail in the next section.

Turbulence

We have seen above that over small scales of space or time the local velocity fluctuates because of the basically discontinuous nature of matter but, in practice, measurements are generally carried out over sufficiently long durations, so these effects are not observed. However, the conservation equations as expressed above in fact apply for the instantaneous, local values of the variables. In previous developments, it has thus been implicitly assumed possible to extend the validity of these equations to the nonfluctuating (averaged) part of the different variables. In fact only the linear terms remain unchanged after averaging, and thus the mass conservation remains valid whatever the importance of fluctuating terms. This is nevertheless not the case for the second (inertia) term on the left-hand side of the momentum equation.

Let us express a local variable in the form of the sum of its average and fluctuating values. For the local velocity this gives $\mathbf{v} = \langle \mathbf{v} \rangle + \mathbf{v}'$, in which $\langle \mathbf{v} \rangle$ is the time-averaged value of $\mathbf{v}$ and $\mathbf{v}'$ is its fluctuating part. By definition, we have $\langle \mathbf{v}' \rangle = 0$. In equations (1.13) and (1.15), the different terms, and possibly space and time derivatives of all the local and instantaneous variables, transform into similar terms involving the averaged values, except for the inertia term since $\langle \mathbf{v} \cdot \nabla \mathbf{v} \rangle = \langle \mathbf{v} \rangle \cdot \langle \nabla \mathbf{v} \rangle + \langle \mathbf{v}' \cdot \nabla \mathbf{v}' \rangle$. As a consequence, the local and averaged expressions for the momentum equation are similar only if the additional term $\langle \mathbf{v}' \cdot \nabla \mathbf{v}' \rangle$ is negligible. This situation corresponds to what we call *laminar flows*. On the contrary, when this term is significant, we are dealing with *turbulent flows*. Note that for turbulent flows, it is also possible to have similar local and average forms for the momentum equation by including the term $\rho \langle \mathbf{v}' \cdot \nabla \mathbf{v}' \rangle$ in the stress expression so as to obtain a turbulent stress tensor. However, in contrast to laminar flows, the relationship between this turbulent stress tensor and flow history depends on boundary and initial conditions.

Rheometry is relevant only in the case of laminar flows because, despite some local fluctuations, this is solely the average resistance to displacement of the different elements relative to each other, which gives rise to stresses. Thus, if we measure this resistance under some specific flow conditions but for all types and rates of relative displacements, we can use the corresponding data for predicting flow characteristics under any other conditions in the laminar regime. In the turbulent regime, a significant fraction of stresses also results from the transport of momentum due to motion fluctuations, but in contrast with laminar flows, the local expressions of these "turbulent" stress terms depend on the boundary and initial conditions of the flow in a complex manner. As a consequence, a simple rheometrical study under specific conditions cannot provide sufficient information for predicting turbulent flows under other conditions. The prediction of turbulent flow characteristics requires sophisticated modeling developments [5].

1.3 CONSTITUTIVE EQUATION

Here, for the sake of simplicity, we will consider mainly the constitutive equation in mechanical terms, leaving aside the effects of temperature changes on flow characteristics. This in particular means that the effect, in the constitutive equation, of the variations of fluctuating motions of elementary volumes during flow can be neglected.

1.3.1 Physical Origin

As soon as the elements in a material are submitted to a stress and/or displace relative to each other as a result of a macroscopic force or deformation applied to it, the local forces acting between the material elements can vary in time and space. Then the distribution of forces within the material evolves as a function of the configuration and state of these elements (e.g., the deformations of polymer molecules or bubbles). Since the actual configuration depends on the complete history of configuration and state changes, the actual force distribution is a function of both flow history and intrinsic properties of the material. Considering the problem on the local scale and accounting for each elementary force and its time evolution as a result of configuration changes would lead to an extremely complex set of equations, and for most materials such an approach is not realistic. This, added to the already mentioned difficulty in precisely expressing the local surface forces in real materials composed of various particle dimensions and types, led us to assume that a *constitutive equation* (i.e., a function relating the local stress tensor over an elementary volume to some average of the history of motions experienced by the material elements in this volume) applies for that material. This is in fact another, more physical, way to reach our conclusion in Section 1.2 regarding the need for a specific relation between stress and velocity fields, which in that case followed from a direct, continuous description of the mechanical properties of the material.

1.3.2 General Characteristics

Although the constitutive equation depends on the specific physical properties of each material, some general, basic trends of the constitutive equations have been derived from a *rational* approach of mechanics [6]. First it can reasonably be assumed that the actual stress tensor ($\mathbf{\Sigma}(\mathbf{x}, t)$) is given in a unique way as a function of the previous motions of material elements in a given sample; this is the *determinism principle*. We thus expect a relationship of the type $\mathbf{\Sigma}(\mathbf{x}, t) = \mathsf{H}_{0 < \vartheta < t; \mathbf{y} \in \Omega}(\chi_t(\mathbf{y}, \vartheta))$ in which Ω is the sample volume and H is a function depending on material properties. In this expression we assume that the observation of the system begins at the instant $\vartheta = 0$ for which its configuration is $\chi_t(\mathbf{y}, 0)$.

Since the stress tensor directly derives from the interactions of neighboring elements in contact (see Section 1.2), the configuration history has an impact on the local stress tensor only through the configuration changes that occur near the small material volume under consideration. This is the *principle of local action*. A possible way to express this in practice consists in considering that the stress tensor $\mathbf{\Sigma}$ depends only on the history of the local, relative configuration gradient, $\mathbf{\Sigma}(\mathbf{x}, t) = \mathsf{H}_{0 < \vartheta < t}(\chi_t(\mathbf{x}, 0), \mathbf{F}_t(\vartheta))$, in which H is another function depending on material properties. Note that as soon as we consider solely configuration changes, we have to keep some information concerning the initial state; this is the origin of the presence of $\chi_t(\mathbf{x}, 0)$ in the function. Using the stress tensor at the initial time $\mathbf{\Sigma}_0 = \mathbf{\Sigma}(\mathbf{x}, 0) = \mathsf{H}_{\vartheta=0}(\chi_0(\mathbf{x}, 0))$ and after inversion of this relation, we can rewrite the preceding expression in the general form:

$$\mathbf{\Sigma}(\mathbf{x}, t) = \mathsf{H}_{0 < \vartheta < t}(\mathbf{\Sigma}_0, \mathbf{F}_t(\vartheta)) \tag{1.21}$$

These behavior characteristics are those of a *simple material*. A final rational, assumption is that the material behavior does not vary with a change in frame of observation; this is the principle of *material objectivity* or *frame indifference*. The implications of the principle of material objectivity on the characteristics of the constitutive equation will be considered later, after we have examined the effects on stresses and deformations of a change of frame of observation.

1.3.3 Effect of Change in Frame of Observation

If we view the system in another frame of observation (while keeping the same referential), from the current position we obtain

$$\mathbf{x}^* = \mathbf{c}(t) + \mathbf{Q}(t)(\mathbf{x} - \mathbf{q}) \tag{1.22}$$

where $\mathbf{q}$ is a constant vector. The first term of the right-hand side, the vector $\mathbf{c}(t)$, expresses the displacement of the origin of the frame of observation while the second term expresses the frame rotation. It is necessary that $\mathbf{Q}$ be an orthogonal tensor, such that for any t, we obtain

$$\mathbf{Q}(t)\mathbf{Q}(t)^T = 1 \tag{1.23}$$

Obviously we also have $\boldsymbol{\xi}^* = \mathbf{c}(\vartheta) + \mathbf{Q}(\vartheta)(\boldsymbol{\xi} - \mathbf{q})$ since (1.22) is valid at any time and any point within the material. The laws of composition of derivatives reveal that the relative configuration gradient (1.2) in the new frame of observation can be expressed as

$$\mathbf{F}_t^*(\vartheta) = \nabla\chi_t^*(\mathbf{x}^*, \vartheta) = \nabla_{(\mathbf{x}^*)}\boldsymbol{\xi}^* = [\nabla_{(\boldsymbol{\xi})}\boldsymbol{\xi}^*][\nabla_{(\mathbf{x})}\boldsymbol{\xi}(\mathbf{x})]\lfloor\nabla_{(\mathbf{x}^*)}\mathbf{x}\rfloor \tag{1.24}$$

in which $\nabla_{(\mathbf{x})}\boldsymbol{\xi}$ is the gradient of $\boldsymbol{\xi}$ relative to $\mathbf{x}$. From (1.2), (1.22), and (1.24) it follows that

$$\mathbf{F}_t^*(\vartheta) = \mathbf{Q}(\vartheta)\mathbf{F}_t(\vartheta)\mathbf{Q}(t)^T \tag{1.25}$$

Let us now consider the stress vector $\mathbf{t}^*$ along a material surface element of normal vector $\mathbf{n}^*$. In the new frame this is expressed as $\mathbf{t}^* = \boldsymbol{\Sigma}^* \cdot \mathbf{n}^*$. From (1.22) we also have $\mathbf{t}^* = \mathbf{Q}(t)\mathbf{t}$ and $\mathbf{n}^* = \mathbf{Q}(t)\mathbf{n}$. From these relations along with (1.11), we find the relation between the stress tensor expressions in the two frames:

$$\boldsymbol{\Sigma}^* = \mathbf{Q}(t)\boldsymbol{\Sigma}\mathbf{Q}(t)^T \tag{1.26}$$

It follows from (1.26) that $\boldsymbol{\Sigma} \cdot \mathbf{u} = \lambda\mathbf{u} \Rightarrow \boldsymbol{\Sigma}^* \cdot (\mathbf{Q} \cdot \mathbf{u}) = \lambda(\mathbf{Q} \cdot \mathbf{u})$, which means that $\boldsymbol{\Sigma}$ and $\boldsymbol{\Sigma}^*$ have the same eigenvalues and the same diagonal tensorial expressions in the corresponding bases of eigenvectors [respectively $(\mathbf{u}_{i=1,2,3})$ and $(\mathbf{Q} \cdot \mathbf{u}_{i=1,2,3})$]. Since they are simple functions of eigenvalues $\mathrm{Tr}(\boldsymbol{\Sigma})$, $\mathrm{Tr}(\boldsymbol{\Sigma}^2)$ and $\det(\boldsymbol{\Sigma})$ are thus independent of the frame of observation and we can define three *invariants* of the stress tensor $\boldsymbol{\Sigma}$ as follows:

$$\Sigma_I = \mathrm{Tr}\boldsymbol{\Sigma}; \quad \Sigma_{II} = \tfrac{1}{2}[(\mathrm{Tr}\boldsymbol{\Sigma}) - \mathrm{Tr}(\boldsymbol{\Sigma}^2)]; \quad \Sigma_{III} = \det(\boldsymbol{\Sigma}) \tag{1.27}$$

From (1.25) we also have

$$\mathbf{L}^* = \frac{\partial}{\partial\vartheta}\left(\mathbf{F}_t^*(\vartheta)\right)_{\vartheta=t} = \mathbf{Q}'(t)\mathbf{Q}(t)^T + \mathbf{Q}(t)\mathbf{L}\mathbf{Q}(t)^T \tag{1.28}$$

and since from (1.23), $\mathbf{Q}'(t)\mathbf{Q}(t)^T + \mathbf{Q}(t)\mathbf{Q}'(t)^T = 0$, we deduce the following from (1.9):

$$\mathbf{D}^* = \mathbf{Q}(t)\mathbf{D}\mathbf{Q}(t)^T \tag{1.29}$$

As a consequence, we may use definitions similar to (1.27) for three invariants $(D_I; D_{II}; D_{III})$ of the strain rate tensor. Note that from the mass conservation (1.14), the first invariant $(D_I = \nabla \cdot \mathbf{u})$ is equal to zero when density variations are negligible. These invariants of the stress and strain rate tensor express global amounts of deformation or stress, and it is practical to use them as parameters of constitutive equations since they provide expressions directly following the frame indifference principle. However, this is relevant only for isotropic materials since the different directions play a similar role in these invariants.

1.3.4 Solids and Fluids

Generalities

From a physical perspective, three types of materials are generally considered [7]:

- *Solids*, in which the molecules, forming a long-range ordered structure, are close to each other and embedded in deep potential wells from which they can barely escape
- *Liquids*, in which the molecules are close to each other but form a disordered structure and can relatively easily escape from their potential wells
- *Gas*, in which the molecules are far from each other and move almost freely relative to each other

Since a gas or a liquid usually can deform at will and retain its physical properties whereas a solid, or more precisely its ordered structure, tends to be irreversibly modified or breaks beyond a critical deformation, it is natural to first separate these materials, from a mechanical perspective, into two corresponding classes, say, "fluids" and "solids." This description applies only for materials with a simple structure (i.e., composed of one or few molecule types), and cannot a priori encompass the wide variety of industrial or natural materials containing more complex element networks.

In mechanics the concept of a simple material as defined above seems sufficiently general to encompass most industrial and natural materials, which might initially be considered either as solids, fluids, or, perhaps, hybrid materials. Then, applying the preceding classification derived from physical arguments, as well as a natural (inherent) classification based on their mathematical properties [6], rational mechanics has suggested the separation of simple materials into two broad classes: *solids* and *(simple) fluids*. According to these definitions, *solids* are materials that have a preferential, natural configuration, on the basis of which their behavior must be defined. More precisely, at any given time the current stress tensor can be determined from knowledge of the current deformation relative to the preferential configuration. *Fluids* are materials that do not have a preferential configuration; they may be deformed at will but then always relax. This means, for example, that if after any type of flow the material remains at rest for a sufficient time, it eventually completely "forgets" its previous deformations, and thus its final configuration does not depend on the previous flow history.

Although these definitions make it possible to propose a straightforward (rheological) analysis of some particular flows (simple shear; see discussion below), they do not encompass the various behavior types of real materials. This is precisely the case for pasty or granular materials, which are, for example, capable of behaving partly as solids, because under some flow conditions they have a preferential configuration, and partly as fluids, because they are capable of definitively forgetting this configuration after a particular deformation history or may recover this initial configuration after a period of rest. It thus appears necessary to distinguish a third class of materials, "jammed systems" (see below), which borrow their behavior partly from fluids and partly from solids.

Fluids

Because of its potentially interesting simplifications of the treatment of constitutive equations, theoretical rheometry (or viscometry) [1] has been dealt with so far within the frame of fluids, which have a vanishing memory. These materials are such that at a sufficient time (lapse) after some temporal point of reference they will have completely forgotten their history prior to this reference time. This may be expressed as follows. For any flow history and two different states of reference (Σ_i and Σ_i^*), there exists a time ϑ_c (possibly negative) in the past such that the actual state of the material is independent of the state (with respect to the two above mentioned states) at this time:

$$H_{\vartheta_c < \vartheta < t}(\Sigma_i, F_t(\vartheta)) = H_{\vartheta_c < \vartheta < t}(\Sigma_i^*, F_t(\vartheta))$$

As a consequence, if there is no previous time limit, the constitutive equation is independent of the initial state, which is expressed formally as

$$\Sigma(\mathbf{x}, t) = H_{\vartheta < t}(F_t(\vartheta)) \tag{1.30}$$

which no longer contains any reference to a specific state at a particular time.

Let us examine some implications of such a constitutive equation. From the principle of objectivity we have $\Sigma^*(\mathbf{x}, t) = H_{\vartheta < t}(F_t^*(\vartheta))$ in any other frame of observation. Using this relation and substituting (1.25) in (1.26), we deduce a relation that must be respected by the constitutive equation of a simple fluid for any history of $F_t(\vartheta)$ and any tensor $Q(\vartheta)$:

$$Q(t)\lfloor H_{\vartheta < t} F_t(\vartheta) \rfloor Q(t)^T = H_{\vartheta < t}\left(Q(\vartheta)F_t(\vartheta)Q(t)^T\right) \tag{1.31}$$

Let us assume, for example, that the material undergoes the *rest history* defined by $F_t(\vartheta) = I(\vartheta) = I$. With a constant rotation of the frame, defined by the fixed orthogonal tensor Q, equation (1.31) becomes

$$Q\lfloor H_{\vartheta < t} I(\vartheta) \rfloor Q^T = H_{\vartheta < t}(I(\vartheta)) \tag{1.32}$$

Since (1.32) is valid for any orthogonal tensor Q, it may be demonstrated mathematically that the function $H_{\vartheta < t}(I(\vartheta))$ is necessarily proportional to I. This means that the stress tensor in a simple fluid that has experienced nothing but rigid rotations is isotropic: $\Sigma = -pI$. This important result follows from application of the principle of material objectivity in the specific case of materials with vanishing memory. A corollary is that such materials cannot support indefinitely a nonisotropic stress tensor "without eventually giving way to it" [1]. Indeed, if the material never flowed, we would have $F_t(\vartheta) = I$, and the stress tensor would necessarily be isotropic.

Solids

According to the definition above, solids appear to have a constitutive equation of the general form (1.21), in which the dependence of the function H on the state at the initial time plays a critical role. In that case the principle of objectivity leads to

$$\mathbf{Q}(t)\lfloor H_{0<\vartheta<t}(\Sigma_0, \mathbf{F}_t(\vartheta))\rfloor\mathbf{Q}(t)^T = H_{0<\vartheta<t}(\mathbf{Q}(0)\Sigma_0\mathbf{Q}(0)^T; \mathbf{Q}(\vartheta)\mathbf{F}(\vartheta)\mathbf{Q}(t)^T)$$
(1.33)

which does not provide further information on the general characteristics of the constitutive equation.

Jammed Systems

The inability to support a nonisotropic stress tensor indefinitely without flowing appears to be a property in disagreement with common observations of pasty or granular materials, which, when submitted to gravity only can maintain indefinitely on our scale of observations a shape significantly different from that expected from a simple hydrostatic pressure for a simple liquid, specifically, a horizontal free surface. This is, for example, the case for various concentrated suspensions, emulsions, or foams such as paints, inks, cement pastes, fresh concrete, muds, purees, toothpaste, and mayonnaise. When these products are for example, stored in closed containers, they can remain undeformed over durations longer that our typical, maximum time of observation (years). All such materials are composed of numerous elements (particles, droplets, bubbles, etc.) interacting by means of different types of surface forces. Since Brownian effects are not sufficient, some elements can remain jammed between other elements as long as the force applied to them remains insufficient. We will refer to them as "jammed materials," a term originating in the physics literature [8]. The corresponding term in rheology is *yield stress fluids*, but this term assumes a specific behavior type, namely, the simple yielding behavior, which does not appear to be sufficiently general, particularly with respect to time-dependent behavior and solid–fluid transition. From a physical perspective these materials are intermediate between liquids and solids since their elements are in disarray (e.g., as molecules in a liquid), but, in a pattern similar to that found in a solid they are embedded in potential wells, from which they can rarely escape in the absence of a sufficient force.

A solid is characterized by the fact that if its actual configuration deviates significantly from the configuration of reference, the solid can never return to this initial reference point under the usual mechanical constraints, and thus will never recover its initial behavior. On the contrary, jammed systems can undergo any type of deformation but can recover their initial configurations (on average) under some specific constraints, for example, after some degree of flow and a sufficient time at rest. Solids are thus usually studied under small deformations, even if some theories do account for their irreversible changes of behavior for large deformations. According to these definitions, jammed systems can appear as "solid" under some limited range of conditions; under small deformations they may behave as solids, that is, their rheological behavior will be similar to that of solid, and when they stray too far from their corresponding preferential configuration, they break and become "fluid," but they can recover their initial configurations and rheological behavior, for example, after a sufficient period of rest or after a similar preparation. An example illustrating this concept is modeling paste. When lying on a table, the paste is slightly elastically deformed like a solid; if squeezed between two plates, it may flow like a liquid. Then, if it is stretched, it may separate into two parts like a fractured solid; however, it will recover its initial "mechanical appearance" if the two parts are mixed back together under sufficient force (are modeled).

Let us now express the general characteristics of jammed systems from a mechanical point of view. Depending on flow history we expect their behavior to be that of either a solid or a simple fluid type. For such materials the constitutive equation should include a "solid regime" described by an equation such as (1.21) for which the state of reference plays a critical role, and a "liquid regime" described by a relation such as (1.30). In this context the set of constitutive relations for such materials must also include an additional relation describing the conditions for the transition from the solid to the liquid regimes as a function of flow characteristics. The general formulation of such constitutive equations appears to be extremely difficult.

Pressure

Now we can comment that for most fluids, in particular those consisting of particles or elements dispersed in a liquid or a gas, the influence of an isotropic stress applied onto the material, namely, a *pressure*, on the flow characteristics is negligible. As a consequence, the stress tensor may be expressed to within an indeterminate isotropic tensor, which can generally be determined from the momentum equation and boundary conditions. By convention we express this in the form

$$\mathbf{\Sigma} = -p\mathbf{I} + \mathbf{T} \tag{1.34}$$

where $p = -\mathrm{Tr}\mathbf{\Sigma}/3$ is the pressure and $\mathbf{T}$ the deviatoric part of $\mathbf{\Sigma}$, which verifies $\mathrm{Tr}\,\mathbf{T} = 0$. In that case this is the deviating part $\mathbf{T}$, which reflects the constitutive equation of the material and can be expressed in the form of equation (1.21) or (1.30).

This approach is useful when the role of the pressure in the second term of the right-hand side of equation (1.34) is effectively negligible. However, this hypothesis appears to fail for granular materials, whose behavior may, under certain circumstances, result from that of a continuous network of grains in contact. In that case, the shear stress in simple shear is approximately proportional to some normal component of the stress tensor (see Section 3.1) which is not independent of the pressure.

1.3.5 Simple Shear and Viscometric Flows

Simple Shear

The constitutive equation in its general form given by equation (1.21) cannot easily be determined experimentally. Indeed, the stress tensor depends on an unknown function of a great number of variables, including all the values of the components of the relative configuration gradient at different times. Determining this relation would require measurement of the values of the deformation and stress components at any given time and position in the material, which, in practice is unrealistic. The solution found in rheometry consists in studying certain aspects of the constitutive equation via specific flows for which the configuration gradient history has a simple form.

In this context, the basic flow is the *simple shear*, which is such that in some Cartesian coordinate system in which the position **x** is defined as a function of its components (x, y, z), the components of the velocity have the form

$$v_x = \dot{\gamma}y, \; v_y = 0, \; v_z = 0 \tag{1.35}$$

in which $\dot{\gamma}$, called the *shear rate*, is a constant. In that case the deformation undergone by the fluid after the time Δt is expressed as $\gamma = \dot{\gamma}\Delta t$. The motion history is given by the solution of the differential equation (1.4) with the initial condition $\boldsymbol{\xi}(t) = \mathbf{x}$, which yields

$$\xi = x + \dot{\gamma}y(\vartheta - t); \eta = y, \zeta = z \tag{1.36}$$

where (ξ, η, ζ) are the coordinates of $\boldsymbol{\xi}$. Then the relative configuration gradient tensor is:

$$\mathbf{F}_t(\vartheta) = \begin{bmatrix} 1 & \dot{\gamma}(\vartheta - t) & 0 \\ 0 & 1 & 0 \\ 0 & 0 & 1 \end{bmatrix}$$

or equivalently

$$\mathbf{F}_t(\vartheta) = \mathbf{I} + (\vartheta - t)\mathbf{M} \tag{1.37}$$

where **M** is the tensor, defined as

$$M = \begin{bmatrix} 0 & \dot{\gamma} & 0 \\ 0 & 0 & 0 \\ 0 & 0 & 0 \end{bmatrix} \tag{1.38}$$

It follows from (1.37) that $\dot{\gamma}$ is the only parameter appearing in the right-hand side of (1.21), which means that the components of $\boldsymbol{\Sigma}$ depend only on $\dot{\gamma}$ and a state of reference. The shear rate may also depend on time (*generalized simple shear*), in which case a similar result is found—the components of $\boldsymbol{\Sigma}$ are functions of the history of $\dot{\gamma}$ and the state of reference. This is the fundamental result for simple shear of any type of material (solids, fluids, jammed systems).

Simple Shear with a Fluid
Let us now assume that the material is a simple fluid. Here, in simple shear, from equations (1.30) and (1.37), the stress tensor is

$$\boldsymbol{\Sigma} = \mathbf{H}_{\vartheta < t}(\mathbf{I} + (\vartheta - t)\mathbf{M}) \tag{1.39}$$

which means that the components of Σ depend only on $\dot{\gamma}$. We can apply the identity (1.31) in that case using

$$Q(\vartheta) = \begin{bmatrix} -1 & 0 & 0 \\ 0 & -1 & 0 \\ 0 & 0 & 1 \end{bmatrix}$$

and find

$$Q\lfloor H_{\vartheta < t}(I + (\vartheta - t)M)\rfloor Q^T = H_{\vartheta < t}\left[Q(I + (\vartheta - t)M)Q^T\right]$$
$$= H_{\vartheta < t}(I + (\vartheta - t)M) \tag{1.40}$$

which means that

$$\Sigma = Q\Sigma Q^T \tag{1.41}$$

Expressing the components of the stress tensor as

$$\Sigma = \begin{bmatrix} \sigma_{xx} & \sigma_{xy} & \sigma_{xz} \\ \sigma_{xy} & \sigma_{yy} & \sigma_{yz} \\ \sigma_{xz} & \sigma_{yz} & \sigma_{zz} \end{bmatrix}$$

we deduce from (1.41) that

$$\sigma_{xz} = 0 = \sigma_{yz} \tag{1.42}$$

Moreover, since the components of Σ only depend on $\dot{\gamma}$, the four following *material functions* completely define the stress tensor in simple shear:

- *The shear stress:*
$$\tau(\dot{\gamma}) = \sigma_{xy} \tag{1.43}$$

- *The first normal stress difference:*
$$N_1(\dot{\gamma}) = \sigma_{xx} - \sigma_{yy} \tag{1.44}$$

- *The second normal stress difference:*
$$N_2(\dot{\gamma}) = \sigma_{yy} - \sigma_{zz} \tag{1.45}$$

- *The pressure:*
$$p = \tfrac{1}{3}(\sigma_{xx} + \sigma_{yy} + \sigma_{zz}) \tag{1.46}$$

The *apparent viscosity* is then defined as follows:

$$\eta(\dot{\gamma}) = \frac{\tau}{\dot{\gamma}} \tag{1.47}$$

In this book we will, however, use the notation μ for fluids assumed to be Newtonian, that is, the viscosity of which is constant.

Viscometric Flows

Now we are interested in *viscometric flows*, motions that at any time are a simple shear in a particular frame that may vary in time. This means that at each time the relation (1.37) is valid with both a frame of observation and a value $\dot{\gamma}$ possibly depending on time: $\mathbf{F}_t^*(\vartheta) = \mathbf{I} + (\vartheta - t)\mathbf{M}(\dot{\gamma}(t))$. Let us consider the corresponding frame at time ϑ seen from the actual frame at time t; it may be deduced from the actual frame by the orthogonal tensor $\mathbf{Q}(\vartheta)$, which is such that $\mathbf{Q}(t) = \mathbf{I}$. Under these conditions the relative configuration gradient at any time in the actual frame is deduced from (1.25):

$$\mathbf{F}_t(\vartheta) = \mathbf{Q}(\vartheta)(\mathbf{I} + (\vartheta - t)\mathbf{M}(\dot{\gamma}(t))) \tag{1.48}$$

Let us now apply the principle of objectivity (1.31) for the flow history $\mathbf{I} + (\vartheta - t)\mathbf{M}$ and a change of frame defined by $\mathbf{Q}$:

$$\mathsf{H}_{0<\vartheta<t}(\Sigma_0, \mathbf{I} + (\vartheta - t)\mathbf{M}) = \mathsf{H}_{0<\vartheta<t}(\mathbf{Q}(0)\Sigma_0\mathbf{Q}(0)^T, \mathbf{Q}(\vartheta)(\mathbf{I} + (\vartheta - t)\mathbf{M}))$$

$$= \mathsf{H}_{0<\vartheta<t}(\mathbf{Q}(0)\Sigma_0\mathbf{Q}(0)^T, \mathbf{F}_t^*(\vartheta)) \tag{1.49}$$

For a fluid, or for a jammed material in its liquid regime, we get

$$\mathsf{H}_{\vartheta<t}(\mathbf{I} + (\vartheta - t)\mathbf{M}) = \mathsf{H}_{\vartheta<t}(\mathbf{F}_t^*(\vartheta)) = \Sigma(\mathbf{x}, t) \tag{1.50}$$

which means that the actual stress tensor is the same as the stress tensor associated with the simpler history $\mathbf{I} + (\vartheta - t)\mathbf{M}$, that is, without factoring in the time changes of the frame in which we have instantaneous simple shear. On the left-hand side (LHS) of (1.50) the function depends only on the different values of $\dot{\gamma}$ in time, which implies that *the stress tensor is simply a function of the history of* $\dot{\gamma}$. This result is in particular valid for a simple shear in a fixed frame but with time variations of $\dot{\gamma}$. On the contrary, for a solid or a jammed material in its solid regime, equation (1.49) does not provide a straightforward result concerning viscometric flows except if $\mathbf{Q} = \mathbf{I}$, which is merely a simple shear with a time-dependent shear rate.

Practical Approach for Simple Shear of Jammed Systems

Because it is difficult to express the constitutive equation of jammed systems in the different possible regimes in a general way, it seems preferable in practice to maintain the dependence on the initial state of the material, even under conditions for which the material widely deforms, like a fluid. This makes it possible to account for the possible solid–liquid transition and the possible nonvanishing memory of pastes or granular materials under some conditions. Under these conditions, in generalized simple shear, the stress tensor of a jammed system may be expressed as

$$\Sigma(t) = \mathsf{H}_{0<\vartheta<t}(\Sigma_0, \dot{\gamma}(\vartheta)) \tag{1.51}$$

In practice it often appears useful to express the actual shear rate as a function of the stress history. By inverting (1.51) we can deduce

$$\dot{\gamma}(t) = H_{0<\vartheta<t}(\Sigma_0, \Sigma(\vartheta)) \qquad (1.52)$$

Note that a specificity of jammed systems is that in steady state, when $\dot{\gamma}$ and Σ are constant, the relationship between these two variables may still depend on the initial state of the material Σ_0. In particular, for the shear stress, we deduce from (1.51) a steady-state dependence of the type

$$\tau = \varsigma_{\Sigma_0}(\dot{\gamma}) \qquad (1.53)$$

in which ς is a function depending on Σ_0.

Note that in some cases it may be assumed that all stress components except the tangential one in the direction of shear play a negligible role in expression (1.51) which, for a constant shear stress, leads to

$$\dot{\gamma}(t) = \chi_{\Sigma_0}(\tau, t) \qquad (1.54)$$

Finally, we note that in the rest of the book, when dealing with flow problems, we will use the generic terms "material" or "fluid" to describe a jammed system under flow.

1.4 VISCOMETRIC FLOWS

The practical interest in the viscometry of flows is that this method in principle makes it possible to determine the resistance to flow of a fluid from simple macroscopic measurements. Indeed, if we have been capable of devising flows with appropriate properties of symmetry such that the fluid layers glide over each other parallel to some boundary, we can expect that a measure of the relative velocity between two solid boundaries in contact with the fluid, and the total force exerted on one of these boundaries, will provide the values of the local material functions (including its apparent viscosity). For simple fluids it has been shown that in that case, which corresponds to that of equation (1.30), the four material functions completely define the stress tensor. In fact, even in that case, there is a direct correspondence between the macroscopic measurements and the local material functions of the fluid only when the shear rate is homogeneous within the fluid volume. However, this situation is expected only for very small gaps [4], but large gaps are often required with pasty or granular materials for the continuum assumption to be valid, at least in terms of the density. Moreover, even for small gaps, the theory of viscometric flows as established by Coleman et al. [1] is valid only for steady-state flows of simple fluids, capable of completely forgetting their initial state. As we have seen above, within the frame of jammed systems (pastes or granular materials), this assumption may not be valid.

In fact, the problem of describing the specific flow characteristics of a material is rather complex because, even for viscometric flows, a sufficient knowledge of the constitutive equation of the material is required. Moreover, it would be necessary to determine the exact boundary conditions, such as possible edge effects and wall slip. Here we will separate the problems and simply assume that the flow a priori possesses the basic, qualitative, ideal characteristics generally assumed for viscometric flows (for simple fluids) and that jamming properties (leading to yielding and thixotropy) or perturbing effects act to modify, to a certain extent, the quantitative characteristics of this ideal flow without changing its qualitative aspects. The possible consequences of these additional effects will be examined in detail in Chapter 3.

To summarize, our approach consists in first assuming some ideal form of the velocity field and deducing the corresponding stress field and then considering the perturbing effects or jamming properties that play a role within this simple frame. Such a procedure obviously constitutes a nonconventional, approximate, mechanical approach. However, with regard to the difficulty of determining a priori the strain–stress field with no knowledge of the material behavior, this appears to be a good method for dealing with viscometric flows of pastes and granular materials in a sufficiently simple way and at the same time taking into account the specificities of these flows.

In the sections that follow we will make the following, ideal assumptions: (1) the material remains homogeneous and constant (there is no irreversible chemical transformation); (2) there are no edge effects; (3) the ambient fluid induces only a uniform pressure p_0 in the fluid located along the interface—that is, the effects of surface tension effects and of the flow of the ambient fluid are neglected; (4) inertia effects are negligible; (5) there are no thermal effects; (6) the flow remains laminar; (7) there is no relative velocity between the fluid and the solid along the rigid boundaries (no wall slip); and (8) there are no flow instabilities, that is, the flow field retains the simplest form, verifying the conservation equations and the boundary conditions. We start by considering the simplest viscometric flow that may be obtained in practice, namely, free surface flow over an inclined plane (Section 1.4.1), then we turn to the viscometric flows that can be obtained from usual laboratory rheometers: flow between parallel disks (Section 1.4.2), between a cone and a plate (Section 1.4.3), between two coaxial cylinders (Section 1.4.4), and through a straight cylindrical conduit (Section 1.4.5).

1.4.1 Free Surface Flow over a Plane

We consider the flow of a material over a solid inclined plane of a slope i with respect to the horizontal and describe it in terms of (x, y, z) coordinates as represented in Figure 1.6. Neglecting edge effects is equivalent to the plane surface and the free surface being infinite. This implies that all positions within each plane parallel to them are equivalent so that the velocity and the stress tensor at each point in the material depends only on their distance y from the plate.

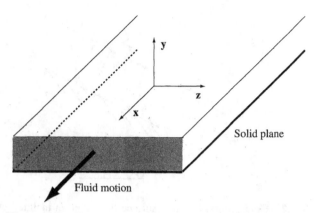

Figure 1.6   Geometric characteristics of inclined plane flow.

Under such conditions the mass conservation (1.14) gives $\partial v_y/\partial y = 0$, which means that the velocity component perpendicular to the planes is uniform, and in particular is equal to the normal velocity along the solid plane, specifically, zero. In addition, for reasons of symmetry, the velocity component in the direction z is zero everywhere; a fluid element at any point in each plane (x, y) "sees" a similar situation on each side of this plane, so that it has no particular reason for moving in one direction rather than in the other, and we have $v_z = 0$. Note that these results concerning the velocity components in the y and z directions are valid only in the absence of mechanical or thermal flow instabilities possibly leading to the formation of vortices with a particular fluid length scale. Finally, for a stable flow the only nonzero component of the velocity is v_x, which depends only on y. This means that the fluid motion takes the form of planar layers sliding over each other in the direction of steepest slope (x).

Let us now apply the momentum equation in integral form (1.15) to a fluid portion limited by the free surface, the planar surface situated at a height y and two sections perpendicular to the solid plane at a distance L (see Figure 1.7). Since the flow is uniform, there are no flux variations between these two sections and the normal forces acting over them balance. The momentum equation finally reduces to the balance of the gravity force acting on the volume and the viscous force resulting from the contact between the two local, adjacent, fluid layers at the height y. In projection along the direction x the momentum equation provides the expression for the shear stress

$$\sigma_{xy} = \tau(y) = \rho g(h - y)\sin i \tag{1.55}$$

in which h is the thickness of the fluid layer. This equation shows that the shear stress varies linearly in the fluid layer and assumes its maximum value

$$\tau_p = \sigma_{xy}(0) = \rho g h \sin i \tag{1.56}$$

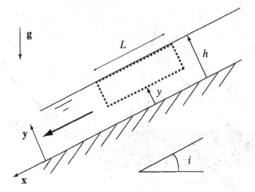

Figure 1.7   Fluid portion in a free surface flow over an inclined plane.

along the interface with the plate. The momentum equation applied to this fluid portion in projection along the y direction yields

$$\sigma_{yy} = p_0 + \rho g(h - y)\cos i \qquad (1.57)$$

Since the flow consists in the relative motion of parallel planar layers of a material along the x direction, we have a local simple shear of intensity possibly varying in time. More precisely, the history of the relative configuration gradient at a given point is a function of the gradient of velocity at a fixed distance from the axis, namely, the shear rate:

$$\dot{\gamma} = \frac{\partial v_x}{\partial y} \qquad (1.58)$$

which is a priori a function of the height, time, and initial state, but for the sake of simplicity we will simply write it $\dot{\gamma}(y, t)$. The stress components of the fluid at a level y above the plane are functions of the history of $\dot{\gamma}(y, t)$ and of the state of the material at the initial time. For this flow the stress distribution, and in particular the shear stress at the wall, is known as soon as the fluid depth has been measured. On the contrary, the corresponding shear rate, which depends entirely on the constitutive equation, varies with the position in the fluid and cannot be determined easily from macroscopic measurements. In practice, two quantities can be measured readily: the flow rate (by unit surface) (q) through a cross section and the velocity of the fluid at the free surface $(\mathbf{U} = U\mathbf{x})$, which is also the maximum velocity of the fluid. These quantities can be related to the velocity distribution in the fluid, and thus to the distribution of shear rate, through simple integrations:

$$q(t) = \int_0^h v_x(y, t)\,dy = \int_0^h (h - y)\dot{\gamma}(y, t)\,dy \qquad (1.59)$$

$$U(t) = v_x(h, t) = \int_0^h \dot{\gamma}(y, t)\,dy \qquad (1.60)$$

These relations are valid at any time in the flow for any fluid type as long as our initial assumptions are valid.

Here we consider the specific case of steady-state flows. From equation (1.55) the shear stress at any height above the plane is fixed. If the other stress components play a negligible role in (1.51), the shear rate and shear stress are linked by a relation of the type

$$\dot{\gamma} = \chi_{\Sigma_0}(\tau) \qquad (1.61)$$

simply derived from equation (1.54). We can then use the shear stress, $\tau = \tau_p(h - y)/h$, instead of the depth in the integrals above, so as to obtain (dropping the index Σ_0 for the sake of clarity)

$$q = \frac{h^2}{\tau_p^2} \int_0^{\tau_p} \tau \chi(\tau)\,d\tau \qquad (1.62)$$

$$U = \frac{h}{\tau_p} \int_0^{\tau_p} \chi(\tau)\,d\tau \qquad (1.63)$$

Thus we find that the flow rate and the fluid velocity at the free surface are simple functions of the fluid depth, the plane slope, and an integral that by derivation may yield the shear rate along the plane. These equations may be differentiated relative to the fluid depth or the plane slope such that,

$$\left(\frac{dq}{dh}\right)_i = h\chi(\tau_p) \qquad (1.64)$$

$$\left(\frac{dU}{dh}\right)_i = \chi(\tau_p) \qquad (1.65)$$

With the help of these equations the constitutive equation of the material in the form of shear rates associated with different shear stresses can be deduced from an ensemble of measurements of the flow rate or the surface velocity as a function of the flow depth. In practice, this generally requires variation of the flow rate starting from the material prepared under fixed conditions and, at each level, waiting for the steady state before measuring the corresponding fluid depth.

1.4.2 Flow between Parallel Disks

We consider a material contained and sheared between two parallel, circular plates in relative rotation at a velocity Ω around their common axis, and separated by

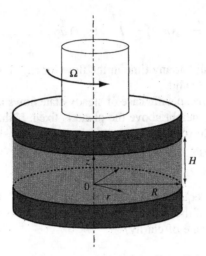

Figure 1.8   Geometric characteristics of parallel-plate flow.

a distance H. We describe the flow in the cylindrical coordinates attached to the lower plate (O, r, θ, z) (see Figure 1.8). We assume that the peripheral interface of the material with the ambient fluid situated at a distance R from the axis is a constant, straight cylinder centered around the disk axis.

Because of the symmetry of the problem by rotation around the central axis, the components of the velocity and the stress tensor a priori do not depend on θ. Moreover, the radial and axial components of the velocity would tend to push the fluid out of its initial volume. If the velocity had a nonzero radial or axial component somewhere, there would be some vortices within the material in order to respect the mass conservation and the constant boundaries. This situation, typical of an instability, will not be considered here. Under these conditions the only nonzero component of velocity is the tangential one, $v_\theta = v_\theta(r, z)$. Now we can state that the boundary conditions are $v_\theta(r, 0) = 0$ (fixed lower plate) and $v_\theta(r, z_0) = \Omega r$. The simplest velocity field satisfying these boundary conditions corresponds to a linear variation of the velocity between 0 and H at any distance r: $v_\theta(r, z) = \Omega r z / H$. This means that the fluid layers parallel to the disk slide over each other by rotating around the central axis. The local shear rate simply corresponds to the spatial rate of variation of the tangential velocity:

$$\dot{\gamma} = \frac{\Omega r}{H} \tag{1.66}$$

It follows that the flow consists in the relative motion of disks of material around their common axis and the history of the relative configuration gradient at a given point is described entirely by the history of the local shear rate.

Now, we can formulate the conservation of torque (see Section 1.2.3) over a fluid portion limited by the upper disk surface and a parallel surface within the fluid (Figure 1.8). We find that the torque applied to the upper disk may be expressed as follows:

$$M = \int_0^R 2\pi r^2 \tau(r) \, dr \tag{1.67}$$

The specificity of this flow is that the shear stress is well known as a function of the rotation velocity but the resulting shear rate is strongly heterogeneous, varying from zero at the axis to the maximum value $\dot{\gamma}_R = \Omega R / H$ at the periphery. It follows that the torque and the rotation velocity provide only global information concerning heterogeneous radial distributions of the shear stress and the shear rate. Apparently a specific technique for interpretation of a series of different torque–rotation velocity values is needed.

Let us consider a transient flow under a given rotation velocity Ω (leading to a constant shear rate distribution) from a given initial state. We can transform (1.67) by a change of variables using (1.53) and (1.66) (again dropping Σ_0):

$$M(t) = \frac{2\pi}{(\Omega/H)^3} \int_0^{\dot{\gamma}_R} \dot{\gamma}^2 \varsigma(\dot{\gamma}, t) \, d\dot{\gamma} \tag{1.68}$$

This equation may be differentiated relative to $\dot{\gamma}_R$:

$$\tau_R = \varsigma(\dot{\gamma}_R, t) = \frac{M(t)}{2\pi R^2} \left[3 + \frac{\dot{\gamma}_R}{M} \frac{\partial M(t)}{\partial \dot{\gamma}_R} \right] \tag{1.69}$$

Thus the relation between the shear stress and the shear rate at the periphery at each time may be obtained through equation (1.69) from a sufficient set of experimental values of the pair $(M(t), \Omega)$. The second term of the sum in brackets in (1.69) is a priori not negligible, since for a Newtonian fluid it is equal to 1 and is equal to n for a power-law fluid, but for yield stress fluids it becomes very small when the maximum stress (at the periphery) is not too far from the yield stress.

A further analysis of the flow equations [1] gives a similar equation for the first normal stress difference:

$$(2\sigma_1 - \sigma_2)\pi R^2 = 2N + \dot{\gamma}_R \frac{dN}{d\dot{\gamma}_R} \tag{1.70}$$

Here, N is the total normal force (including the atmospheric pressure) exerted onto the upper plate and $\sigma_1 = \sigma_{zz} - \sigma_{rr}$ and $\sigma_2 = \sigma_{\theta\theta} - \sigma_{rr}$. Thus the analysis of a series of pairs $(N, \dot{\gamma}_R)$ makes it possible to derive the value of the function $2\sigma_1 - \sigma_2$.

1.4.3 Flow between a Cone and a Plate

We consider the flow between a cone and a plate such that the cone axis is perpendicular to the plate and the cone vertex falls exactly onto the plate surface. We describe the flow in the spherical coordinates attached to the plate (r, θ, φ)

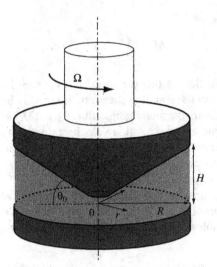

Figure 1.9   Geometric characteristics of cone and plate flow.

(see Figure 1.9). Here the cone is rotating around its axis at a rotation velocity Ω. The angle θ_0 between the cone and the plate is assumed to be small so that the ratio of the fluid thickness to the distance from the axis remains much smaller than 1. The material remains between the cone and the plate and its peripheral free surface is assumed to be constant. Under such conditions, because of the symmetry of the problem, the velocity and the stress tensor do not depend on φ. A simple velocity field with only a lateral component (v_φ) is the simplest situation that may be expected. As for the parallel disks, some radial or tangential motion somewhere within the fluid would indicate the occurrence of some flow instability. Moreover, the flows in all such fluid portions are similar since edge effects are negligible and the boundary conditions are similar; the geometry is similar (with constant cone angle) and both the material thickness and the relative velocity of the solid surfaces are proportional to the distance. It results that the material velocity is simply proportional to the distance r from the cone submit. Finally the only nonzero component of the velocity is v_φ, which may be expressed as: $v_\varphi = r\omega(\theta)$, where ω is an unknown function. The flow associated with such a velocity field consists in the relative motion of conical material layers rotating around the central axis and sliding one over each other. At a given point the history of the relative configuration gradient is thus given entirely by the history of the relative velocity of the local conical layers or equivalently the gradient of the tangential velocity at a fixed distance from the submit, that is, the shear rate:

$$\dot{\gamma} = \frac{1}{r}\frac{dv_\varphi}{d\theta} = \omega'(\theta) \tag{1.71}$$

If the flow is homogeneous, we obtain the relation $\dot{\gamma} = \Omega/\theta_0$ from (1.71).

Let us now apply the torque conservation in a conical portion of fluid with one interface falling along the cone and the other inside the fluid. It follows that if the peripheral free surface is situated at a distance R, the total torque applied by the cone onto the material is expressed as follows:

$$M = \int_0^R 2\pi r^2 (\cos\theta)^2 \tau \, dr = \frac{2}{3}\pi R^3 (\cos\theta)^2 \tau(\dot{\gamma}) \qquad (1.72)$$

Thus for a given torque value the shear stress varies in material thickness, and the maximum relative difference is equal to

$$\frac{\Delta\tau}{\tau} = \frac{\tau(\theta_0) - \tau(0)}{\tau(0)} = (\tan\theta_0)^2$$

For a cone angle smaller than $4°$, this relative difference is slightly smaller than 0.5%, so that the impact of these variations on the determination of the constitutive equation will often be negligible as compared to that of perturbing effects or time changes of material behavior.

In addition, a detailed description of flow characteristics leads [1] to the following equation for the normal stress difference:

$$N = K + \frac{\pi R^2 \sin\theta_0}{2}[\sigma_1 - \sigma_2] \qquad (1.73)$$

in which $\sigma_1 = \sigma_{\theta\theta} - \sigma_{rr}$, $\sigma_2 = \sigma_{\varphi\varphi} - \sigma_{rr}$ and $K = \pi R^2 \sin\theta_0 \sigma_{\varphi\varphi}(R) + \int_0^R 2\pi r \sin\theta_0 p(r) \, dr$. Note that K may be considered as constant for a series of tests within the same sample so that the value of the function $\sigma_1 - \sigma_2$ can be directly found from equation (1.73) by observing the evolution of N with the rotation velocity.

Note that in practice the cone should preferably be truncated in order to avoid any contact between the cone and the plate leading to an additional and probably uncontrolled frictional force. This truncation implies that the flow is not as assumed in the preceding theory at the approach of the central axis. The resulting effect is even smaller as the truncation diameter is small. Nevertheless, in order to avoid particle jamming with suspensions, this truncation must be sufficiently large for the central gap to be several times larger than the diameter of the largest particles.

1.4.4 Flow between Two Coaxial Cylinders

Let us consider the flow of a material contained and sheared between two concentric cylinders. Such a flow is often called a "Couette flow." We describe the fluid motion in the cylindrical coordinates (O, r, θ, z) attached to the outer cylinder (see Figure 1.10). Here the inner cylinder rotates around its axis at a velocity Ω. Assuming negligible edge effects is equivalent to assuming that the length of the cylinders is infinite, and in that case the variables of the flow do not depend

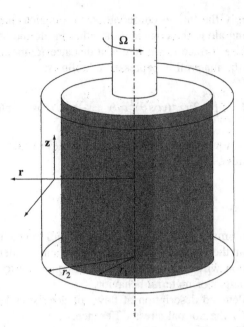

Figure 1.10   Geometric characteristics of concentric coaxial cylinders.

on z. Moreover, the flow is symmetric by rotation around the axis; that is, the variables do not depend on θ. Under these conditions the simplest velocity field corresponds to radial and axial velocity components equal to zero. If the radial or axial component of the velocity deviated from zero somewhere, there would be a net outward, radial or vertical, flow, associated with some flow instability. For a stable flow it follows that only the tangential velocity component is different from zero, and it depends only on r. We will express this in the following form

$$v_\theta = r\omega(r) \tag{1.74}$$

in which ω is the local rotation velocity of the fluid layer situated at a distance r from the central axis. A velocity field as given by equation (1.74) shows that the flow consists in the relative motion of concentric cylindrical layers of material rotating around the central axis. At a given point the history of the relative configuration gradient is given entirely by the history of the relative velocity of the local concentric layers or equivalently the gradient of the relative velocity of these fictive cylinders: $\omega'(r)$. Here, the shear rate corresponds to the local velocity induced by this gradient:

$$\dot{\gamma} = r\omega'(r) \tag{1.75}$$

In the absence of inertia effects, the torque applied by the inner cylinder onto the material may be found by integration of (1.17) over any cylindrical surface of radius r:

$$M = 2\pi h r^2 \tau(r) \tag{1.76}$$

This means that the stress in the gap ($\delta = r_2 - r_1$) separating the cylinders is not uniform; it depends on the distance from the cylinder axis. The amplitude of this heterogeneity depends on the ratio of the gap to one of the cylinder radii since the ratio of the maximum to the minimum stresses is $(r_2/r_1)^2 = (1 + \delta/r_1)^2$.

We can compute the expression for the rotation velocity as a function of the material motion along a radius:

$$\Omega = \int_{r_i}^{r_e} \frac{\dot{\gamma}}{r} \, dr \tag{1.77}$$

This means that there is no straightforward relation between the mean shear rate and the rotation velocity. In the limit of a small gap to radius ratio, the shear stress is almost constant throughout the gap, and we can expect the shear rate to be approximately constant so that we have

$$\Omega \approx \dot{\gamma} \frac{r_2 - r_1}{r_1} \tag{1.78}$$

which provides a simple relation between the rotation velocity and the shear rate in the fluid. In that case the shear stress–shear rate data are directly found from torque–rotation velocity data with the help of equations (1.76) and (1.78). In other cases one needs to interpret a set of torque–rotation velocity data in terms of a relation between the shear stress and the shear rate.

Assume that we applied a constant torque M to a material for which the actual shear rate depends only on the tangential stress component [equation (1.54)]. In the case of Couette flow this is in fact a very reasonable hypothesis with regard to boundary conditions (see Section 4.3.1). Using this relation and (1.76), we can transform (1.77) into

$$\Omega = \int_{\tau_1}^{\tau_2} \frac{\chi_{\Sigma_0}(\tau, t)}{\tau} \, d\tau \tag{1.79}$$

in which $\tau_1 = M/2\pi h r_1^2$ and $\tau_2 = M/2\pi h r_2^2$. Dropping Σ_0 for the sake of clarity and differentiating (1.79) relative to M, we get

$$2M \frac{\partial \Omega(t)}{\partial M} = \chi(\tau_1, t) - \chi(\tau_2, t) \tag{1.80}$$

As a consequence, the shear rate at the wall may be obtained by summing a series of such relations with successive decreasing torques in a ratio $\beta = (r_1/r_2)^2$. Since a zero shear rate is necessarily associated with a zero torque we eventually obtain

$$\dot{\gamma}(\tau_1, t) = \sum_{p=0}^{\infty} \left(2M \frac{\partial \Omega}{\partial M} \right)_{\beta^p \tau_1, t} \tag{1.81}$$

The shear rate can thus be deduced by determining the different terms of this series [the right-hand side (RHS) term of (1.81)], associated with the different values of $\beta^p \tau_i$ by varying either (1) the radii of the cylinders and computing the derivative for the same torque value or (2) the torque and keeping the same geometry. In any case it is necessary to determine the function $\Omega(M, t)$ in a generally wide range in order to accurately determine the slope of this function for specific values of M. Note that for an infinitely wide outer cylinder ($r_2 \to \infty$) we have $\dot{\gamma}(\tau_2 = 0) = 0$, so from equation (1.80) we can derive the simpler shear rate expression:

$$\dot{\gamma}(\tau_1, t) = 2M \left(\frac{\partial \Omega}{\partial M} \right)_t \tag{1.82}$$

1.4.5 Flow in a Cylindrical Conduit (Poiseuille Flow)

We consider the uniform flow of a material in cylindrical coordinates (r, θ, z) such that z is in the direction of the conduit axis (see Figure 1.11). We assume that the flow is uniform, which means that the flow characteristics are identical in each cross section, that is, the velocity field is independent of z. It is also

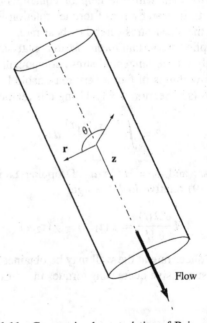

Figure 1.11 Geometric characteristics of Poiseuille flow.

symmetric around the conduit axis, so that all variables are independent of θ. However, some stress components may depend on z because the driving force is the pressure difference between the extremities of the conduit. In the absence of any flow instability only the longitudinal component of the velocity deviates from zero and depends only on r:

$$v_z = f(r) \tag{1.83}$$

A velocity field of this type means that the flow consists in the relative motion of concentric, cylindrical, material layers parallel to the central axis. At a given point the history of the relative configuration gradient is given entirely by the history of the relative velocity of the local cylindrical layers, namely, the shear rate:

$$\dot{\gamma} = f'(r) \tag{1.84}$$

Let us consider a fluid portion limited by two sections perpendicular to the cylinder axis and separated by a distance L, and a cylinder of radius r within the fluid. Along the air–fluid interfaces at the boundaries ($z = 0$ and $z = L$) the stress tensor reduces to an isotropic pressure independent of r. It is thus natural to assume that within the fluid the normal stress σ_{zz} does not depend on r but only on the distance from the boundaries (z). Under these conditions, in the absence of inertia, the momentum equation (1.15) applied to such a fluid portion and in projection along the axial direction yields

$$(\sigma_{zz}(L) - \sigma_{zz}(0))\pi r^2 + 2\pi r L \sigma_{rz}(r) = 0 \tag{1.85}$$

from which we deduce

$$\tau = \sigma_{rz} = \frac{A}{2} r \tag{1.86}$$

with $A = (\sigma_{zz}(0) - \sigma_{zz}(L))/L$. Since τ depends only on r, we can conclude from equation (1.83) that A, which could depend only on z, is in fact constant.

This shear rate distribution is far from being homogeneous in a conduit flow, as the fluid is submitted to a stress history that significantly varies from the wall to the conduit axis. As a consequence, there is no simple relation between the applied stress, via the pressure drop by unit length A, and some specific or mean shear rate in the material. We can nevertheless compute the flow rate (total fluid discharge through a cross section):

$$Q = 2\pi \int_0^R r v_z \, dr = -2\pi \int_0^R \frac{r^2}{2} \frac{dv_z}{dr} dr \tag{1.87}$$

Assuming again a behavior of the type (1.54), we can further analyze the flow under constant pressure drop A. Equation (1.87) becomes

$$Q(t) = \frac{8\pi}{A^3} \int_0^{RA/2} \chi_{\Sigma_0}(\tau, t) \tau^2 \, d\tau \tag{1.88}$$

Differentiating (1.88) relative to A, we get the expression for the shear rate at the wall as a function of the variation of the flow rate with the pressure drop by unit length:

$$\dot{\gamma}_R = \chi_{\Sigma_0}\left(\frac{-RA}{2}, t\right) = \frac{1}{\pi R^3 A^2}\left(\frac{d(QA^3)}{dA}\right)_t \tag{1.89}$$

REFERENCES

1. B. D. Coleman, H. Markowitz, and W. Noll, *Viscometric Flows of Non-Newtonian Fluids*, Springer-Verlag, Berlin, 1966.
2. W. B. Russel, D. A. Saville, and W. R. Schowalter, *Colloidal Dispersions*, Cambridge Univ. Press, Cambridge, UK, 1989.
3. G. K. Batchelor, *An Introduction to Fluid Dynamics*, Cambridge Univ. Press, Cambridge, UK, 1967.
4. R. B. Bird, W. E. Stewart, and E. N. Lightfoot, *Transport Phenomena*, Wiley, New York, 2001.
5. M. Lesieur, *Turbulence in Fluids*, Kluwer Academic Publishers, Dordrecht, 1997.
6. C. Truesdell, *Introduction to Rational Mechanics of Continuous Medium*, Masson, Paris, 1974 (in French).
7. D. Tabor, *Gases, Liquids and Solids*, Cambridge Univ. Press, Cambridge, UK, 1991.
8. A. J. Liu and S. R. Nagel, *Jamming and Rheology*, Taylor & Francis, New York, 2001.

CHAPTER 2

RHEOPHYSICS OF PASTES AND GRANULAR MATERIALS

Industrial or natural pastes, slurries, or granular materials are general composed of a large number of elements of various types such as droplets, bubbles, polymers, or particles of different sizes and shapes and at different volume fractions in a liquid or a gas. One might fear that this variety of characteristics would lead to an infinite number of different rheological behavior types. This is effectively the case from a quantitative perspective; as for any other type of material, the absolute value of the apparent viscosity under given flow conditions depends on the detailed characteristics of the material, such as the physical properties and relative fractions of its components. However, from a qualitative perspective, these systems exhibit a limited number of rheological behavior types. This is due to the qualitative similarities between the interactions that occur among the elements on a local scale for different systems. In particular, all "jammed" systems are composed of objects that interact with their neighbors and form a continuous network of interactions from one fluid boundary to another. The finite force necessary to break this network and induce some flow confers such materials their "jamming" character. Then the typical rheological characteristics of pastes or granular materials, such as elasticity, plasticity, and thixotropy, are imparted from the temporal evolution of the spatial configuration of this network.

In this chapter we review the main trends of the rheological behavior of the systems with components as described above, and as far as possible we consider these behavior types in relation to the interactions between these components

Rheometry of Pastes, Suspensions, and Granular Materials: Applications in Industry and Environment
By Philippe Coussot   Copyright © 2005 John Wiley & Sons, Inc.

and their evolution during flow. First, in order to distinguish the main classes of materials, we analyze the different types of mutual interactions between the basic elements of pastes or granular materials. These interactions appear to have close similarities and, apart from hydrodynamic interactions, may be separated into two main classes—hard and soft interactions—depending on whether the elements are solid particles in contact or not; in this context the soft elements in any case include bubbles, droplets, polymer chains, and colloidal particles. Then we focus on the rheology of *soft jammed materials*—pastes—composed mainly of elements developing soft interactions. These systems are basically yield stress fluids exhibiting a viscoelastic solid regime, a liquid regime, and thixotropy. Afterward we turn to *granular materials*, composed mainly of solid grains in direct interaction. One fundamental characteristic of these systems is that the local rheological behavior depends not only on the local flow history but also on boundary conditions, which as yet makes it difficult to identify an intrinsic material behavior. Finally, we consider intermediate materials—*granular pastes*—composed of elements developing hydrodynamic or soft interactions and a large volume fraction of solid grains such that, depending on flow conditions, either soft (or hydrodynamic) or hard interactions may play a major role in the rheological behavior of the mixture.

2.1   INTERACTIONS BETWEEN MATERIAL ELEMENTS

Here we consider as *elements* of the material the material's different, basic, components that can undergo only finite, reversible deformations during flow. There are obviously a wide variety of elements in industrial or natural pastes or granular materials, but we can distinguish a limited number of basic classes: long molecular chains, bubbles, droplets, colloidal particles, and coarse (noncolloidal) grains, and between these elements, the molecules of the interstitial fluid, either a gas or a liquid. Here we review the interactions that develop between the elements within the same category when immersed in a liquid. Besides these main categories there are also many specific elements such as blood cells, snowflakes, and vegetable fibers, which generally have properties intermediate between those of two of the classes mentioned above. It is beyond the scope of this book to review in detail the interactions between such elements, but we postulate that they belong to one of the two basic classes described in the Introduction.

In order to propose a generic description of the main Interactions without useless details about the specific characteristics of each material, we will use the following, simplified guidelines for characterizing the material structure (Figure 2.1):

- We assume that the material is composed of discrete elements dispersed in a continuous phase, a Newtonian, a liquid or gas of density ρ_0 and viscosity μ_0 (see discussion below), which will be referred as the *interstitial fluid*; the distribution of the elements in the fluid will be referred to as the *configuration*.

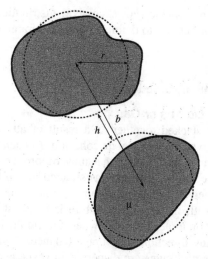

Figure 2.1 Two-dimensional scheme of some physical characteristics of the elements (gray) in the interstitial fluid (white).

- The *volume fraction* of a class of elements in the interstitial fluid ϕ is defined as the ratio of the volume of elements to the total volume of the sample; we can also define the maximum packing fraction ϕ_m as the highest possible volume fraction of elements; the latter parameter is unfortunately not well defined since it depends on the element configuration, which in turn depends on the flow history or equivalently on the procedure for determining it; note that for deformable elements the maximum packing fraction is equal to 1.

- The element size and shape are described by a single parameter, the characteristic element length $d = 2r$, which can be estimated as the diameter of the sphere of volume equal to the average volume $\langle \Omega \rangle$ of the elements [i.e., $d = (6\langle \Omega \rangle / \pi)^{1/3}$]; under these conditions the volume fraction of elements is simply related to d via $\phi = N\pi d^3 / 6$, where N is the number of elements per unit volume.

- The characteristic separation distance between the center of two elements b may be estimated by assuming that if the characteristic length of the elements were equal to b, the maximum packing fraction would be reached ($\phi_m = N\pi b^3/6$), which implies $b = d(\phi_m/\phi)^{1/3}$. Note that if the average shear rate within the material is $\dot{\gamma}$, the characteristic, relative velocity between two neighboring elements is $b\dot{\gamma}$.

- The typical separation distance between the surface of two neighbouring elements h is simply defined as

$$h = b - d = 2r \lfloor (\phi_m/\phi)^{1/3} - 1 \rfloor \tag{2.1}$$

These guidelines are in fact quite relevant only for elements with an aspect ratio, the ratio of the maximum to the minimum lengths of the elements, of the order of 1 and for a reasonable (not excessive) distribution of element dimensions. Otherwise it is necessary to define additional parameters for characterizing the material.

2.1.1 Hydrodynamic Interactions

Interactions within Liquids or Gas

Liquid molecules are stacked together as a result of attraction due to van der Waals forces, which gives the material its cohesion. At each instant the molecules form an apparently jammed structure that may be broken only if some energy is supplied; each molecule may be seen as situated in a potential well resulting from the forces exerted by all its neighbours, and some energy is needed to extract it from this well. In fact this structure is constantly broken because of the thermal agitation of molecules, and when the liquid is macroscopically at rest, the motions in any direction on average balance. A similar result is found with a gas even if gas molecules are dispersed in space and move, more or less freely, over distances much greater than their lengths before colliding with other molecules. Momentum transfers via these collisions are the source of the gas viscosity when it is sheared at a sufficiently low rate.

Now, when a force, as weak as it may be, is applied to a liquid or a gas, this balance of motions is broken and locally the molecules on average move relative to each other in some specific direction, at a rate approximately proportional to the stress level. For a simple liquid or a simple gas, that is, with relatively low-molecular-weight molecules, the resulting rheological behavior is generally Newtonian; thus the shear stress in simple shear is simply proportional to the shear rate:

$$\tau = \mu_0 \dot{\gamma} \qquad (2.2)$$

The corresponding usual tensorial expression of this constitutive equation is

$$\mathbf{\Sigma} = -p\mathbf{I} + 2\mu_0 \mathbf{D} \qquad (2.3)$$

The fluids as defined by (2.3) have three ideal characteristics, which render them as simple fluids (see Section 1.3.3) but cause them to strongly differ from any of the material types described below:

- Their apparent viscosity is constant ($\eta = \tau/\dot{\gamma} = \mu_0$).
- They flow as soon as they are submitted to a nonisotropic stress of any value.
- They immediately "forget" their previous history, since only the actual rate of deformation plays a role in the stress field.

In practice obviously no fluid exactly obeys these principles under any flow conditions. For example, during the first instants of a rapid solicitation, simple liquids may exhibit some elasticity, or under extremely low rates of flow the thermal agitation of molecules can induce significant, additional, viscous dissipation. Nevertheless, these deviations from the ideal behavior occur under extreme flow regimes that seldom play a role in usual paste or granular flows. Within the frame of our description of main interactions between elements, we will assume that the rheological behavior of the interstitial liquid or gas is simply given by an equation of the type (2.3).

In fact, the viscosity of gas is generally so small, usually much smaller than that of a liquid, that gas flows are often turbulent (see Section 1.2.2). In the following text, for the sake of simplicity, we will neglect this aspect, which could have an impact only in the case of rapid dry granular flows. However, most of the complex behavior of materials with a gas as the interstitial fluid comes from the absence of cohesion between the elements; the overall volume fraction of elements may significantly vary because the ambient fluid (beyond the boundaries of the material) is generally the same gas. This contrasts with the case of materials with a liquid as the interstitial fluid, where the liquid forms a continuous network throughout the material and the molecules within it are sufficiently attracted to each other and thus maintain all the elements in an almost constant, overall volume as long as the material does not fracture. The average volume fraction of elements in the paste sample thus remains constant.

Hydrodynamic Interactions between Elements

In general the liquid or the gas within pasty or granular materials constitutes an interstitial fluid in which elements that are much larger than the liquid molecules are immersed. This is, for example, the case for emulsions, foams, suspensions, powders, or polymer gels. Thanks to the small ratio of liquid or gas molecule to element dimensions, the interstitial fluid may be considered as a continuum medium at the scale of the elements. The relative motion of two elements implies some flow of the interstitial fluid and thus requires some force. This is called *hydrodynamic interaction* between the elements.

Let us first consider the displacement of a single element through a fluid macroscopically at rest. This element undergoes a resistant hydrodynamic force due to the flow of the surrounding fluid that it induces. For a Newtonian fluid the corresponding drag force (F_D) depends on the shape of the element and is proportional to its characteristic length, to the fluid viscosity, and to the relative velocity between the element and the fluid (see Section 5.1.3). When a rigid element is embedded in a fluid that is macroscopically in simple shear flow, an analogous effect occurs—the presence of the element induces a perturbation to the flow characteristics that the fluid would experience in the absence of the element. This finally induces additional energy dissipations [see equation (1.19)] so that the apparent viscosity of the fluid plus the elements is greater than that of the fluid alone. Thus a *suspension*, that is, a mixture of liquid and rigid elements, will have an apparent viscosity that increases with the volume fraction of elements.

This effect may be easily understood by noting that the rigid elements take the place of some fluid volumes that can no longer flow. Let us consider a

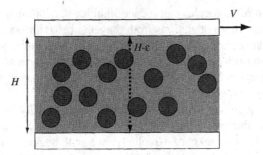

Figure 2.2 Schematic representation of a rough theoretical estimation of the apparent viscosity of a suspension (see text); the effective, average, sheared thickness of the fluid is smaller than the separating distance between the tools.

simple shear between two parallel plates with a gap H and without slip along the solid–fluid interfaces. Roughly speaking, the presence of the elements has an impact equivalent to reducing the effective thickness of the sheared fluid to $H - \varepsilon$ (see Figure 2.2), in which ε depends on the amount of elements in the fluid. As a consequence, under a given apparent shear rate ($\dot{\gamma}_{app}$), associated with a relative velocity of the plates V (i.e., $\dot{\gamma}_{app} = V/H$), the average local (effective) shear rate ($\dot{\gamma}_{eff}$) is greater: $\dot{\gamma}_{eff} = V/(H - \varepsilon)$. The resulting shear stress applied along the plates is approximately $\mu_0 \dot{\gamma}_{eff} = \mu_0 V/(H - \varepsilon)$ so that the apparent viscosity of the suspension, equal to the ratio of the stress to the apparent shear rate, is equal to $\mu_0/(1 - \varepsilon/H)$. The ratio ε/H is an increasing function of the volume fraction of elements in the fluid, and more precisely of the ratio of the solid fraction to the maximum packing fraction. Thus our (rough) estimation of the suspension viscosity shows that it should increase with the volume fraction of elements (ϕ), and tend to infinity when $\varepsilon \to H$, which in fact means that the solid fraction tends toward the maximum packing fraction: $\phi \to \phi_m$. Any empirical or theoretical approach of the viscosity of suspension predicts these two basic qualitative effects.

Suspension Viscosity
The exact determination of suspension viscosity requires a more detailed description of the flow field, which is virtually impossible to describe in terms of localized simple shear as above. In fact, the local strain field is affected by the presence of all the elements in a complex manner, and only in limiting cases has it been possible to find analytical expressions. Before looking at these results it is worth noting two important, general, theoretical properties of suspensions that result from a statistical approach of the problem [1]:

1. As long as the configuration of elements in the liquid remains isotropic and constant, the suspension remains Newtonian.
2. In that case the suspension viscosity is proportional to the viscosity of the interstitial fluid.

For low values of ϕ, say, smaller than 4%, the elements are far from each other and theoretical arguments assuming negligible mutual interactions lead [2] to the following expression for the (Newtonian) viscosity (μ) of a suspension of spherical elements of viscosity $\overline{\mu}$ in a Newtonian liquid of viscosity μ_0:

$$\frac{\mu}{\mu_0} = 1 + \phi \left(\frac{\mu_0 + 2.5\overline{\mu}}{\mu_0 + \overline{\mu}} \right) \tag{2.4}$$

In this approach it is assumed that at any given time the inclusions remain spherical and there is a continuity of the velocity along the interface between the elements and the liquid. When the elements are rigid, we have $\mu_0/\overline{\mu} \to 0$ so that the coefficient in factor of ϕ tends toward 2.5.

The expression (2.4) also provides the viscosity of a suspension of droplets or bubbles in a liquid as long as they remain spherical. For example, for bubbles ($\overline{\mu} \approx 0$), we get $\mu/\mu_0 = 1 + \phi$. Here the suspension viscosity increases with ϕ, whereas the apparent "average viscosity" ($\phi\overline{\mu} + (1 - \phi)\mu_0$) of the medium decreases; this observation is derived from the assumption of constant shape of the inclusions. The simple presence of undeformable objects induces some perturbation of the flow field, which, leading to additional viscous energy dissipation, implies that the viscosity of the suspension exceeds that of the interstitial liquid. The more gradual increase in viscosity with ϕ than with solid inclusions results from the no-slip condition along the interface between the inclusion and the liquid; the viscous dissipation in the liquid is larger for solid inclusions, in which the velocity along the interface results from a rigid-body motion, than for fluid inclusions, in which the velocity along the interface results from the coupling between the flow inside and outside the inclusion. On the contrary, when bubbles or droplets can be freely deformed as the suspension flows, the suspension viscosity decreases with ϕ. The latter regime is obtained when the ratio of (1) the characteristic stress due to interfacial tension (γ_{LG}/r) (see Section 3.4) to (2) the characteristic viscous stress ($\mu_0\dot{\gamma}$) (i.e., $\gamma_{LG}/\mu_0\dot{\gamma}r$) is much smaller than 1, while for $\gamma_{LG}/\mu_0\dot{\gamma}r$, inclusions are not expected to deform during flow. For sufficiently large values of ϕ, direct interactions (see Section 2.1.3) between such inclusions may play a significant role in the suspension viscosity (see Section 2.2).

For larger values of ϕ there is no analytical solution to the problem, and various empirical expressions have been proposed [3], which all predict that the viscosity will tend toward infinity when $\phi \to \phi_m$, in which ϕ_m is the maximum packing fraction of the elements. In most of these expressions the ratio of the suspension to the interstitial fluid viscosity is expressed as a complex function of the ratio ϕ/ϕ_m only. For spheres of uniform size and moderate volume fractions, and as long as the particle configuration remains isotropic and constant, this approach seems reasonable since ϕ/ϕ_m is directly related via equation (2.1) to a global geometric characteristic of the particle configuration, specifically, some average ratio of the separating distance to particle size. For elements of aspect ratio (A) significantly different from 1 the suspension viscosity also depends on A [4,5]. Moreover, these expressions should be corrected to account for possible nonuniform grain size distribution. However, as a first approximation, this effect may be included without correction by simply considering that it occurs as if it basically affected the value of ϕ_m. This appears consistent with the general observation that for ϕ not too close to ϕ_m a suspension will have a greater viscosity than will another suspension of similar volume fraction ϕ but with a wider grain size distribution, because ϕ_m increases as the grain size distribution spreads.

Within this frame, the different expressions proposed in the literature and attempting to predict the viscosity of suspensions over a wide range of volume fractions including values approaching ϕ_m must be seen as only rough approximations of reality. Here we retain two typical expressions of this type that provide estimations of suspension viscosity in reasonable, apparent agreement with various experimental data. The form of the first one [6] was derived from theoretical arguments related to homogenization principles (which describe the properties of heterogeneous materials):

$$\mu = \mu_0 (1 - \phi/\phi_m)^{-2.5\phi_m} \tag{2.5}$$

Here the power has been chosen so as to predict the expected trends for $\phi \to 0$ and $\phi \to \phi_m$. The second equation [7] is essentially empirical, it has been fitted to a set of data obtained for different volume fractions and grain size distributions:

$$\mu = \mu_0 \left(1 + \frac{3}{4} \frac{\phi}{(\phi_m - \phi)} \right)^2 \tag{2.6}$$

In fact, there remain various problems with these approaches:

- As the volume fraction of elements increases, the hypothesis of constant isotropic distribution is more likely to fail, due to collective orientation or migration effects (see Section 3.6), which at least implies that the suspension behavior might deviate from the Newtonian model when $\phi \to \phi_m$.
- When the volume fraction of elements is sufficiently large, the separating distance is very small, and thus the elements may develop mutual interactions with specific characteristics (see next paragraph) not taken into account in the approaches described above.
- As already mentioned, the maximum packing fraction ϕ_m is not a single, intrinsic parameter of the material; rather, it strongly depends on the way the packing has been obtained. For example, assume that the maximum packing fraction of a granular material corresponds to the volume fraction of a certain amount of this material lying on a solid surface under the action of its own weight. The volume fraction of uniform solid spheres progressively poured in a dry vessel without any further action may be as low as 55%, whereas it can reach 64% if they are mixed and well vibrated [8]; this shows that the apparent maximum packing fraction significantly depends on the way it was obtained, which obviously constitutes a strong limitation of the formulas, especially for large solid fractions for which the suspension viscosity is very

sensitive to the exact value of ϕ_m (either a finite or an infinite viscosity may be found depending on the value of ϕ_m used).

2.1.2 Colloidal Interactions

Colloidal interactions (or equivalently forces) appear between small elements, typically with a characteristic length between 1 nm and 1 μm, immersed in a liquid. The basic property of these forces is that they act at distance between the elements through the liquid even at rest. A detailed description of the different force types and their characteristics may be found in the book by Russel et al. [9]. Here we summarize some basic aspects of importance in view of our simple overview and analysis of the rheological behavior of pastes and slurries. In the following text we will consider these forces ($\mathbf{F}_i$) via the potential energy of interaction (Φ_i) to which they are associated through $\mathbf{F}_i = -\nabla\Phi_i$.

At very short distances the elements repel each other because of the impossibility of interpenetration of the electronic clouds; this is the *Born repulsion*, which gives rise to Φ_b tending to infinity when $h \to 0$. At greater distances the elements interact via *van der Waals forces* (see Section 2.1.1), which also act between elements much larger than atoms or molecules because of thermodynamic fluctuations of the electromagnetic field inside and around the elements. Thus there is a mutual, in general attractive, force acting between two very close elements even if they are not in contact. The corresponding potential energy (Φ_a) roughly decreases with the inverse of the square distance between elements:

$$\Phi_a \propto -\frac{r^2}{h^2} \tag{2.7}$$

There are also *electrostatic forces* that result from the presence of ions adsorbed at the surface of the elements. In a liquid these ions tend to slightly diffuse, which leads to the formation of what we refer to as a *double layer* because successive, positive and negative ions layers are believed to form around the element. As a first approximation, the corresponding potential energy (Φ_e) decreases exponentially with the distance between the elements

$$\Phi_e \propto \exp(-\kappa h) \tag{2.8}$$

in which κ^{-1} is the Debye length, which in particular increases with the ion concentration in the liquid. Because of the particular form of this potential energy, the distance κ^{-1} has a physical meaning; specifically, the force is finite and approximately constant up to this distance and rapidly decreases to zero beyond it.

The coupling of these different actions leads to a net force between two elements that in particular depends on their separating distance. In general the total potential energy has a large maximum in the limit of small h and tends rapidly towards zero at long distances. In some cases the potential energy has one or several minima between these two limits, which provides preferential distances at which the particles tend to remain. When the particles fall in a deep minimum at short separation distances we speak of *aggregation*, a phenomenon that is often irreversible under usual flow conditions [9]. Within a colloidal suspension the interaction potential between two close particles can seldom be predicted from the preceding considerations concerning two particles alone in the liquid; it at least depends on the orientations and separating distances of all the neighbors of these particles. It remains that, on average, the above mentioned qualitative results remain valid; the potential energy between two neighboring particles mainly increases, possibly with some minima, as their separating distance decreases, and the particles aggregate if they fall in a deep minimum at short distance. If this does not occur, for example, as a result of appropriate flow conditions or absence of a deep first minimum [10], we say that the colloidal suspension is *stable*. This also assumes that negligible sedimentation occurs in the suspension.

Polymers are added in many industrial pasty materials. A polymer molecule is generally intermediate in size between a liquid molecule and a colloidal particle and is a deformable object. Polymer molecules may be adsorbed or anchored on the particle surface and tend to form a thin layer of long molecular chains surrounding the particle. The general resulting effect is a repulsive force between the elements at a short distance because the chains cannot interpenetrate. Thus the addition of polymers in the suspension adds a repulsive potential energy. Some attraction may also occur at certain distance. The presence of nonabsorbed polymer chains dissolved in the liquid may lead to attractive forces because of some osmotic pressure effect between two neighboring elements [11]. However it may be considered that the adsorbed polymer molecules will generally tend to ' foster the stability of colloidal suspensions.

The material may also be essentially composed of long polymer chains in suspension in a liquid. These chains interact by steric effects; since they tend to occupy the largest volume, approaching two such chains will result in storage of more potential energy and thus will necessitate the application of some force. The resulting polymeric suspension may be seen as consisting of soft objects (polymer chain + surrounding liquid) interacting via repulsive forces, with some similarity to concentrated emulsions or foams.

Because of the thermal agitation of the surrounding liquid, in the absence of other forces, colloidal particles can undergo random (Brownian) motions, causing them to diffuse through the liquid. The corresponding diffusion coefficient for spheres in a Newtonian liquid is $D = kT/6\pi\mu_0 R$, in which k is the Boltzmann constant, which means that the mean square displacement of the particle due to Brownian motion in a time t is $2Dt$. When they are also submitted to significant colloidal interactions as described above, the Brownian diffusion can be limited by the potential barrier that the particles have to overcome because of their interactions with the surrounding elements. Such effects have as yet not been described in detail, but we can expect that the Brownian agitation at least enhances restructuration effects at the origin of thixotropy, since they allow particles to explore different relative positions with different potential energies.

To summarize, colloidal interactions are complex, but as soon as the stability of the system is ensured by any means (electrostatic repulsion, adsorbed polymers, etc.), the particles tend to remain at some distance from each other. Only if some force is exerted on them can they approach each other, which means that the distribution of forces is on average equivalent to a set of repulsive potential energies between neighboring particles.

2.1.3 Interactions between Bubbles or Droplets

Bubbles or droplets are inclusions of a gas or a liquid in a liquid. Here we consider the interactions between bubbles or droplets of sufficiently large size for colloidal effects to be negligible, but we must keep in mind that in practice such a distinction may not be so easy. When such large objects are submitted only to a uniform pressure, they have a spherical shape, which minimizes their surface energy. The deformation of a sphere with its volume held constant leads to an increase in the surface of the gas–liquid or the liquid–liquid interface, which increases the surface energy (see Section 3.4). This implies that a force is necessary to deform such an object (a bubble or a droplet), a force that increases with amplitude of the deformation. This process is reversible; when the force is released, the object returns to its initial spherical shape. Thus bubbles or droplets are basically elastic objects. This approach obviously neglects the flow of the liquid inside a droplet, which seems realistic as a first approximation. However, the situation is more complex when the concentration of these objects in the liquid is high. In that case they are separated by thin liquid layers covered by surfactant molecules positioned along the interfaces with the droplets or bubbles, and the flow of these interstitial liquid films may play a significant role during deformation of the material. The resulting, detailed, physical effects are complex [12,13], but ultimately, roughly speaking, it can be noted that an approach of the centers of two such objects initially close to each other requires application of a force that progressively increases as the distance between the centers decreases. According to our assumption concerning the maximum packing fraction (equal to 1) of deformable objects, equation (2.1) defines a theoretical separation distance between such objects that progressively decreases to zero as the elements deform further while remaining in contact. Note that this deformation is limited by the presence of the other elements. The potential energy of interaction between such deformable elements thus increases continuously as their separation distance decreases.

2.1.4 Interactions between Two Solid Particles

Lubricated Contacts

When they remain at some reasonable distance, say, of the order of r, noncolloidal solid particles interact only "hydrodynamically"; that is, as described in Section 2.1.1, their presence basically induces an increase of the viscosity of the fluid, assumed to remain Newtonian. The scenario significantly changes when the particles are able to approach each other at a short distance, specifically, at a distance h much less than r and of the order of the particle roughness ε. There are thus two limiting cases: (1) if h is equal to or smaller than ε, the particles are generally in direct contact; and (2) if h is larger than ε, there is still some interstitial fluid between any pair of points of the two particles. In case 2 the particles still interact hydrodynamically, but the force required to induce some relative motion is much larger than when h is of the order of the particle size (see Figure 2.3a). Considering the small ratio of the fluid thickness to the characteristic length of the solid surface, the flow in the gap may, under some circumstances, be described within the frame of the "lubrication assumption"—the longitudinal components (in the direction of the solid surface) of the velocity are much larger than the velocity component in the normal direction, and the dominant variations of the longitudinal components of the velocity are those in the normal direction. We thus will speak of a *lubricated contact*.

For example, let us represent the particle surfaces as two parallel planes. The viscous force ($\mu_0\dot{\gamma}$) for a given relative velocity will obviously diverge when $h \to 0$ since the shear rate ($\dot{\gamma} = V/h$) tends to infinity. In reality, because of the complex shape of the particle surface on the scale of its roughness, the viscous force for such a tangential, relative motion does not fully diverge before frictional processes occur (see discussion below), because there generally remains a continuous (thin) flow path for the fluid between the two particles even when they are in contact at some point. However, within a concentrated suspension under shear flow, two particles in adjacent layers also tend to approach and move away from each other; so that, h decreases and then increases. During the first stage

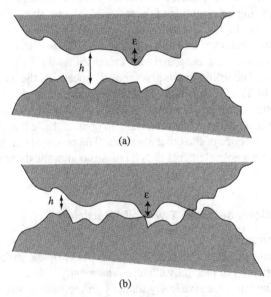

Figure 2.3 Schematic representation of two different types of direct interaction between solid particles: (a) lubricated contact; (b) direct contact.

of this process there is a normal, repulsive force between the particles, which diverges as $1/h$ when $h \to 0$. These few elements already show that complex additional hydrodynamic effects may occur within concentrated suspensions at the approach of ϕ_m, which may in particular induce non-Newtonian phenomena such as nonzero normal stress differences. However, despite some attempts [14–17] to at least theoretically approach the resulting rheological behavior in the specific regime for which lubricated contacts dominate, current knowledge of the effective behavior in that case remains poor. Here we will simply neglect these specific effects induced by lubricated contacts and consider that when lubricated contacts predominate, the suspension remains Newtonian with a viscosity tending to infinity as $\phi \to \phi_m$.

Direct Contacts: Collision and Friction

Here we consider the case of direct contact between the particles, where some elements of the solid surfaces of the two particles touch each other (see Figure 2.3b). Here we assume that the role of the interstitial fluid is negligible. In that case we expect new types of interactions associated with the transmission of energy through the solid structures of the particles. Although their geometric characteristics on a macroscopic scale are simple, these interactions are complex; on a local scale, they may at the same time involve elastic, plastic, or rupture effects, the importance of which will depend on various parameters such as the particle velocities, rotation velocities, shapes, and material characteristics. In fact, it is possible to distinguish important classes of direct interactions (viz., *collision* and *friction*) by considering two specific cases: purely normal or purely tangential (with sustained contact) relative motion between the particles.

Let us consider for the sake of simplicity (qualitatively similar results are obtained with other geometries) two cubic, perfectly elastic particles (of side a) coming into contact along one of their faces as the result of a normal, relative motion. After the first instant of contact the particles deform as a result of the transformation of their kinetic energy in elastic energy. In the limit of small deformations ($\delta = \Delta a/a$), the resulting force between the particles is denoted as $E\delta$, where E is the elastic modulus of the particle. In a frame of reference linked to the surface of contact between the particles, the velocity of one particle is $V = a\dot{\delta}$, and its kinetic energy (i.e., $mV^2/2$), is transformed into elastic energy (i.e., $E\delta^2 a^3/2$), associated with its deformation. The maximum deformation (δ_m) is reached when the initial kinetic energy is entirely transformed in elastic energy, that is, when the particle velocity becomes zero. As a consequence, from the conservation of energy, we get

$$m\dot{\delta}^2 = Ea(\delta_m^2 - \delta^2) \tag{2.9}$$

which shows that the particle velocity progressively decreases up to zero and then increases again. At the end of the contact, when δ again reaches zero, its value V_f is equal in amplitude (but in the opposite direction) to the initial velocity V_i:

$$V_f = -V_i \tag{2.10}$$

Equation (2.9) can also be used to compute the contact duration (i.e., $t_c = 2 \int_0^{\delta_m} d\delta/\dot{\delta}$), and we deduce that t_c is proportional to $\sqrt{m/Ea}$. More complex results may be obtained for other particle shapes, but they also show that the contact duration will be larger for softer particles (with low values of the elastic modulus E). In practice, such a process may also involve some irreversible, or plastic, deformations of the particles, which implies some energy loss during this process. As a result, in that case, the final relative velocity is smaller than the initial one

$$V_f = -eV_i \qquad (2.11)$$

where e is the coefficient of restitution ($e < 1$), which in fact may slightly vary with the velocity. Other effects such as rotation or inertia could also be taken into account, but this is beyond the scope of this book.

Let us now consider two particles in contact and in relative tangential motion. In that case some force appears to be required to maintain the relative motion. This seems to result [18–20] from elastoplastic deformations of a very limited number of regions in contact at the top of the particle surfaces (see Figure 2.3b). Despite the complexity of the physical processes on the local scale, it appears that a simple law can describe the macroscopic properties in a number of situations. This is the Coulomb model, in which the frictional (tangential) force (T), required to move two particles in direct contact tangentially to each other, is simply proportional to the normal force (N) maintaining them in contact:

$$T = fN \qquad (2.12)$$

On the contrary, the relative velocity is equal to zero when $T \leq fN$. In practice, it is often difficult to measure the friction coefficient, in particular under a given relative velocity, because stick–slip may occur; although it may remain constant on average, the force needed to maintain the motion at a low, constant rate strongly fluctuates with time. Moreover, it appears that the force necessary to maintain the motion is slightly less than the force required to induce the motion.

At first sight these two processes (normal and tangential motion) respectively correspond to a collision and to a friction. However, in practice direct contacts generally involve relative motion with both normal and tangential components, so they likely involve both collisional and frictional processes. A basic, qualitative distinction is possible by comparing the effective duration of contact (T_c) to that expected for a pure collision (t_c). If T_c is of the order of t_c, mainly collisional processes as described above probably have time to take place; thus, we can speak of a *collision*. If, on the contrary, T_c is much larger than t_c, mainly frictional processes as described above take place, and we can speak of *friction*.

Let us now consider the potential energy of interaction between two solid (noncolloidal) particles at rest in a fluid. This energy is equal to zero as the separating distance is greater than the particle roughness, or more precisely, as the particles are not in contact; the particles do not interact. Below a small, critical value of h, say, ε, the particles are in contact and may be slightly deformed elastically or plastically as the relative normal force exerted onto the particles increases. As the separating distance decreases below ε, the potential energy of the particles rapidly increases as a result of their elastic deformation, and tends toward infinity. Since the solid particles are in general weakly deformable, they can store a high energy under small deformations and as a first approximation, all proceeds as if their potential energy of interaction could indefinitely increase while their separation distance remains equal to zero.

2.1.5 Classification of Forces

Materials in which hydrodynamic interactions prevail can be considered as Newtonian fluids, the viscosity of which in general increases with the volume fraction of elements. However, hydrodynamic interactions develop only if there is some flow. This contrasts with the other possible interactions between elements that can develop within the interstitial fluid in the material at rest. Considering the potential energy as a function of the separation distance between neighboring elements as described above, two classes of interactions (apart from hydrodynamic interactions) may be distinguished (see Figure 2.4):

1. *Soft Interactions.* Interactions between colloidal particles, bubbles, droplets, or long molecules in a liquid, for example, such that, on average (owing to the stability of the system for colloidal suspensions), the potential energy continuously varies (mainly increases) as the separation distance between the centers of the elements decreases.

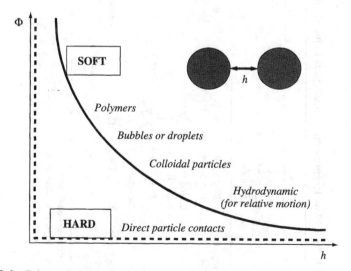

Figure 2.4 Schematic aspects of the potential energy of interaction between two types of elements as a function of their separation distance.

TABLE 2.1 Flow Regimes and Rheological Behavior of Pastes and Granular Materials

	Paste[a]	Granular Material[b]	Granular Paste[c]
	Polymer gels, Colloids, Concentrated foams, Emulsions	Sand, Powders, Seeds	Saturated soils, Highly concentrated suspensions
Pasty regime: $\Gamma < 1$	Solid regime (from rest): $\gamma < \gamma_c \Rightarrow \tau = g(\gamma) + h(\dot{\gamma})$ Liquid regime (steady state): $\sqrt{-T_{II}} > \tau_c \Rightarrow$ $$\Sigma = -p\mathbf{I} + \tau_c \frac{\mathbf{D}}{\sqrt{-D_{II}}} + F(D_{II})$$ Thixotropy	Frictional regime: $\dot{\gamma} \neq 0 \Rightarrow \tau = \sigma \tan\varphi$ $Ba \ll 1$	Newtonian interstitial liquid: $\Sigma = -p\mathbf{I} + \Sigma_g$ Pasty interstitial fluid: $\dot{\gamma} \neq 0 \Rightarrow$ $\tau = \tau_c(\phi) + \sigma \tan\varphi$ Frictional regime: $Le \gg 1$
Hydrodynamic regime: $\Gamma \gg 1$	$\Sigma = -p\mathbf{I} + 2\mu(\phi)\mathbf{D}$	Collisional regime: $\tau \propto \dot{\gamma}^2$ $Ba \gg 1$	Lubricational regime: $Le \ll 1$ $\Sigma = \Sigma_f$

[a] Characterized by soft interactions.
[b] Characterized by hard interactions.
[c] Characterized by soft (or hydrodynamic) or hard interactions.

2. *Hard Interactions.* Interactions between noncolloidal solid particles, which are strongly nonlinear. The potential energy abruptly increases from zero when the particles come into contact.

We can expect that all materials in which one of these interaction types prevail will have similar qualitative, rheological properties since they appear to be composed of an ensemble of elements interacting in a similar way. Thus we can distinguish two very different types of materials: the *soft jammed systems* (*pastes*), composed mainly of elements between which soft interactions predominate, and *granular materials*, in which hard interactions prevail. A third, intermediate class, *granular pastes* (or *slurries*), may be distinguished, in which either soft (or hydrodynamic) or hard interactions may play a major role depending on flow conditions. In the following section we discuss the behavior of these different classes of materials. The results are summarized in Table 2.1.

2.2 RHEOLOGY OF SOFT JAMMED SYSTEMS (PASTES)

Here we consider *soft jammed systems*, in which soft interactions predominate. We are interested in systems with at least one flow regime in which these interactions significantly affect the rheological behavior of the system. This means that there exists a continuous network of soft interactions of significant intensity throughout the fluid. This in particular implies that the fraction of and/or the mutual interaction energy between the elements in the liquid is sufficiently high. On the contrary, when the fraction of elements and their mutual potential energy are too low, the material behavior is governed by hydrodynamic interactions and as a first approximation can be considered as Newtonian, with an apparent viscosity exceeding that of the interstitial liquid according to formulas given, for example, in Section 2.1.1. Thus, as the fraction of elements increases, the material should continuously evolve from a simple, Newtonian liquid (at least for elements with a very low volume fraction) to a soft jammed system.

The potential energy of mutual interactions depends primarily on the spatial distribution of elements and thus, as a first approximation, should not vary much with shear rate, since, except in some specific cases (e.g., ordering), a flow does not significantly affect the average properties of the element configuration and in particular the distribution of separating distances. On the contrary, viscous dissipations originating from hydrodynamic interactions significantly increase with the rate of flow. This implies that the relative importance of soft interactions and hydrodynamic dissipations depends not only on the material characteristics (particularly the volume fraction of elements) but also on the flow regime. Roughly speaking, this may be appreciated from the ratio of the typical shear stress $\mu \dot{\gamma}$ within an equivalent suspension of analogous elements interacting only hydrodynamically in simple shear at a rate $\dot{\gamma}$, to the yield stress of the soft jammed system (i.e., τ_c). In this context the viscosity μ is a function of μ_0 and ϕ as described in Section 2.1.1. The yield stress τ_c reflects the critical force required for motion in the limit of low shear rates, when viscous dissipations are negligible and only soft interactions occur, and, as we will see in Section 2.2.2, can be related to the average potential energy of interaction. We have:

$$\Gamma = \frac{\mu\dot{\gamma}}{\tau_c} \qquad (2.13)$$

For small values of Γ, with large soft interactions or low shear rate, soft interactions predominate, while for larger values of Γ, with low soft interactions (e.g., dilute colloidal suspensions) or high shear rate, hydrodynamic dissipations prevail [21]. In the following text we will focus mainly on the rheological behavior of soft jammed systems in the limit of low Γ values (i.e., the *pasty regime*), for which they exhibit their specific properties, while in the limit of large Γ values (i.e., the *hydrodynamic regime*), they approximately behave as the equivalent suspension of noninteracting elements, or, in other words, are approximately Newtonian. Note that since $\tau/\tau_c \rightarrow 1$ when $\Gamma \rightarrow 0$ (see Section 2.2.3) and $\tau/\tau_c \approx \Gamma$ when $\Gamma \gg 1$, we expect that τ/τ_c is a function of Γ only. This provides a master flow curve independent of volume fraction, from which the behavior of a similar material under any volume fraction may be deduced. Existing data for different types of materials seem to confirm the validity of this approach [21–23].

Each element within soft jammed systems interacts with several other elements surrounding it. It thus undergoes the action of different (on average repulsive) forces that, collectively—assuming that they remain constant—determine a potential well for the element (see Figure 2.5). When the material is macroscopically at rest, each element should thus tend to progressively reach its position of minimum energy between its neighbors. However, since the other elements are in an analogous situation, this position always evolves, and the whole set of elements constantly tends to evolve toward a configuration of equilibrium. When the fluid is macroscopically at rest, the material evolves autonomously toward equilibrium. On the contrary, when the fluid flows, the spatial distribution of

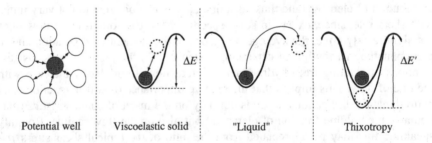

| Potential well | Viscoelastic solid | "Liquid" | Thixotropy |

Figure 2.5 Potential well in a soft jammed system and correlation between different motions within it and macroscopic rheological characteristics: solid regime when the element does not overcome the potential barrier ΔE, liquid regime when it overcomes this barrier, and thixotropy when the depth ($\Delta E'$) of the potential well varies with time.

elements in space may be significantly altered at a generally high rate. In that case the system remains far from equilibrium. The common, basic, rheological trends of pasty materials derive from these qualitative aspects of their structure. In the following text we review these effects according to this simple scheme and leave aside various other effects such as antithixotropy or shear thickening [24,25] effects, which have sometimes been observed with pasty or granular materials but, in the frame of actual knowledge, cannot easily be accounted for in the simple modeling approach or explained by natural physical considerations.

2.2.1 Solid Regime: Viscoelasticity

Let us consider the set of elements in equilibrium and assume that we slightly displace one element relative to its neighbors. As it moves slightly from its position of minimum potential energy, the element stores some potential energy so that its displacement requires some force, but as soon as the force is released, the element rapidly returns to its initial position of minimum energy (see Figure 2.5). For a very slight motion the mathematical problem can be linearized and the corresponding force is proportional to the displacement. Thus, a soft jammed system, consisting of a network of elements undergoing such processes, is linearly elastic under small deformations. Since any relative motion of the elements also involves hydrodynamic interactions, a deformation of the paste also implies some viscous dissipation. Therefore, a relatively large viscous component should be present with the behavior observed under small deformations. Nevertheless, as long as the initial configuration of the elements is recovered when the load is released, the material can be considered as remaining in a solid state, associated with this configuration of reference. Once an element overcomes the energy barrier of its potential well, it tends to fall into another potential well (see Figure 2.5). As soon as one such event occurs somewhere in the material as a result of a critical, macroscopic deformation (γ_c), the initial configuration is irreversibly broken. Indeed, the local structure of pasty materials is generally in a highly disordered state (with a wide range of sizes, shapes, and interaction types), and a reverse deformation would be unlikely to yield exactly the reverse local motion of one or several elements [26]. Finally, as long as the (macroscopic) deformation remains below a critical value, there is no irreversible configuration change and a paste can be considered as a *viscoelastic solid*.

The simplest model for representing viscoelasticity in the solid regime assumes that the total stress is the sum of an elastic and a viscous components. This is the simple Kelvin–Voigt model, which, in simple shear, is expressed as

$$\tau = G\gamma + \mu\dot{\gamma}; \ \gamma < \gamma_c \tag{2.14}$$

in which G is the elastic modulus and μ the viscosity of the fluid in this regime. Slightly more generally, in the solid regime, we can expect a constitutive equation of the form

$$\tau = g(\gamma) + h(\dot{\gamma}) \tag{2.15}$$

which makes it possible to account for possible nonlinear elasticity and viscous effects.[1] In equation (2.15), g and h are two increasing functions such that $g(0) = 0$ and $h(0) = 0$. More complex forms of viscoelastic constitutive equations may be used [27] but, considering the limited practical interest of the constitutive equation in the solid regime of pastes along with the poor phenomenological knowledge of this field, it seems that using an equation of the type (2.15) is sufficient for expressing basic quantitative aspects of the behavior in this regime.

As soon as the configuration of elements is irreversibly broken, the material can be considered as falling in the *liquid regime* (see discussion below). However, in this regime the material can still exhibit some elastic properties. This is, for example, a critical aspect of the behavior of polymer melts or polymer solutions, due to the ability of polymer molecules to transiently stretch as a result of rotations around the carbon links; this results in at least transient storage of elastic energy. For pasty materials, since the dominant interactions result from the relative motion of the mesoscopic elements, elastic effects might result from the fact that during flow there subsist locally some parts of the initial solid configuration that have not been broken but have deformed and stored some elastic energy. However, from a general point of view, it is not clear (theoretically or experimentally) whether elastic effects can play a significant role in the behavior in the liquid regime, and in the following discussion we will neglect this aspect.

A critical question concerns the way a material can return to its solid regime. Existing data do not make it possible to clarify this and here we will simply assume that after some flow in the liquid regime a material returns to its solid regime only if it is left at rest for a very short time. Although such a tendency might not be completely consistent with certain observations, it facilitates a complete, consistent, rheological modeling of the material. Note that this assumption plays a critical role in paste rheometry because one needs to control the material state at any time during the tests, and it is in particularly critical to decide whether the material is in its solid or liquid regime (see Section 3.1).

2.2.2 Solid–Liquid Transition: Yielding

Yielding occurs when the initial configuration is irreversibly broken because at least one element has jumped out of its potential well somewhere in the material. This phenomenon strongly resembles the dislocations that may occur in a crystalline solid structure. In that case the macroscopic behavior of the material is generally considered to be basically *plastic*, possibly with some elastic or viscous effects. However, for such solid materials the mechanical properties are also generally irreversibly affected by the deformation because no other deformation makes it possible to recover a structure with analogous properties on average. This obviously contrasts with simple fluids but also fundamentally contrasts with jammed systems such as pastes and granular materials, for which the apparent

---

[1] Throughout this chapter, for the sake of simplicity, we only consider the shear stress and shear rate amplitudes (positive by definition) when we deal with simple shear.

basically plastic behavior at the transition beyond their solid and their liquid regimes is usually associated with reversible changes of the internal structure.

As for usual plastic solid materials, the yielding of soft jammed systems in simple shear is naturally associated with a *critical deformation* (γ_c), in which one or several elements are definitively extracted from their potential wells, and a minimum stress (τ_c), making it possible to reach this critical deformation. This critical stress will be referred to as the *yield stress*. It may be associated with some minimum, average stress necessary to extract the elements from an average potential well. The depth of this average potential well may be related to the yield stress; it is equal to the average energy stored during a flow at an infinitely low shear rate up to a deformation γ_c, which, assuming a linear elastic behavior below the yield stress, is expressed as $\tau_c b^3 \gamma_c / 2$. In view of a mechanical approach, these characteristics are more generally described with the help of a *yielding criterion*, which gives the critical stress conditions beyond which plastic effects occur. In simple shear the yielding criterion expresses the fact that only finite deformations smaller than γ_c are expected for a shear stress smaller than the yield stress, so that no steady flow at a finite shear rate is expected:

$$\tau < \tau_c \Leftrightarrow \dot{\gamma} = 0 \qquad (2.16)$$

Note that (2.16) is equivalent to $\tau > \tau_c \Leftrightarrow \dot{\gamma} \neq 0$. Moreover, for a fluid following a constitutive equation of the type (2.15) in the solid regime, we have

$$g(\gamma_c) = \tau_c \qquad (2.17)$$

This relation ensures the continuity between the solid and liquid regimes.

To describe the yielding condition for more complex flow conditions, it is necessary to use a three-dimensional yielding criterion taking into account the three-dimensional stress field [28]. Two such criteria are of current use, the Tresca and the Von Mises criteria, but we shall discuss only the second one, which has generally been used [29,30] for describing yield stress fluids. Its major advantage is to provide a straightforward tensorial expression for describing the solid–liquid transition, which recovers the expression (2.16) in single shear. The Von Mises yielding criterion is expressed as

$$\sqrt{-T_{\mathrm{II}}} < \tau_c \Leftrightarrow \mathbf{D} = 0 \qquad (2.18)$$

It has been suggested that yield stress materials could flow at low shear rates for stresses smaller than the yield value, thus exhibiting a Newtonian plateau at low stress levels [31]. Indeed, as for solids, even when Brownian motion is negligible, the number of jumps per unit time is statistically very small under low stress levels, but we can still expect occasional escape of elements from their potential wells, which would ultimately lead to a significant though extremely slow configuration change. However, this possible effect has almost no practical implications since the shear rate of the corresponding expected flows under small stresses is typically of the order of 10^{-6} s^{-1}, which induces a significant deformation, say, larger than 100%, only over observation times of the order of several weeks, a time during which other effects (chemical reactions, drying, phase separation, etc.) may play a major role. Moreover, there is no clear evidence that homogeneous flows at extremely low shear rates can effectively occur. Indeed, pasty materials frequently withstand years of storage without experiencing any of these phenomena and without any apparent deformation [32–36]. Finally, under such low, apparent shear rates, a severe shear banding likely develops, so that the flow is in the "discrete regime" (see Section 3.3). In the following text we shall neglect this possible characteristic (Newtonian behavior under low shear stress).

2.2.3   Liquid Regime: Flow

Yielding corresponds to the transition from the solid regime to the liquid regime, associated with the beginning of irreversible changes in the element configuration. Immediately following breakup of the initial configuration, another jammed configuration presumably forms, since each element escaping from its initial potential well immediately falls into another potential well formed by its new neighboring elements. Thus, in the limit of extremely slow flows, we can expect that a network of interactions on average similar to the initial one has time to re-form after any macroscopic deformation. Under these conditions the shear stress (τ) needed to maintain the flow (i.e., continuously breaking up the current network) remains equal to that (τ_c) needed to break the initial network. Such a behavior is a perfectly plastic one. At higher flow rates we can expect two new effects: (1) the network of interactions usually does not have enough time to reform before again being broken during the next step of deformation, and (2) hydrodynamic effects due to the relative motion of elements can play a role. As a consequence, the stress tensor now also depends on the flow rate. In order to represent both these effects in a single model, the stress tensor of pastes in the liquid regime is generally expressed as the sum of a rate-independent term and a rate-dependent term. Such a model makes it possible to recover the pure plastic behavior at low flow rates and the viscous behavior for larger flow rates. Such materials are generally called *viscoplastic fluids*.

In pure plasticity, as soon as the stress tensor reaches the critical stress value (yield stress) appearing in the yielding criterion, the material flows, with a deformation localized along the surfaces where this critical value is reached. Within the frame of fluid mechanics, it is generally assumed that in regions where yield stress has been overcome the material flows at a finite rate, which increases with the difference between the local stress and the yield stress. This is in agreement with the simple physical approach described above. Different models have been proposed in this context [37] but, in simple shear, most of them can be summarized by the following, generic form in steady state

$$\tau > \tau_c \Rightarrow \tau = \tau_c + f(\dot{\gamma}) \tag{2.19}$$

in which f is a positive function of $\dot{\gamma}$. The Bingham model corresponds to $f(\dot{\gamma}) = \mu_B \dot{\gamma}$, in which μ_B is a material parameter, and the Herschel–Bulkley model to $f(\dot{\gamma}) = K \dot{\gamma}^n$, in which K and n are two material parameters. For the Casson model we have $f(\dot{\gamma}) = K\dot{\gamma} + 2\sqrt{K\tau_c \dot{\gamma}}$. This model is often used in industry because it provides a more progressive increase of the stress with shear rate than does the Bingham model, although still with only two parameters (τ_c and K). In tensorial form, these models may be expressed as

$$\sqrt{-T_{\mathrm{II}}} > \tau_c \Rightarrow \Sigma = -p\mathbf{I} + \tau_c \frac{\mathbf{D}}{\sqrt{-D_{\mathrm{II}}}} + F(D_{\mathrm{II}})\mathbf{D} \qquad (2.20)$$

in which F is a positive function and D_{II} is the second invariant of $\mathbf{D}$. For example, for the Bingham model we have $F = 2\mu_B$, where μ_B is a constant, and for the Herschel–Bulkley model we have $F(D_{\mathrm{II}}) = 2^n K / \left(\sqrt{-D_{\mathrm{II}}}\right)^{1-n}$.

In (2.19) and (2.20) f and F are in general assumed to be increasing functions, tending to zero as $\dot{\gamma}$ or D_{II} tends toward zero. This effectively corresponds to the expected simple yielding behavior, where the flow rate progressively increases from zero as the difference between the applied stress and the yield stress increases. A practical consequence is that in simple shear experiments for which the stress distribution is approximately homogeneous, beyond the yield stress, the fluid should be approximately homogeneously sheared even at low shear rate values. In fact, various recent experimental observations (see Section 3.3) show that the yielding behavior of pasty materials is more complex; at sufficiently low apparent flow rate, the deformation localizes in the regions of higher stresses. In fact, all occur as if there were a *critical shear rate* ($\dot{\gamma}_c$) that prevented steady flow of the material. Thus, when an apparent shear rate lower than $\dot{\gamma}_c$ is imposed (due to the relative motion of boundaries), the strain field evolves toward a situation in which the material remains rigid in some regions and is sheared at $\dot{\gamma}_c$ elsewhere.

An appropriate form of the constitutive equation for describing such a behavior, which appeared to be in agreement with MRI data (see Section 4.2), is similar to (2.19) but with $f(\dot{\gamma}) \to 0$ when $\dot{\gamma} \to \dot{\gamma}_c$. In a more straightforward way, this may be expressed as

$$\tau > \tau_c \Rightarrow \tau = f_0(\dot{\gamma}) \qquad (2.21)$$

where f_0 is now an increasing function of $\dot{\gamma}$ such that $\tau_c = f_0(\dot{\gamma}_c)$. The continuity between the solid and the liquid regimes is still ensured. It is nevertheless worth noting that even if the local behavior of a material exhibiting such a constitutive equation may strongly differ from that resulting from a simple yielding behavior (2.20), its apparent behavior under simple flow conditions, such as, for example, it may be deduced from the torque–rotation velocity curve in a rotational rheometer, is often still very close to that predicted by the usual forms (2.20) (see Section 3.3). Indeed, roughly speaking, below the yield stress there is no flow and, as a result of stress heterogeneity, just beyond the yield stress the flow generally localizes in a thin layer. The thickness of this layer increases as the applied stress increases, and ultimately, at sufficiently high stress levels, the flow invades the entire sample. Here we assume that the discrete regime is not reached, that is, that the sheared thickness remains much greater than the element size (see Section 3.2.8). Thus, even if the flow characteristics are not as expected from the simple yielding models, the yield stress remains a basic characteristic of the material behavior, and flow is always expected only beyond this yield stress. Consequently, in most practical cases involving nonviscometric flows (Chapters 5 and 7), we will, as a first approximation, neglect this peculiar aspect of the behavior and simply retain the simple yielding behavior as approximately expressed by an equation such as (2.20).

2.2.4 Time Effects: Thixotropy

The preceding observations regarding the liquid regime pertained to steady-state flows. It has also been assumed that the solid–liquid transition is instantaneous, that is, that beyond its critical deformation, the material suddenly becomes a liquid, the constitutive equation of which may be described by a model such as (2.20). In fact, after having left their initial potential wells, the elements usually need some time before reaching a stable position within their neighborhood. Also, at (macroscopic) rest, the elements may progressively fall into deeper and deeper potential wells resulting from slight rearrangements of the configuration under the coupled action of thermal agitation and colloidal forces (see Figure 2.5). Thus, in contrast to the simplifying assumption at the origin of our physical explanation of the yielding process, we are dealing with out-of-equilibrium systems [38]. The consequence of these effects is that the apparent viscosity evolves in time due to the progressive destructuring or restructuring of the fluid [39–41]. Let us recall that these effects are reversible, since the initial configuration may be recovered on average after some appropriate flow history [42]. They thus must be distinguished from some irreversible aging processes, which occur as a result of chemical reactions in some systems (cement hardening, gel formation, etc). Similar thixotropic processes occur with any type of material, even for simple liquids, but in that case the characteristic times for reaching equilibrium are generally much shorter than the typical time of observation, and their thixotropic character can be neglected in practice.

The simplest way to account for these thixotropic effects consists in introducing a dependence of the stress on the instantaneous (average) configuration of elements that can, as a first step, be represented by a single parameter, λ. This parameter might, for example, represent the average size of (weak) aggregates [43,44], the fraction of particles belonging to structural units [45], or the number of bonds or links between particles [46]. Within the context of our general conceptual presentation, this *structure parameter* represents the instantaneous, average depth of potential wells. For the sake of simplicity, we will assume that the overall scheme concerning the mechanical regimes (Sections 2.2.1–2.2.3) remains valid and that thixotropy affects the solid and the liquid regimes separately.

In the solid regime, since the potential wells become deeper and deeper in time when at rest, the network of interactions strengthens, and thus the elastic modulus G of a Kelvin–Voigt material increases with the time of preliminary rest. More generally, the thixotropic constitutive equation in the solid regime will retain a form of the type (2.15), with g and h now also increasing with λ. The depth of potential wells is governed by the exact local arrangement of the elements but, on average, the distribution of element centers in space remains almost constant because the volume fraction of elements in the liquid is constant. As a consequence, we would expect that, unlike the elastic modulus, which is related to the depth of the potential well, the critical deformation, which is related mainly to the average distribution of particles in space, does not change much with the state of the structure. Under these conditions the yield stress will evolve as

$$\tau_c = g(\gamma_c, \lambda) \tag{2.22}$$

or, for a Kelvin–Voigt model, as $\tau_c = G(\lambda)\gamma_c$.

In the liquid regime, we may now write [47,48]

$$\tau = \tau(\dot{\gamma}, \lambda) = \eta(\dot{\gamma}, \lambda)\dot{\gamma} \tag{2.23}$$

A simple tensorial formulation of this behavior can be derived: $\boldsymbol{\Sigma} = \boldsymbol{\Sigma}(\mathbf{D}, \lambda)$. Various forms of this function have been proposed in the literature [49–52]; typically, a dependence on λ is introduced in the different terms of the simple yield stress models (2.20). However, according to the present state of knowledge, there is no clear argument in favor of one specific form. A further step would be to use a tensorial parameter for describing the state of structure to account for a possible anisotropy of the element configuration, but for most materials the current physical knowledge in this field is severely insufficient to allow a relevant approach of this type.

Now, to be thorough, we obviously need to describe the evolution of the value of λ as a function of flow history. A typical approach [47,53] consists in reducing this dependence to a relationship between the instantaneous rate of variation of λ and the current flow parameters: $d\lambda/dt = k^*(\lambda, \dot{\gamma})$. Various forms for the function k^* have been suggested in the literature [see the reviews in Refs. 54 and 55), but the most simple and common one is based on the assumption that the current rate of variation of λ results from the competition between two effects: the tendency of the system to return to its equilibrium configuration (*restructuring*) and its continuous breakage as a result of flow (*destructuring*). Physically, this may be understood by considering limiting cases. Under extremely low shear rates the material has time to restructure between two successive flow steps, leading to jumps of elements out of their potential wells, so that the potential wells are deep; under high shear rates the material does not have enough time to restructure and the potential wells are much shallower. It is logical to assume that the corresponding rate of destructuring is proportional to the flow rate (i.e., proportional to $\dot{\gamma}$), which gives $d\lambda/dt = f^*(\lambda) - g^*(\lambda)\dot{\gamma}$, in which f^* and g^*

are two functions depending on material characteristics. In fact, by an appropriate change of variables [i.e., $\Lambda = f(\lambda)$ such that $f^*f' = 1/\theta$ and $g^*f' = K(\Lambda)$], the second term of the kinetic equation becomes constant. In the absence of any specific physical definition of λ, such a change of variables is fully justified. Now, using the symbol λ instead of Λ, we find a general expression for the kinetic equation

$$\frac{d\lambda}{dt} = \frac{1}{\theta} - k(\lambda)\dot{\gamma} \qquad (2.24)$$

in which k is a function of material characteristics and θ is a characteristic time of restructuring of the material [in fact associated with the (new) exact form for $\eta(\lambda)$]. In this context the constitutive equation of thixotropic materials is given by the set of equations (2.15) and (2.22)–(2.24).

The rheological behavior in steady state is found by solving equation (2.24) with $d\lambda/dt = 0$, then introducing the corresponding value of $\lambda_S(\dot{\gamma})$ in (2.23) to obtain $\tau(\dot{\gamma})$. In general [43,54–55], the predicted flow curve is monotonous: τ is an increasing function of $\dot{\gamma}$. However, a few authors [45,50,57–59] have proposed constitutive equations for which the shear stress in steady state appears to be a decreasing function of $\dot{\gamma}$ for $\dot{\gamma} < \dot{\gamma}_c$. The resulting transient behavior [45,60,61] is in agreement with various observations (see Section 3.3) and in particular the apparent *viscosity bifurcation* of pastes, an effect not considered by conventional thixotropy models. When one applies a stress greater than a critical value (τ_c), the fluid flows in steady state at a relatively high rate ($\dot{\gamma} > \dot{\gamma}_c$); when the applied stress is lower than this critical value, the fluid progressively evolves toward complete stoppage. Thus no steady flows can be obtained at a shear rate below a critical, finite value ($\dot{\gamma}_c$).

This viscosity bifurcation effect is certainly intimately linked to the yielding character of pastes. A yield stress fluid has a certain internal structure that breaks when flow occurs. At rest this structure generally recovers rapidly. Under any flow condition or at rest, there is competition between the restructuring and the destructuring processes, as described by equation (2.24). However, whereas the rate of restructuration is almost independent of the current flow rate, the rate of destructuring is basically proportional to the shear rate. Under controlled stress, if the flow rate reached is sufficiently high, the destructuring predominates, resulting in decreased viscosity and thus increased flow rate, further increasing the destructuring, and so on; if the flow rate reached is insufficient, the restructuration prevails, resulting in decreased flow rate, which in turn increases the restructuring, and so on. In fact, this description inherently derives from the general physical scheme proposed to explain the thixotropy of soft jammed systems discussed at the beginning of this section, which means that an appropriate theoretical approach should not separate thixotropy and yielding effects.

In practice, the typical rheological properties of thixotropic fluids are a viscosity decrease in time under flow and an increase of the apparent yield stress with the time spent at rest. Both effects are related to the competition between the destructuring and restructuring effects described above. The increase in apparent yield stress is the simplest effect to appreciate within the frame of usual thixotropic models. Indeed, after some time (Δt) at rest $(\dot{\gamma} = 0)$, the structure parameter is $\lambda = \lambda_0 + \Delta t/\theta$, where λ_0 is the initial state of the material (before rest). From (2.23) and (2.24) it appears that under controlled stress the structure parameter can decrease, a situation likely leading to flow and in general associated with yielding, only if the stress exceeds a critical value [i.e., $\eta(\lambda, \dot{\gamma})/\theta k(\lambda)$]. In agreement with experimental observations, this expression should increase with λ, which imposes some constraints on the functions η and k. One of the simplest approaches in this context consists in assuming viscosity variations in the form $\eta = \eta_0(1 + \lambda^n)$ (in which n is a material parameter >1) and $k(\lambda) = \alpha\lambda$. In that case it may be shown [62] that the flow effectively evolves toward either (1) stoppage, if the stress is lower than a critical stress value, or (2) flow at a shear rate larger than a finite shear rate value otherwise. For $\lambda \gg 1$ this critical stress, $\tau_c(\lambda)$, corresponds to the largest root of the RHS of (2.24), in which the shear rate expression following from (2.23) has been introduced:

$$\tau_c(\lambda) = \frac{\eta_0(1 + \lambda^n)}{\alpha\theta\lambda} \tag{2.25}$$

For sufficiently long times of rest $\Delta t \gg \theta$, this apparent yield stress scales as Δt^{n-1}, a power-law dependence that appears to agree with existing data in this field [39,61,62].

2.2.5 Synthesis

In previous paragraphs we highlighted different aspects of the rheological behavior of pasty materials under different circumstances. The tools chosen to represent these characteristics are specifically appropriate for each of these aspects (viscoelastic solid regime, viscoplastic behavior in the liquid regime, thixotropy). A more appropriate, but also more difficult, approach would consist in using a single model capable of predicting different aspects of rheological behavior. Some models of this type have been proposed [53,55,57]. In the following text since our basic motivation is the mechanical characterization of the material, we will not attempt to combine these aspects in a single model. Rather, in accordance with conventional rheometry, we will focus on the different regimes and try to extrapolate the parameters of models as proposed in previous paragraphs from data obtained under appropriate tests. In the context of nonviscometric flows, it becomes much more difficult to obtain straightforward information about the rheological behavior of the material, and we will only focus on the solid–liquid transition and determination of the material yield stress.

2.3 RHEOLOGY OF GRANULAR MATERIALS

Here we consider materials in which hard interactions between elements predominate under various flow conditions. This means that these materials are composed mainly of solid grains in an interstitial fluid, which plays a negligible role; in general, the interstitial fluid is a gas, but intuitively we expect that in a mixture of sufficiently large particles such as boulders with a low-viscosity liquid (such as water), hard interactions will also prevail. More precisely, in a given material, the relative motion of solid particles is not affected by the interstitial fluid when hard interactions predominate over soft or hydrodynamic interactions. To determine under which conditions this occurs, a logical approach consists in comparing the energies associated with hard and hydrodynamic interactions (here, for the sake of simplicity, we neglect possible soft interactions). However, for granular materials, the energy associated with hard interactions may depend strongly on external constraints such as the pressure and the local configuration of particles, and as a consequence cannot be estimated easily from macroscopic flow characteristics. Here we will simply discuss the typical energies associated with a pure collision, which corresponds to the particle inertia, and a pure friction between two particles within a simple shear flow of the material. By comparing them to the typical viscous dissipation induced by the corresponding relative motion of the particles, we obtain two dimensionless numbers [63]:

- *The Stokes Number (St)*. This is the ratio of the characteristic kinetic energy of the particles [i.e., $mV^2 \approx \rho_p r^3 (b\dot{\gamma})^2$, in which ρ_p is the particle density] to the characteristic viscous energy associated with the motion of the particle through the fluid at the velocity $b\dot{\gamma}$ relative to a neighboring particle (i.e., $F_D b = k\mu_0 r b^2 \dot{\gamma}$) (see Section 5.1.3):

$$St = \frac{\rho_p r^2 \dot{\gamma}}{k\mu_0} \qquad (2.26)$$

- *The Leighton Number (Le)*. This is the ratio of the characteristic energy associated with a frictional displacement over the distance b (i.e., fNb) to the characteristic viscous energy:

$$Le = \frac{fN}{k\mu_0 r b\dot{\gamma}} \qquad (2.27)$$

A granular material corresponds to a material composed of solid particles for which either St or Le is much greater than 1. In this context two flow regimes can be distinguished depending on whether frictions or collisions dominate, namely, the *frictional* (Ba $\ll$ 1) and the *collisional* (Ba $\gg$ 1) regimes, in which the Bagnold number Ba = St/Le = $\rho_p b r^3 \dot{\gamma}^2 / fN$.

Since hydrodynamic interactions play a negligible role in a granular material, momentum transfers occur only via particle contacts. As a consequence, the rheological behavior of a granular material depends generally on the configuration of particles, and more precisely on the spatial distribution of direct contacts, which may change in time as a function of macroscopic flow properties and boundary conditions. In essence, this is not very different from pasty materials, in which the spatial distribution of elements in the liquid can also change as a function of flow conditions and time, and governs their behavior. However, the resulting variety of material properties is as yet not well correlated to flow characteristics and boundary conditions. In this context, we will describe the main rheological trends expected in the basic flow regimes identified above (frictional and collisional), but will describe the flow characteristics (mainly in the frictional regime) found only under specific conditions where the boundaries are assumed to be fixed (see Chapter 6). Here we will retain the basic characteristics of granular material flows as they may be seen from a macroscopic approach; a more in-depth physical analysis may be found in Ref. 64.

2.3.1 Frictional Regime

In this regime the particles must remain in contact for sufficiently long times for the frictional interactions to be dominant (see Section 2.1.4). As a consequence, at each instant there is a continuous network of particles in contact, which gives these materials their specific behavior in this regime (see Figure 2.6). In contrast with pastes, purely elastic effects are unlikely to appear because most of the relative motion between two particles in contact induces irreversible, frictional, energy dissipation. Moreover, negligible thixotropic effects are expected because the particles are non-Brownian, which implies that they a priori do not tend to recover any equilibrium configuration under macroscopic rest. The basic rheological behavior of such material is thus Coulomb-like, but the corresponding coefficient of friction also depends on the flow history.

Coulomb-like Behavior
Locally the force needed to tangentially move two particles in contact is proportional to the normal force acting between them by a factor f [see equation (2.12)]. This is the same for an assembly of particles in contact; roughly speaking, the tangential stress needed to shear the material is proportional to the normal stress, specifically, $\tau = \mu\sigma$, in which μ is the *coefficient of friction* of the material.

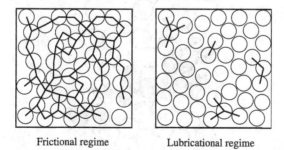

Frictional regime   Lubricational regime

Figure 2.6 Schematic representation of instantaneous distribution of particles in a granular material in frictional regime (left) and in lubricated or collisional regime (right). The short continuous lines represent direct contacts between the particles.

This coefficient is often expressed in the form $\mu = \tan\varphi$, which introduces the *angle of friction* φ, which plays a particular role in some granular flows [see equation (6.2)]. It is worth noting that, although this behavior clearly finds a qualitative explanation in the local properties of particle interactions, no clear quantitative relationship between μ and f has been established. In fact, the precise, instantaneous, macroscopic characteristics of the behavior of a granular assembly are not only the result of the basic properties of the local interactions but also the specific particle configuration leading to a specific distribution of contacts and normal force intensities. In particular, it was shown from numerical simulations [65] that the (theoretical) coefficient of friction of a granular assembly composed of nonfrictional particles [$f = 0$ in (2.12)] differs from zero. This means that the purely geometric jammed state of the granular phase can play a fundamental role in the macroscopic frictional processes.

Generalizing these results to any flow conditions, it was suggested that the apparent behavior of a granular material in the frictional regime is such that flow takes place (see Figure 2.7) along the surfaces where the *Coulomb criterion* is reached:

$$\dot{\gamma} \neq 0 \Rightarrow \tau = \mu\sigma \tag{2.28}$$

in which τ and σ are the amplitudes of respectively the tangential and the normal stress vectors along the surface under consideration. Such a description of the rheological behavior of a material is more complex than are the usual formulations of constitutive equation for fluids. Indeed, although in practice the use of (2.28) requires a clear view of the stress field so as to determine the value of the ratio τ/σ along each surface, such a formulation does not express in a straightforward manner the different components of the stress tensor as a function of the kinematics. As a consequence, it is usually necessary to make additional, generally arbitrary, assumptions concerning some stress components before attempting to solve a flow problem (see Chapter 6). Only in some very simple cases is a rigorous approach possible (see the Mohr–Coulomb approach in Section 6.4).

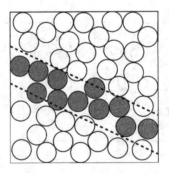

Figure 2.7 Localization of shear along the surface where the Coulomb criterion is reached in a granular material; the particles move in mass (i.e., without relative motion), and shear occurs mainly in the region shown between the dotted lines.

Flow History Dependence

In the field of soil mechanics, flow history dependence has been extensively studied in the frictional regime [66]. For example, it has been observed that in a triaxial test (see Section 6.4) the shape of the curve of the first normal stress difference as a function of the induced strain depends on the initial particle configuration. For an initially loose configuration, the normal stress difference tends to continuously increase with the strain, while the particle configuration becomes denser; for an initially dense configuration, it increases up to a critical value, making it possible to break and slightly disperse the particles, and then slowly decreases. In both cases it is expected that a critical state [67] and, correspondingly, an asymptotic stress value, independent of strain history, will be reached after a sufficient deformation. Other experiments, consisting of push–pull maneuvers at a fixed velocity of a granular column through a conduit, showed that stick–slip effects occurred under low velocities (typically <1 μm/s) [68], with the force fluctuating in time between two significantly different values. This phenomenon is reminiscent of stick–slip effects [69] in solid friction, but here it likely reflects some collective rearrangements of, rather than local frictional properties of, grains.

These observations illustrate some of the various possible consequences of the dependence of the apparent coefficient of friction on the instantaneous particle configuration and its evolutions in time. It is difficult to give a general frame for these variations because they also depend on flow conditions. Here we suggest remembering that, during the startup flow of a dense granular material at a sufficient (not too low) velocity the apparent coefficient of friction usually assumes a value that, on average, decreases in time so as to reach an approximately constant value (μ) associated with some *critical state* of the material, a critical state that nevertheless depends on flow conditions. For a given material and a given flow type, the apparent coefficient of friction thus generally ranges between a maximum value (μ_{start}) associated with flow start and a minimum value (μ) corresponding to a well-developed flow. Note that, because of the lack of systematic experimental comparisons, it is not yet clear whether the corresponding values found under different specific flow conditions are consistent.

The coefficient of friction should also increase with the rate of flow, as for any material, since the energy dissipation a priori increases, so that the value μ should be used in (2.28) mainly to describe the material behavior in the limit of well-developed slow flows. Current knowledge does not provide general behavior expressions including the influence of flow rate. More recent systematic simulations and experiments [65] have nevertheless shown that in homogeneous simple shear flows the apparent rheological behavior (see Figure 2.8) of the material may be expressed in the following form

$$\mu(\dot{\gamma}) = \mu + K\left(\frac{\dot{\gamma}}{\sqrt{\sigma}}\right) \tag{2.29}$$

in which K is a material parameter. Note that it was shown that for such flows the volume fraction slightly decreases with shear rate, which again indicates that

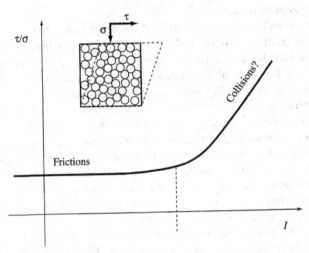

Figure 2.8   Friction coefficient (τ/σ) as a function of the variable $I = \dot{\gamma}/\sqrt{\sigma}$ for a granular material undergoing simple shear (in logarithmic scale).

the coefficients in (2.29) depend on flow conditions. For conduit flows the mean coefficient of friction was shown to decrease with velocity at very low velocity levels (in the stick–slip regime), then it remains approximately constant over two decades of velocities and finally increases slightly at higher velocities [68]. These results are reminiscent of phenomena occurring in solid friction [70], which led to the development of the Dieterich–Ruina model [71] for the coefficient of friction as a function of velocity, in agreement with these observations.

Finally, the coefficient of friction of a granular material, such as that for solids in contact, increases in time, aging as a result of contact creeps or evolutions of slight capillary effects resulting from the air humidity [68,72,73]. The evolution typically observed is of the type

$$\mu(t) = \mu + b \ln\left(\frac{t}{\theta}\right) \tag{2.30}$$

2.3.2   Collisional Regime

Here we consider rapid flows for which the effects of particle collisions predominate in the resistance to flow. Since the contact duration is generally much smaller than the characteristic time for relative motion of two neighboring particles, only a few contacts can exist at each instant and an instantaneous view of the material in this regime would show the particles dispersed with almost no contact, in contrast with the material aspect at low flow rates (see Figure 2.6). Thus these flows have some characteristics similar to those of gas flows. This similarity has led to a number of theoretical developments ("kinetic theory") aimed at predicting the flow properties of granular materials in this regime [74–78].

The basic physical concept underlying these approaches is as follows. The relative motion of two parallel layers of materials induces a shear stress that results from the fluctuations of particle motion around their average path and thus is equal to the rate of momentum transfer between the layers per unit time (see Chapter 1). Under these conditions the shear stress is proportional to both (1) the frequency of collisions between particles of two adjacent layers and (2) the momentum transfer per collision. Term 1 directly derives from the motion fluctuations and is proportional to the square root of the "granular temperature" $T_{gran}^{1/2}$. The second term is simply proportional to the relative velocity of colliding particles, which is proportional to the rate of shear ($\dot{\gamma}$). For gas flows the temperature is often independent of flow properties, so for perfect gases the shear stress is simply proportional to the shear rate (Newtonian behavior). On the contrary, for granular materials the fluctuations result from macroscopic flow. More precisely, energy considerations generally suggest that the granular temperature is proportional to $\dot{\gamma}^2$, so we have $T_g^{1/2} \propto \dot{\gamma}$, and we finally get

$$\tau \propto \dot{\gamma}^2 \tag{2.31}$$

but the coefficient of proportionality still remains hard to predict, even with a complete knowledge of the physical characteristics of the material components. We finally remark that in simple shear, for the same reasons, the momentum transfers in the normal direction also tend to induce a normal force proportional to $\dot{\gamma}^2$, and thus proportional to τ.

2.3.3   Frictional–Collisional Regime Transition

Let us assume that starting from the frictional regime, we progressively increase the rate of the granular flow. The number and intensity of collisions between particles will increase, and beyond a certain flow rate we expect that the collisional regime will be reached. The complete transition from the frictional to the collisional regimes theoretically occurs when Le/St decreases far below 1. The rheological behavior in the intermediate regime, for which both collisions and frictions play a significant role (say, when Le/St is of the order of 1), is more difficult to cope with than in the two extreme regimes, and as yet we do not have clear information about it (see, however, Ref. 65).

2.4   RHEOLOGY OF GRANULAR PASTES

Here we consider materials in which both soft (or hydrodynamic) and hard interactions may play a major role depending on flow regime. More precisely, we focus on highly concentrated suspensions of noncolloidal particles in a liquid or a paste (soft jammed system). Such materials are composed of a continuous matrix of interstitial fluid, in which a high fraction of grains, possibly in contact, are dispersed. Note that now the interstitial fluid (the matrix) can be a simple liquid or a soft jammed system. On a local scale the interactions between the different components of a granular paste correspond to interactions via either the interstitial fluid or direct contact between the particles. Since the local stress is some average of all these interactions over a representative surface element (see Section 1.2.2), we can decompose it in the sum of two terms corresponding to the average over this surface of respectively the first and the second types of interaction. It follows that the stress tensor associated with the flow of the complete system, including both phases, may be decomposed in the sum of two terms, one associated with the flow of the matrix (taking into account the presence of the particles) and another needed to induce the relative motions of the granular phase assumed to be in an inviscid fluid (so that only direct contacts can occur):

$$\Sigma = \Sigma_f(\phi) + \Sigma_g \qquad (2.32)$$

It is worth noting that this decomposition has been used in both fluid mechanics for concentrated suspensions of noncolloidal particles [1] and soil mechanics [79], which illustrates that the behavior of granular pastes is of common interest in these two fields.

The first term (Σ_f) corresponds to the constitutive equation of a paste or a simple suspension containing a large fraction (ϕ) of solid inclusions and in which only hydrodynamic interactions occur. Since the grain size distribution of industrial or natural materials is generally wide, it is likely that, in contrast with uniform sphere suspensions for which ordered configurations may develop, under any flow condition the particle configuration is in a similar state of disorder on average (in time and space). As a consequence, we expect that hydrodynamic dissipations do not depend much on the fraction of direct contact between the particles. In this context it is likely that Σ_f, which reflects the hydrodynamic interactions within the matrix around these particles, a priori does not significantly vary with flow conditions. On the contrary, the second term (Σ_g), which corresponds to the constitutive equation of the granular phase, strongly varies with the fraction of direct contacts, which gives rise to different flow regimes for the granular paste. At low flow rates contacts of long duration giving rise to friction may dominate within the granular phase. At larger flow rates the grain interactions can be lubricated by the interstitial fluid. At even higher flow rates we can expect collisional effects, but in the following text we shall not consider this regime.

2.4.1   Frictional Regime

In this regime, which corresponds to slow flow conditions, there is a continuous network of interactions via direct contacts between the particles and significant frictional interactions develop, but we can distinguish two very different cases depending on the rheological behavior of the interstitial fluid.

Granular Paste with a Newtonian Interstitial Liquid

When the interstitial fluid is a Newtonian liquid, the first stress term of (2.32) is negligible at low velocities ($\mathbf{D} \to 0 \Rightarrow \mathbf{\Sigma}_f \to -p\mathbf{I}$) and the total stress tensor is almost equal to that of the granular phase with an additional term due to the liquid pressure ($\mathbf{\Sigma} = -p\mathbf{I} + \mathbf{\Sigma}_g$). Finally, the material behavior is qualitatively equal to that of a granular material in the well-developed frictional regime, specifically, plastic with a critical shear stress proportional to the normal stress [see equation (2.28)]. The normal stress may now account for the liquid pressure term, but in the absence of gravity effects, the pressure is uniform in the liquid and the interstitial liquid does not impact the local normal stresses between grains. In that case the coefficient of friction ($\tan \varphi$) is similar for dry and wet granular materials.

However, in most cases, the normal stress amplitude depends significantly on gravity effects and, because of the presence of the interstitial liquid, the coefficient of friction of a granular paste is different from that of the equivalent, dry granular material. The physical origin of this effect may be understood as follows. The apparent weight of particles embedded in a liquid phase depends on the pressure distribution in the liquid. Let us consider two particles stacked one above the other with their centers of gravity along the same vertical line. The normal force between them is equal to the weight of the upper particle $\rho_p \Omega g$, in which Ω is the particle volume, and the minimum tangential force required to move the upper one relative to the lower one is $f \rho_p \Omega g$. If the two particles are now embedded in a liquid of density ρ_0, the apparent weight of the upper particle is decreased of its buoyancy force: $\rho_0 \Omega g$ (see Section 5.1). For a Newtonian interstitial liquid, the minimum tangential force (in the limit of extremely slow motion) needed to move the upper particle results solely from the friction at the point of contact between the two particles and is now $T = f(\rho_p - \rho_0)\Omega g$. In the limit of very slow flows this force is also that required to move a material element, including one particle and some fluid surrounding it, relative to another material element. Let us now estimate the net normal force between two such elements. This normal force results from the weight of the liquid and the particle and is equal to $\rho_p \Omega g + \rho_0 \Omega^* g$, in which Ω^* is the average volume of liquid around each particle ($\Omega^* = \Omega(1 - \phi)/\phi$). It follows that the apparent coefficient of friction of the two material elements, which is equal to the ratio of the minimum tangential force to the net normal force, is $f(\rho_p - \rho_0)/(\rho_p + \rho_0 \Omega^*/\Omega)$, which is smaller than f. It may be demonstrated, by directly using the preceding macroscopic expression for the stress tensor of the granular paste, that a similar result is obtained for the macroscopic coefficient of friction of a granular paste, $\tan \varphi^*$, when the normal stress results solely from the weight of material elements:

$$\tan \varphi^* = \tan \varphi \frac{\rho_p - \rho_0}{\rho_p + \rho_0(1 - \phi)/\phi} \tag{2.33}$$

Let us now consider that the liquid surrounding the particle is in motion relative to the particles that are still supposedly at rest; there is a fluid flow in the vertical direction. The apparent weight of the upper particle is now again diminished by the drag force exerted by the fluid (see Section 5.1.3). It follows that the minimum tangential force required to induce some relative motion between the particles is now

$$T = f\lfloor(\rho_p - \rho_0)\Omega g - kL\mu V\rfloor \tag{2.34}$$

The apparent friction coefficient is obviously even smaller in that case, and it can even tend toward zero (when $T \to 0$) when the liquid velocity approaches a critical value. This result can again be extrapolated to a granular assembly. This is at the origin of the "liquefaction" of granular packings immersed in a moving liquid such as occurs at the bottom of an earthdam.

Granular Paste with a Pasty Interstitial Fluid

When the interstitial fluid is a paste exhibiting a yield stress, the resistance to flow should now include both the "granular" and the "pasty" effects expressed in (2.32). Here we assume that the paste elements are much smaller than the particles. According to (2.20), in the limit of low flow rates the stress tensor of the pasty phase is $\Sigma_f \approx -p\mathbf{I} + \tau_c\mathbf{D}/\sqrt{-D_{\mathrm{II}}}$, which leads, for the granular paste, to a modified yielding criterion of the type

$$\dot{\gamma} \neq 0 \Rightarrow \tau = \tau_c + \sigma \tan\varphi \tag{2.35}$$

Such a result is, for example, found in soils formed from a high concentration of clay, sand, and gravel in water [79] or for fresh concrete when the volume fraction of coarse particles added in the cement paste is sufficiently large [80].

In (2.35) the yield stress τ_c is now a function of the fraction of solid inclusions in the paste. From dimensional arguments analogous to those used in the case of a suspension of noncolloidal particles in a Newtonian fluid in Section 2.1.1, it may be shown that, as long as there is no interaction between the grains and the elements of the paste

$$\tau_c(\phi) = \tau_c(0)f(\phi) \tag{2.36}$$

in which f is a function of the solid fraction of grains alone. Unlike the scenario for suspension viscosity, there is as yet no theoretical argument, making it possible to predict the shape of the function f. Some data for clay suspensions in which coarse grains were added [81–83] have shown that f is almost constant as long as ϕ is not too close to the maximum packing fraction of grains in the paste (ϕ_m), and then diverges when $\phi \to \phi_m$. The coefficient of friction in (2.35) should be equal to that of the granular phase alone provided there is no interaction between the two phases, but this is seldom the case with industrial or natural materials containing a continuous grain size distribution ranging from colloidal to coarse grains (typical examples are muds or fresh concrete); in such cases it may be difficult to distinguish the interstitial pasty fluid from the solid inclusions interacting via direct contacts.

2.4.2 Lubricational Regime

The continuous network of direct contacts in granular pastes may disappear at sufficiently high flow rates. This is probably so because the relative tangential motion of two close particles separated by a thin layer of fluid induces a normal force that tends to move the particles away from each other (see Section 2.1.4) and increases with the rate of shear. In this *lubricational* regime the granular stress term becomes negligible and the behavior of the material is governed by that of the interstitial fluid: $\Sigma = \Sigma_f$. The value of the latter tensor was discussed in Section 2.1.1, but clear data on this phenomenon for solid fractions approaching the maximum packing fraction are scarce.

2.4.3 Frictional–Lubricational Regime Transition

The transition from the frictional to the lubricational regimes occurs when the energy dissipated by relative motions of particles separated by fluid layers exceeds the energy dissipated by frictional contacts. It is expected that the lubricational regime will be reached when the Leighton number [see equation (2.27)] falls far below 1.

In practice, this transition may, for example, be observed from rheometrical tests in a Couette system with different sheared heights [84]. Under low rotation velocities (likely in the frictional regime) the applied torque appears to increase with the square height whereas under higher rotations velocities (likely in the lubricational regime) it is proportional only to the sample height, as expected in the case of a simple fluid behavior [see equation (1.76) and Figure 2.9]. This probably results from the fact that the normal force between the grains, here due mainly to the particle weight, does not play any role in the lubricational

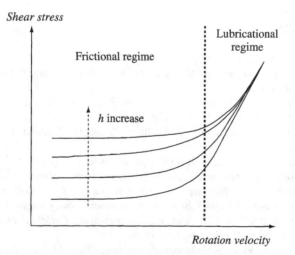

Figure 2.9 Frictional–lubricational transition in a granular paste: torque as a function of rotation velocity in a Couette system for different heights of fluid in the cup.

regime but plays a major role in the frictional regime. Since the normal (vertical) force in a granular packing over sufficiently short distances (see Chapter 6), is approximately proportional to the depth, the average normal force increases with the sample height. Thus in the frictional regime the average shear stress due to the granular phase also increases in this way and the required torque now effectively appears to increase with the square height of the sample. On the contrary, at higher flow rates the granular shear stress becomes negligible and the shear stress due to the interstitial fluid, which is independent of the height of material, becomes dominant. The torque in this regime is thus expected to be simply proportional to the fluid height. This experiment in fact constitutes a very practical means to determine whether we are dealing with a granular paste, capable of flow in both regimes depending on flow rate; if the shear stress depends on material height, the flow regime is frictional but becomes lubricational when the shear stress becomes independent of material height. Also remark that the shear stress in the frictional regime is almost independent of flow rate, in contrast with the results in the lubricational regime (see Figure 2.9).

More recent experiments with very concentrated suspensions in a Newtonian liquid [85] showed that in these systems slight heterogeneities in volume fraction develop during flow that may govern the apparent rheological behavior. In particular, the apparent frictional–lubricational transition appears to in fact correspond to a transition from a regime in which the shear is localized to a regime in which the material in the gap is completely sheared. Thus the apparent stress at the plateau level is almost constant in the first regime because the stress along the interface between the sheared and unsheared regions is almost constant. This suggests that this stress is associated more with a jamming transition than a frictional–lubricational transition.

REFERENCES

1. G. K. Batchelor, *J. Fluid Mech.* **41**, 545 (1970).
2. G. K. Batchelor, *An Introduction to Fluid Mechanics*, Cambridge Univ. Press, Cambridge, UK, 1967.
3. M. R. Kamal and A. Mutel, *J. Polymer Eng.* **5**, 293 (1985).
4. A. B. Metzner, *J. Rheol.* **29**, 739 (1985).
5. R. B. Bird, W. Stewart, and E. N. Lightfoot, *Transport Phenomena*, 2nd ed., Wiley, New York, 2001.
6. D. Quemada, *J. Mécan. Théor. Appl.* (special issue) 267 (1985).
7. J. S. Chong, E. B. Christiansen, and A. D. Baer, *J. Appl. Polymer Sci.* **15**, 2007 (1971).
8. L. A. Utracki, in *Rheometrical Measurements*, A. A. Collyer and D. W. Clegg, eds., Elsevier Applied Science, Amsterdam, 1988, Chapter 15.
9. W. B. Russel, D. A. Saville, and W. R. Schowalter, *Colloidal Dispersions*, Cambridge Univ. Press, Cambridge, UK, 1989.
10. G. R. Zeichner and W. R. Schowalter, *Am. Inst. Chem. Eng.* **23**, 243 (1977).
11. P. N. Pusey, in *Liquids, Freezing and Glass Transition*, J. P. Hansen, D. Levesque, and J. Zinn-Justin, eds., 763–942 (Elsevier Science Publ., Amsterdam, 1989.

12. D. Weaire and S. Hutzler, *The Physics of Foams*, Oxford Univ. Press, Oxford, UK, 2001.
13. D. M. A. Buzza, C.-Y. D. Lu, and M. E. Cates, *J. Phys. II France* **5**, 37 (1995).
14. N. A. Frankel and A. Acrivos, *Chem. Eng. Sci.* **22**, 847 (1967).
15. J. D. Goddard, *J. Non-Newtonian Fluid Mech.* **2**, 169 (1977).
16. G. Marrucci and M. M. Denn, *Rheol. Acta* **24**, 317 (1985).
17. M. L. Wang and T. C. Cheau, *Rheol. Acta* **27**, 596 (1988).
18. D. Tabor, *J. Lubr.* **103**, 169 (1981).
19. B. Bushan, *Appl. Mech. Rev.* **49**, 275 (1996).
20. D. François, A. Pineau, and A. Zaoui, *Mechanical Behavior of Materials*, Hermes, Paris, 1993 (in French).
21. P. Coussot, *Phys. Rev. Lett.* **74**, 3971 (1995).
22. M. Dorget, *Rheological Properties of Silica/Silicone Composites*, Ph.D. thesis, Univ. J. Fourier, Grenoble, 1995 (in French).
23. M. Cloitre, R. Borrega, F. Monti, and L. Leibler, *Phys. Rev. Lett.* **90**, 068303 (2003).
24. R. L. Hoffman, *J. Colloid Interface Sci.* **46**, 491 (1974).
25. H. A. Barnes, *J. Rheol.* **33**, 329 (1989).
26. P. Hébraud, F. Lequeux, J. P. Munch, and D. J. Pine, *Phys. Rev. Lett.* **24**, 4657 (1997).
27. C. Truesdell, *Introduction to Rational Mechanics of Continuous Medium*, Masson, Paris, 1974 (in French).
28. W. Prager, *Introduction to Mechanics of Continua*, Ginn, Boston, 1961.
29. J. M. Piau, *Techniques de l'Ingénieur* **A710**, **A711** (1979) (in French).
30. J. G. Oldroyd, *Proc. Cambridge Phil. Soc.* **43**, 100 (1947).
31. H. A. Barnes and K. Walters, *Rheol. Acta* **24**, 323 (1985).
32. J. P. Hartnett and R. Y. Z. Hu, *J. Rheol.* **33**, 671 (1989).
33. G. Astarita, *J. Rheol.* **34**, 275 (1990).
34. J. Schurz, *J. Rheol.* **36**, 1319 (1992).
35. I. D. Evans, *J. Rheol.* **36**, 1313 (1992).
36. R. D. Spaans and M. C. Williams, *J. Rheol.* **39**, 241 (1995).
37. R. B. Bird, G. C. Dai, and B. J. Yarusso, *Rev. Chem. Eng.* **1**, 1 (1982).
38. B. Abou, D. Bonn, and J. Meunier, *Phys. Rev. E* **64**, 021510 (2001).
39. N. J. Alderman, G. H. Meeten, and J. D. Sherwood, *J. Non-Newtonian Fluid Mech.* **39**, 291 (1991).
40. N. Willenbacher, *J. Colloid Interface Sci.* **182**, 501 (1996).
41. D. Bonn, S. Tanase, B. Abou, H. Tanaka, and J. Meunier, *Phys. Rev. Lett.* **89**, 015701 (2002).
42. J. Mewis, *J. Non-Newtonian Fluid Mech.* **6**, 1 (1979).
43. H. Usui, *J. Non-Newtonian Fluid Mech.* **60**, 259 (1995).
44. H. D. Weymann, M. C. Chuang, and R. A. Ross, *Phys. Fluids* **16**, 775 (1973).
45. D. Quemada, *Rhéologie* **6**, 1 (2004).
46. C. F. Goodeve, *Trans. Faraday Soc.* **35**, 342 (1939).
47. D. C.-H. Cheng, *Nature* **216**, 1039 (1967).
48. D. C.-H. Cheng and F. Evans, *Br. J. Appl. Phys.* **16**, 1599 (1965).
49. F. Moore, *Trans. Br. Ceram. Soc.* **58**, 470 (1959).
50. J. Billingham and J. W. J. Ferguson, *J. Non-Newtonian Fluid Mech.* **47**, 21 (1993).
51. C. F. Chan Man Fong, G. Turcotte, and D. De Kee, *J. Food Eng.* **27**, 63 (1996).
52. C. Tiu, and D. V. Boger, *J. Texture Stud.* **5**, 329 (1974).
53. D. Quemada, *Eur. Phys. J. Appl. Phys.* **5**, 191 (1999).

54. C. R. Huang, in *Encyclopedia of Fluid Mechanics*, N. Cherimisinoff, ed., Gulf, Houston, 1986, Chapter 7.
55. A. Mujumdar, A. N. Beris, and A. B. Metzner, *J. Non-Newtonian Fluid Mech.* **102**, 157 (2002).
56. D. De Kee and C. F. Chan Man Fong, *Polymer Eng. Sci.* **34**, 438 (1994).
57. P. Coussot, A. I. Leonov, and J. M. Piau, *J. Non-Newtonian Fluid Mech.* **46**, 179 (1993).
58. J. R. A. Pearson, *J. Rheol.* **38**, 309 (1994).
59. D. C.-H. Cheng, *Rheol. Acta* **42**, 372 (2003).
60. P. Coussot, Q. D. Nguyen, H. T. Huynh, and D. Bonn, *J. Rheol.* **46**, 573–589 (2002).
61. P. Coussot, Q. D. Nguyen, H. T. Huynh, and D. Bonn, *Phys. Rev. Lett.* **88**, 175501 (2002).
62. N. Roussel, R. Le Roy, and P. Coussot, *J. Non-Newtonian Fluid Mech.* **117**, 85 (2004).
63. P. Coussot and C. Ancey, *Phys. Rev. E* **59**, 4445 (1999).
64. G. D. R. Midi, *Eur. Phys. J. E.*, **14**, 367 (2004).
65. F. Da Cruz, *Flow of Dry Grains—Friction and Jamming*, Ph.D. thesis, ENPC, Marne la Vallée, 2004 (in French)
66. I. Vardoulakis and J. Sulem, *Bifurcation Analysis in Geomechanics*, Blackie Academic & Professional, Glasgow, 1995.
67. M. A. Schofield and C. P. Wroth, *Critical State of Soil Mechanics*, McGraw-Hill, London, 1968.
68. G. Ovarlez and E. Clément, *Phys. Rev. E* **68**, 031302 (2003).
69. C. A. Brockley, R. Cameron, and A. F. Potter, *J. Lubr. Technol.* **89**, 101 (1967).
70. F. Heslot, T. Baumberger, B. Perrin, B. Caroli, and C. Caroli, *Phys. Rev. E* **49**, 4973 (1994).
71. A. Ruina, *J. Geophys. Res.* **88**, 10359 (1983).
72. L. Bocquet, E. Charlaix, S. Ciliberto, and J. Crassous, *Nature* **396**, 735 (1998).
73. F. Restagno, L. Bocquet, T. Biben, and E. Charlaix, *J. Phys. Condens. Matter* **12**, A419 (2000).
74. R. A. Bagnold, *Proc. Roy. Soc. Lond.* **A225**, 49 (1954).
75. J. T. Jenkins and S. B. Savage, *J. Fluid Mech.* **130**, 187 (1983).
76. C. S. Campbell, *Ann. Rev. Fluid Mech.* **22**, 57 (1990).
77. C. K. K. Lun, *J. Fluid Mech.* **233**, 539 (1991).
78. M. Babic, H. H. Shen, and H. T. Shen, *J. Fluid Mech.* **219**, 81 (1990).
79. K. Terzaghi, *Theoretical Soil Mechanics*, Wiley, New York, 1943.
80. H. Van Damme, S. Mansoutre, P. Colombet, C. Lesaffre, and D. Picart, *C. R. (Comptes Rendus) Phys.* **3**, 229 (2002).
81. P. Coussot, *Mudflow Rheology and Dynamics*, Balkema, Amsterdam, 1997.
82. C. Ancey and H. Jorrot, *J. Rheol.* **45**, 297 (2001).
83. C. Ancey, *J. Rheol.* **45**, 1421 (2001).
84. C. Ancey and P. Coussot, *C. R. Acad. Sci. Paris* **327**, 515 (1999).
85. N. Huang, G. Ovarlez, F. Bertrand, S. Rodts, P. Coussot, and D. Bonn, *Phys. Rev. Lett.* **94**, 028301 (2005).

CHAPTER 3

EXPERIMENTAL PROCEDURES AND PROBLEMS IN PASTE VISCOMETRY

Although the theory of viscometric flows provides more or less straightforward relations between some macroscopic characteristics and the material functions of the material, in practice it is not so easy to interpret experimental results in rheological terms, that is, in terms of constitutive equations in a wide range of flow regimes. First the rheological behavior of the material cannot be directly deduced from data, in particular for pastes because their typical properties (viscoelasticity, yielding, thixotropy) are associated with changes or transitions in behavior regimes. As a consequence, it is necessary to develop specific experimental strategies for determining the rheological properties of such materials in relation to their basic properties. Here we start by reviewing (Section 3.1) the experimental procedures appropriate for a pasty material following the typical, idealized, rheological features distinguished in Chapter 2. A set of experimental data can also be the result of various additional effects such as flow heterogeneities directly resulting from the geometric characteristics of the rheometer, nonideal boundary conditions (edge deformation, surface tension effects, wall slip), flow heterogeneities (fracture, shear localization), or material heterogeneities (due to migration, segregation, drying, etc.). All these effects contribute to deviation of flow from the ideal flow of a homogeneous and constant material and should be

Rheometry of Pastes, Suspensions, and Granular Materials: Applications in Industry and Environment
By Philippe Coussot   Copyright © 2005 John Wiley & Sons, Inc.

considered in the rheological interpretation of data. In this context we successively review wall slip (Section 3.2), surface tension effects (Section 3.3), shear localization (Section 3.4), fracture (Section 3.5), drying (Section 3.6), migration (Section 3.7), inertia (Section 3.8), and temperature (Section 3.9) effects. In each case we discuss the physical origins of these effects and their impact on the procedures for determining the basic rheological properties of pastes and the possible countermeasures.

3.1 EXPERIMENTAL PROCEDURES

A straightforward rheological interpretation of rheometrical data is a priori not possible for pastes because of transitions or changes in behavior regimes. In practice it is necessary to first assume some specific type of rheological behavior for each flow regime. Then rheometrical data can be used to determine the values of the parameters of this rheological behavior, generally expressed in terms of a mathematical model. Here we assume that the behavior of pastes corresponds to that described in Chapter 2 and summarized in Table 2.1. In this context the appropriate experimental procedures for determining the basic rheological trends of pastes in the pasty regime (viscoelasticity in the solid regime, solid–liquid transition, behavior in the liquid regime, thixotropy) are successively reviewed. Obviously this ideal scheme cannot pretend to cover the variety of specific rheological trends observed for such materials in reality but is intended to provide basic practical tools, making it possible to tackle problems encountered with paste rheometry.

3.1.1 Setup of the Material

For any rheometrical test it is critical to initially ensure that the material is prepared in a clearly identified state. This ensures not only a good reproducibility of measurements but also a relevant interpretation of rheometrical data in terms of behavior regime transitions or changes. Considering the general form of the constitutive equation of pastes [equation (1.51)], the ideal approach, in view of the rheological interpretation of data, would consist of preparing the material with a controlled stress distribution, relative to a controlled initial state of the material's internal structure. Unfortunately, in practice this is rather difficult without precise knowledge of the material behavior. In particular, when it is set up between the tools of the rheometer, the material undergoes some uncontrolled deformation history significantly depending on the particular procedure followed by the experimenter, which may leave it in a nonreproducible, uncontrolled state.

In order to set up the material in a controlled state independent of its previous history, the usual technique consists in imposing a very rapid flow (*preshear*) and then leaving the material at rest for some time (*time of rest*). A rapid preshear indeed makes it possible to reach the steady state of the liquid regime in a relatively short time, so the internal structure of the material should reach a given state depending only on the intensity of this preshear. The time of rest following preshear can play several roles: (1) possible inertia effects resulting from the previous rapid rotation of the rheometrical tools and the material can be avoided; (2) a viscoelastic material has time to return to its solid structure; and (3) in the case of thixotropy, different times of rest, in principle provide different homogeneous states of restructuration of the material.

3.1.2 Viscoelasticity in the Solid Regime

The appropriate procedures in this case are based on the fact that this regime is related to the material in its "solid state"; the material must be in a solid state initially and remain in this state during testing. This in particular implies that the material behavior must be studied after a short period of rest following setup in the rheometer and subsequent preshear, and that the stress to which the material is submitted must not overcome the critical value (τ_c) at which the liquid regime may be reached.

Creep Tests
In a creep test the material is submitted to a constant stress (τ) and the material behavior is monitored from the resulting strain–time curve. Typically, with pastes, two types of curves are observed (see Figure 3.1) depending on the relative value of the applied stress and the yield stress (τ_c); for $\tau < \tau_c$, the curve remains concave with a slope continuously decreasing in time, and exhibits an apparent, horizontal asymptote, whereas for $\tau > \tau_c$, the initial shape of the curve is similar to that under smaller stresses but after some time there is generally an inflection point and the curve tends to reach an inclined straight line. In the first case ($\tau < \tau_c$) the fact that the material apparently stops flowing after a certain time indicates that the critical deformation has not yet been reached, which corresponds to the solid, viscoelastic regime. In the second case ($\tau > \tau_c$) after a certain time the material tends to flow steadily at a finite rate of shear (the slope of the strain–time straight line), which indicates that the liquid regime has been reached. In this section we focus on the creep curves in the solid regime and we consider the solid–liquid transition and the liquid regime in further detail in Sections 3.1.3 and 3.1.4.

In the solid regime, since the creep curves contain information concerning both viscous and elastic properties of the material, it is relatively difficult to interpret them in rheological terms. In order to better understand the origin of the trends of creep curves for pastes, let us first consider the response of the basic, solid, viscoelastic model [see equation (2.14)] to a given stress ($\tau_0 < \tau_c$) starting from rest [$\gamma(t = 0) = 0$]. The solution to the differential equation (2.14) is:

$$\gamma = \frac{\tau_0}{G}\left(1 - \exp -\frac{Gt}{\mu}\right) \tag{3.1}$$

from which we deduce that $\gamma \rightarrow \tau_0/G$ when $t \rightarrow \infty$. The asymptotic deformation thus provides the value of the elastic modulus. In addition, the time derivative

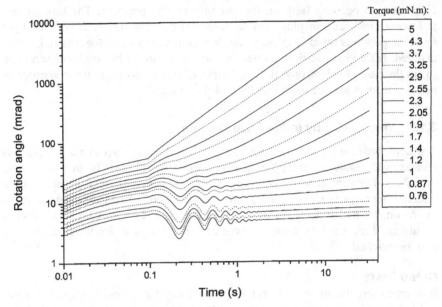

Figure 3.1 Typical set of creep curves under different stress values with a polymer hair gel ("May gel-normal," manufactured by Yplon S.A., Belgium) consisting of water and Carbopol and extremely stable under various conditions. Rheometer geometry: coaxial cylinder geometry, inner diameter 24.5 mm, outer diameter 100 mm, length 35 mm. Before each of these elementary tests (under a given torque) the material was presheared at a high velocity and then left at rest. Here the large gap geometry precludes a straightforward analysis of data in terms of shear stress versus shear rate. As a consequence, we present the data in terms of torque versus rotation angle, which may be roughly interpreted in terms of apparent shear stress versus shear rate, but that exhibits similar, qualitative trends. Note the still unexplained oscillations between 0.1 and 1 s under small torque levels, which might be due to some coupling between elastic and inertia effects at the solid–liquid transition.

of the deformation at the initial instant is $\dot{\gamma}(t = 0) = \tau_0/\mu$, which means that the viscosity of the material in the solid regime may be deduced by measuring the initial shear rate. Thus, for a material following a Kelvin–Voigt model, the rheological parameters can be found from a single creep test under a given stress level, by measuring the initial slope and the final strain of the strain–time curve (see Figure 3.2).

Let us now assume that the material follows a more general model of the form (2.15). In that case it is not possible to determine the material behavior from a single creep test since the functions involved in the constitutive equation are a priori nonlinear. We can nevertheless mention that, since in the limit of long times, $\dot{\gamma} \to 0$ in the solid regime, $h(\dot{\gamma})_{t\to\infty} = 0$, and the asymptotic value of the strain under a given stress value (τ_0) is equal to $g^{-1}(\tau_0)$. Moreover, since $(g(\gamma))_{t=0} = 0$, the initial slope of the strain–time curve under a given stress

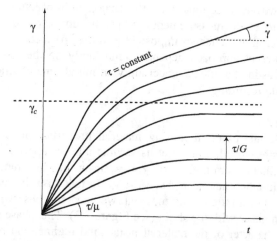

Figure 3.2   Schematic aspects of the creep curves for a (nonthixotropic) pasty material for different (increasing from bottom to top) values of applied stress: viscoelastic solid regime below the critical deformation γ_c; liquid regime beyond γ_c.

τ_0 corresponds to $h^{-1}(\tau_0)$. A set of creep tests under different stress values thus provides the function g^{-1}, through the corresponding values of asymptotic strains, and the function h^{-1}, through the corresponding values of the initial slope of the creep curves. Note that in practice the appropriate procedure consists in imposing different stress levels to the material in the same given initial state, which requires us to impose a given stress, follow the strain in time, and then set up the material in the same initial state before applying another stress value.

It is also possible to observe the behavior of the material in the viscoelastic solid regime from creep tests under stresses higher than the critical value associated with the regime transition ($\tau_0 > \tau_c$). In that case the fluid remains in the solid regime as long as the critical deformation has not been reached. The slope of the creep curve at the initial instant still provides $h^{-1}(\tau_0)$, but, since the shear rate is finite when the critical deformation is reached, the term $h(\dot{\gamma})_{\gamma=\gamma_c}$ a priori differs from zero and we cannot obtain straightforward information about the function g of the material from (2.15).

Deformation Step Tests

Here, from the initial instant, we impose on the material a constant shear rate $\dot{\gamma}_0$ during a time Δt such that $\dot{\gamma}_0 \Delta t = \gamma_0$ and then allow it to relax for $t > \Delta t$ and measure the corresponding stress–time curve. We note that the material remains in its solid regime during the test if $\gamma_0 < \gamma_c$. For a material following a constitutive equation of the type (2.15), since $\gamma(0) = 0$, it follows that $g(\gamma)_{t=0^+} = 0$, and $h(\dot{\gamma}_0)$ is equal to the initial stress value. Moreover, immediately after stoppage ($t = \Delta t^+$), $h(\dot{\gamma})_{\Delta t^+} = 0$, so that $g(\gamma_0)$ is equal to the stress level $\tau(\Delta t^+)$. It is thus possible to determine the two functions of the constitutive equation from a series of deformation step tests of different durations and under different controlled shear rates. However, in contrast with the analogous procedure for creep tests, this procedure relies on measurements at specific times, $t = 0^+$ and $t = \Delta t^+$, immediately after abrupt changes imposed by strains. As a consequence, the precision of the present approach depends significantly on the limitations of the rheometer such as inertia effects, uncertainty on initial time of strain imposition, and delay in data recording.

Dynamic Tests

We can also deduce other interesting material properties in the solid regime from strain or stress oscillations. Let us assume that we apply to the material a periodically oscillating strain given by $\gamma = \gamma_0 \sin \omega t$, in which ω is the frequency and γ_0 the strain amplitude such that $\gamma_0 < \gamma_c$. In that case, in contrast with the response ($\tau \propto \gamma$) of a pure elastic material, we generally observe that the stress does not vary in phase with the strain (see Figure 3.3). This phase shift originates in the viscous character of the material in its solid regime, and may be used to characterize its viscoelastic character.

In order to examine these effects more precisely, let us first consider the case of a Kelvin–Voigt material submitted to such oscillations. The resulting stress is

$$\tau = G\gamma_0 \sin \omega t + \mu\gamma_0\omega \cos \omega t \qquad (3.2)$$

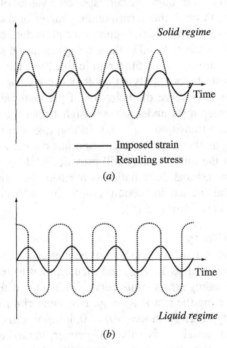

Figure 3.3 Schematic aspects of the stress–time curve for a (nonthixotropic) pasty material under oscillating strain: (a) solid regime (low strain amplitude); (b) liquid regime (large strain amplitude).

which is increasingly shifted from the imposed deformation as the second: first term ratio on the RHS of (3.2) is larger. This effect may be expressed by rewriting the stress in the form

$$\tau = \tau_0 \sin(\omega t + \varphi) \tag{3.3}$$

in which φ is the phase angle. Using (3.2) and (3.3), we can now relate τ_0 and φ to the parameters of the constitutive equation of the material: we have $\tau_0 \cos \varphi = G\gamma_0$ and $\tau_0 \sin \varphi = \mu\omega\gamma_0$, from which we deduce $\tau_0 = \gamma_0 \sqrt{G^2 + \mu^2\omega^2}$ and $\tan \varphi = \mu\omega/G$. In that case the elastic modulus and the viscosity in the solid regime can be determined in a straightforward way from simple measurements of τ_0 and φ.

We can generalize this approach to any kind of viscoelastic solid. The stationary stress response to the oscillating strain is an oscillating signal that is not necessarily sinusoidal but whose amplitude (τ_0) and phase shift (φ) can be estimated from the available data. Then the *elastic modulus* and *viscous (or loss) modulus* of the material may be defined as follows:

$$G' = \frac{\tau_0}{\gamma_0} \cos \varphi; G'' = \frac{\tau_0}{\gamma_0} \sin \varphi \tag{3.4}$$

Similarly, stress oscillations ($\tau = \tau_0 \sin \omega t$) may be imposed and the resulting strain recorded in time. The phase shift and the elastic and loss moduli of the material may also be determined from (3.4) by estimating the amplitude and phase shift of the oscillating strain signal.

Obviously, for a Kelvin–Voigt material G' and G'' are respectively equal to G and $\mu\omega$. For another material type the relative amplitudes of the elastic and viscous moduli give us an idea about the relative weight of elastic and viscous effects in the solid regime. In particular, we deduce the following from (3.4):

$$\tan \varphi = \frac{G''}{G'} \tag{3.5}$$

For a Kelvin–Voigt material G' and G'' follow simple variations with frequency as long as the strain amplitude γ_0 remains smaller than the critical deformation, but this is not the case for other types of materials, in which the elastic and loss moduli may depend on the strain (or stress) amplitude and frequency. In this context two basic tests may be carried out to determine these functions: imposing (1) the frequency and varying the deformation or stress amplitude or (2) the stress or deformation amplitude and varying the frequency. The resulting variations of the elastic and viscous moduli as a function of frequency, deformation, or stress amplitude provide elements for appreciating the material behavior. For example, for polymers, certain theoretical approaches relate these variations to the physical characteristics of the internal structure [1]. However, such general theories do not exist for pasty materials, so experiments of this type cannot yet provide information that may be easily interpreted in physical terms, except by comparison with some specific knowledge in the field of polymers; thus the relevance of this analogy remains to be proved.

3.1.3   Yielding: Solid–Liquid Transition

Here we wish to determine the characteristics of the solid–liquid transition for pastes that, within the frame of our simplified description (see Section 2.2), is described by two material parameters: the critical deformation beyond which the material starts to flow in the liquid regime and the minimum stress (yield stress) required to reach this deformation. Here we neglect thixotropic effects and assume that a constitutive equation of the type (2.15)–(2.19) completely describes the material behavior in the flow range under consideration. This in particular means that we consider the transition from the solid to the liquid regimes as a continuous and perfectly reversible phenomenon. We will see that the basic difficulty for determining the characteristics of the solid–liquid transition lies in the fact that, if the material is effectively viscoelastic in its solid regime, an infinite period of time is in theory needed to reach the critical deformation when the yield stress is applied. This is why several alternative, but also more approximate, techniques and procedures have been developed for estimating the yield stress.

Creep Tests

Let us impose on the material a series of creep tests (starting at each level from the material in the same state) under stress values both lower and higher than the yield stress. In principle, we can distinguish these two regions of stress as follows. Below the yield stress the deformation saturates after an infinite time; beyond the yield stress, the rate of increase of the deformation tends to a constant, finite value (see Figure 3.2). If the material is a simple yield stress fluid, following a constitutive equation of the type (2.15)–(2.19), the slope of the strain–time curve should tend to zero as the stress value decreases relative to the yield stress. This implies that in practice it should be extremely difficult to distinguish the last (experimental) creep curve corresponding to the solid regime from the first curve, corresponding to the liquid regime, since both curves should have a slope tending to zero in the limit of long time periods. As a consequence, the values of the yield stress and critical deformation determined from systematic creep tests under different stress values are approximations, the precision of which depends on the number of stress values tested and the time taken for appreciating whether the strain is finite. Nevertheless, in reality, for many pasty materials the distinction between the solid and the liquid regimes from creep tests seems quite clear because of the viscosity bifurcation effect (see Section 3.3). Beyond a critical stress level, the material becomes capable of flowing at a finite shear rate and no stable flows can be obtained at shear rates intermediate between this value and zero. This implies that in a series of creep curves a gap appears between the shear rate–time curves corresponding to the solid and the liquid regimes in the limit of long time periods. The corresponding strain–time curves exhibit a slope that turns abruptly from zero for applied stresses below the critical stress to a finite value for applied stresses beyond the critical stress.

Stress or Shear Rate Ramp

In order to obtain a more rapid, although obviously more approximate, result, it is possible to impose a "ramp" of steps (i.e., incrementation) of increasing stress levels, without preshear or rest between two successive levels, and determine the yield stress from the critical stress at which the material apparently begins to flow. Let us recall that the deformation rapidly saturates below the yield stress, so that the shear rate is almost equal to zero, while beyond the yield stress the shear rate increases quite rapidly from zero. However, we can note that, when one progressively increases the stress, at each stress level below the yield stress the material seldom has enough time to reach the asymptotic deformation. Thus it is still in its solid regime and the current, apparent, shear rate differs from zero. More generally, since the material undergoes a complex flow history, the critical deformation is not necessarily reached at the critical stress, which implies that the shear rate reached at that level differs from zero. As a consequence, one does not obtain a simple horizontal plateau from such tests and the transition between the solid and the liquid regime cannot be easily identified.

A typical flow curve (see Section 3.1.4) obtained under such conditions is shown in Figure 3.4. The initial part of the curve with a steep slope generally corresponds to the predominance of viscoelastic effects in the solid regime (see discussion above). Then there is a generally marked plateau that, roughly speaking, corresponds to the yielding transition but, because of the abovementioned effects, the level of this plateau may not exactly correspond to the yield stress value; in fact, this level is closer to the yield stress as the rate of increase of the ramp is slower. In the extreme case of steps of infinite duration, the shear rate has time to reach its finite, steady-state value for a stress greater than τ_c and reaches zero for a smaller stress, so that the shear stress–shear rate curve would apparently directly commence from a plateau at the yield stress level.

Another possibility consists in analyzing the material characteristics when the material turns from its liquid state to its solid state. This transition may be observed from progressively decreasing applied stresses starting from steady-state flow at a high rate; for a nonthixotropic yield stress fluid in the liquid regime, a steady-state flow is immediately reached when a smaller shear rate is imposed, so that in theory the exact yield stress value corresponds to the asymptotic stress under infinitely low shear rates (see Figure 3.4).

Increasing or decreasing ramps of shear rates may also be imposed on the material. In principle, for an ideal, nonthixotropic, yielding constitutive equation, phenomena analogous to those described above in the case of stress ramps occur at different times (or equivalently at different shear rates) and the solid–liquid transition may also be roughly characterized. For a thixotropic paste there may be a significant difference. Under controlled shear rate the fluid is forced to destructure so that a steady flow can be rapidly reached; under controlled stress the fluid starts to flow only when the critical stress corresponding to the current state of structure has been overcome.

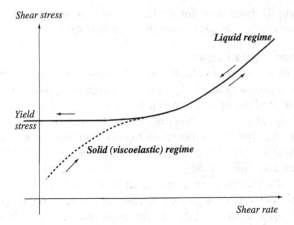

Figure 3.4 Schematic aspects of the shear stress–shear rate curve obtained when an increasing (dashed line)–decreasing (continuous line) stress ramp is applied to a (non-thixotropic) pasty material initially at rest.

Dynamic Tests

Since the relative amplitudes of elastic and loss moduli provide an idea of the relative amounts of elastic energy and viscous dissipation in the material during flow, we can expect these parameters to evolve in a specific way at the transition from the solid to the liquid regimes. Here we consider only controlled strain dynamic tests under varying frequency or strain amplitude. According to the constitutive equation assumed for pasty materials, they remain in the solid regime as long as the strain amplitude is smaller than the critical strain, independently of the frequency. As soon as the critical strain has been overcome, the material falls in its liquid regime and never returns to its solid regime as long as oscillations (or any type of flow) are imposed.

In the solid regime ($\gamma_0 < \gamma_c$), the stress expression (2.15) generalized to the real value of the stress (and not its amplitude) is expressed as $\tau = g(|\gamma|)| \sin \omega t|/ \sin \omega t + h(|\dot{\gamma}|)| \cos \omega t|/ \cos \omega t$. In this expression the value of the first term is limited by the value of the strain amplitude, while the second term increases with the frequency. The elastic behavior predominates under low frequencies, while under sufficiently large frequencies the viscous behavior prevails. This means that in practice G' should tend toward a constant value in the limit of low frequencies but should significantly decrease at high frequencies while G'' increases.

In the liquid regime ($\gamma_0 > \gamma_c$), for a nonthixotropic paste, the stress is solely a function of the actual shear rate (see Section 2.2.3), which implies that elastic effects are negligible and the elastic modulus should drop to zero. For example, for a fluid behavior represented by equation (2.20) the shear stress obtained under controlled strain oscillations is

$$\tau = \frac{|\cos \omega t|}{\cos \omega t}[\tau_c + F(|\gamma_0 \omega \cos \omega t|)] \tag{3.6}$$

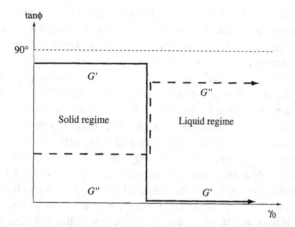

Figure 3.5   Theoretical aspects of G' and G'' variations as a function of strain amplitude for a (nonthixotropic) paste homogeneously sheared; the solid–liquid transition corresponds to the abrupt drop of G'.

This equation shows that the shear stress amplitude assumes its maximum value for the largest shear rate amplitude and crosses the abscissa (horizontal) axis when the sign of the shear rate changes (see Figure 3.3). The stress is thus in phase with the shear rate variations, which implies that $G' = 0$ in the liquid regime.

Let us now assume that we impose oscillations with a ramp of increasing strain amplitudes. Below the critical deformation the material in principle remains in the solid regime, so the elastic modulus should keep a value significantly different from zero; beyond the critical deformation the material should fall in the liquid regime and remain in this regime as long as it is not left at rest any time, which is the case here since it is alternatively sheared in one direction and then in the other direction. The transition between the solid and liquid regimes should thus correspond to the abrupt drop to zero in the G' curve (see Figure 3.5). The strain amplitude and the corresponding applied stress associated with this drop are the critical deformation and the critical stress. Note that for a sufficiently low frequency this drop may also correspond to a rapid change in the evolution of G'' since the constitutive equation from which it derives abruptly changes at the critical strain amplitude.

3.1.4   Flow Curve

The *flow curve* is the shear stress–shear rate curve in steady state. In the absence of thixotropy effects, it may be determined from controlled rate or stress tests in the liquid regime. The basic problem in this context is to make sure that these measurements correspond to the liquid regime. In principle it would be sufficient to wait a sufficiently long time under each velocity or stress level to calculate the corresponding steady-state values, even if it is in difficult practice to distinguish between very slow flows in the liquid regime and residual viscous motion in the solid regime (see Section 3.1.3). As for the yielding transition, this has led people to estimate the complete flow curve from stress or velocity ramps, in which successive, increasing or decreasing steps of stress or shear rate are imposed. For the reasons described above, the result provides only a rough estimate of the effective flow curve, especially at low shear rates for increasing stress or shear rate, because the critical strain may not have been reached yet. On the contrary, a decreasing ramp in principle provides relevant results in terms of the material flow curve in the liquid regime since elastic effects no longer play a role.

To obtain clear information on the material flow curve, it is critical to carry out experiments over a wide range of (apparent) shear rates and plot the data in a shear stress–shear rate curve in logarithmic scale. Indeed, a representation in linear scale focuses on a specific range of shear rates, typically of one order of magnitude of width, from which one cannot reasonably extract a complete, effective information about the rheological behavior of the material. For example, the typical flow curve of a paste in linear scale is a curve with a slope progressively decreasing with shear rate and tending to intersect the abscissa axis at a finite stress value. The latter is naturally associated with the yield stress of the paste, but it appears that its value significantly varies with the range of shear rates chosen for representing the data (Figure 3.6). Moreover the slope of the apparent flow

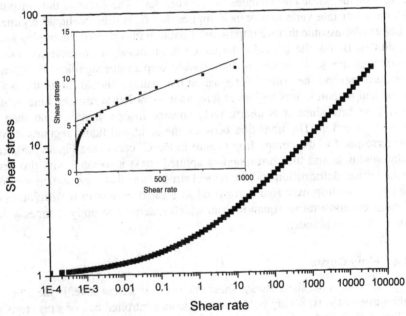

Figure 3.6 Typical aspects of flow curve for a pasty material (following a Herschel–Bulkley model). On logarithmic scale the yield stress is 1, but on linear scale (insert) this value depends on the range of shear rate chosen; in our example it is apparently around 3, but if a simple model (straight line) is fitted to data (here a Bingham model), it is found to be around 5.

curve varies with the observed shear rate range. Typically, fitting a model to the experimental flow curve data in linear scale may thus provide a model capable of accurately representing the data in a range such as $[\dot{\gamma}_1; 10\dot{\gamma}_1]$ with $\dot{\gamma}_1 \neq 0$, but uncapable of predicting flow properties under shear rate values outside this range. Such an approach is nevertheless justified when one seeks a model for predicting flow characteristics under shear rates precisely falling within this specific range. However, in practice flows generally involve sheared and unsheared regions, so the range of shear rates is an interval including zero, and it is not so obvious to determine a priori which range of the type $[\dot{\gamma}_1; 10\dot{\gamma}_1]$ plays a major role on the flow properties under study.

3.1.5 Thixotropy

The thixotropic character of pasty materials adds a degree of complexity to their rheometrical characterization. Since we have distinguished the straightforward procedures for studying the behavior of pastes in the absence of thixotropy, the first step in studying thixotropy consists in observing the effects of time on the results obtained from experiments with these simple procedures.

Solid Regime

According to our assumption concerning the constitutive equation of thixotropic pastes in the solid regime (see Section 2.2.4), a series of creep curves starting from the material after different times of rest should be similar to that shown in Figure 3.2, except that the solid–liquid transition should occur at a yield stress value increasing with the time of rest. The effects of thixotropy can thus be observed from the increase of the functions g and h and the yield stress τ_c as a function of the time of rest, following the techniques described in Sections 3.1.2 and 3.1.3. As a basic requirement for this approach to be valid, the duration of the test must be much shorter than the characteristic time for significant restructuring. Since long times are often required to obtain clear results from creep tests, this condition may be difficult to satisfy.

To determine the evolution of the elastic and viscous moduli in time as a result of thixotropic effects, it may be preferable to carry out dynamic tests, the characteristic time of which $(2\pi/\omega)$ can be very short. The ideal procedure involves imposing successive oscillations under a given amplitude after different times of rest with the same previous preparation of the material. Another, more approximate procedure consists in imposing continuous oscillations to the same sample and following the variation of elastic and loss moduli in time. This procedure is relevant only if we can consider that formation of the solid structure in time is not affected by a flow under a strain lower than the critical strain.

Liquid Regime

In order to follow material thixotropy in the liquid regime, the basic test would consist in imposing a flow at a constant shear rate and following the stress in time or, conversely, imposing the stress and following the shear rate in time. Such a procedure is relevant only if the material is effectively in the liquid state, a condition that is not easy to achieve precisely and may be a source of confusion. The material is in the liquid state only if it is already flowing at a finite shear rate and has overcome its critical strain. Thus appropriate tests consist in imposing first a constant shear rate during a significant time period, followed by a sudden change (increase or decrease) of the shear rate level and following the corresponding stress (or conversely imposing the stress and following the shear rate). The typical aspects of stress–time curves for such tests are shown in Figure 3.7. Since it has been more destructured than expected under the new conditions, for an abrupt decrease in shear rate, the shear stress initially decreases below its final steady-state value under the new shear rate level; since the material is initially less destructured than expected under the new conditions, for an abrupt increase in shear rate the shear stress initially increases beyond the final steady value. Such data thus reflect the time changes of the state of structure with shear rate level.

Besides this ideal procedure the thixotropy of materials is usually studied by suddenly imposing a given shear rate level ($\dot{\gamma}$) on the material initially at rest and following the stress in time (see Figure 3.7). In general the resulting stress almost immediately increases to a large value and then progressively decreases in time toward the steady-state value, an effect referred to as *stress overshoot*. It is also generally observed that the amplitude of this overshoot increases with the time of preliminary rest. The complete stress–time curve in that case is assumed to provide information concerning both the restructuring during the time of rest, through the level of the overshoot, and the destructuring during flow, through the progressive decrease toward the steady-state value in time. Let us examine more precisely the response of a paste following the ideal constitutive equation (2.15)–(2.19). Initially the material is a priori in its solid state

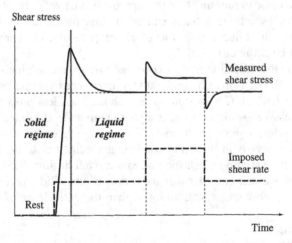

Figure 3.7 Schematic aspects of the shear stress (continuous line)–time variations for a pasty material submitted to sudden changes of imposed shear rate (dark dashed line). The vertical dark dotted line shows the limit between the solid and the liquid regimes.

and the stress increase reflects the (viscoelastic) resistance to flow of the material submitted to a deformation step as described in Section 3.1.2. Assuming that thixotropy does not play any role in such a short duration flow in the solid regime (neither destructuring nor restructuring effects), the material will remain in its solid regime up to the time $\gamma_c/\dot{\gamma}$ for which the deformation reaches γ_c, and the corresponding, maximum stress level reached at that time will be equal to $g(\gamma_c, \lambda_0) + h(\dot{\gamma}, \lambda_0)$, in which λ_0 is the structure parameter of the material at the initial instant. Then the material falls in the liquid regime, and its initial stress value is $g(\gamma_c, \lambda_0) + f(\dot{\gamma}, \lambda_0)$, which a priori differs from the abovementioned value. Such a difference would imply an abrupt increase or decrease in stress after the initial rapid increase, which contrasts with the usual observations; this reflects the limitations of the model. Then the stress $\tau(\dot{\gamma}, \lambda)$, in which the evolution of λ is given by a kinetic equation of the type (2.25), decreases in time. Leaving aside the abovementioned discrepancy, one can analyze the corresponding data by considering that the maximum stress value reflects the yield stress level reached after the preliminary time of rest; restructuring of the material may thus be characterized by plotting this stress as a function of the time of rest. The characteristic time of stress decrease in the liquid regime gives an indication of the destructuration rate. Note that the overshoot level may significantly depend on the precision of the rheometer for recording transient flow characteristics.

Another, more "compact," procedure is still sometimes used; it consists in imposing on the material, after different times of rest, an increasing–decreasing stress (or shear rate) ramp and recording the corresponding shear rate (or stress)–time curve. Since the apparent viscosity of the material progressively decreases in time as this one flows and destructures, the result is a *hysteresis loop*, which differs from the curve obtained for nonthixotropic materials (Figure 3.4) in that the decreasing stress curve does not superimpose the increasing stress curve in the liquid regime (Figure 3.8). In that case the initial part of the increasing curve still corresponds to the solid regime but now with larger elastic and viscous parameters. This implies that, after a long time of rest, the stress curve initially reaches higher stress levels than after a short time of rest, and the apparent viscosity of the material for a given shear rate value decreases in time so that the decreasing curve falls below the increasing curve. Under these conditions the hysteresis loop cannot be interpreted easily in terms of characteristics of the constitutive equation; it only provides a very global idea of the thixotropic character of the material.

3.1.6 Effect of Heterogeneity in Shear Rate

The experimental procedures described above a priori concern ideal, viscometric flow conditions for which the shear rate is homogeneous within the sample. In practice this is seldom the case for various reasons either inherent in the rheometer geometry or resulting from the coupling between boundary conditions, geometry characteristics, and material behavior (see Sections 3.2 and Section 3.3). Here we examine how heterogeneities resulting solely from nonuniform stress distributions may affect the results obtained with the experimental procedures described

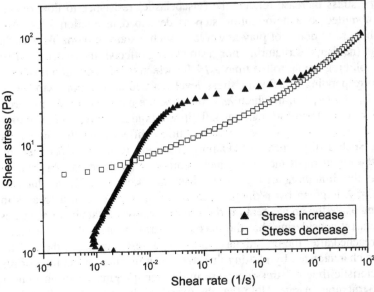

Figure 3.8 Hysteresis loop for an increasing–decreasing stress ramp with an emulsion (mayonnaise) with a (roughened) cone and plate geometry (angle 4°; diameter 4 cm): preshear at 150 Pa during 100 s, then 120 s rest, then stress increase from 1 to 100 Pa in 240 s and decrease in the same time frame. (Reprinted with permission from Ref. 45, Figure 1d. Copyright (2002) by the American Physical Society.)

above. Our frame of reference in this context is Couette flow because the stress heterogeneity is controlled with this geometry.

The basic effect of stress heterogeneities is that the local response of the material to a macroscopic solicitation differs from one point to another. As a consequence, the apparent deformation resulting from the relative rotation of the two cylinders in a Couette system under a given stress along the cylinder differs from some local deformation resulting from this stress value. The cylinder rotation results from the elementary, local rotations of the fluid layers at different distances from the axis (see Section 4.3). It follows that if this stress heterogeneity is not accounted for in calculations of the type presented in Section 1.4, the behavior of the material obtained via a rough analysis of data based on the relationship between the apparent shear rate and shear stress will not reflect the local rheological behavior of the material. Note, however, that this discrepancy will be essentially quantitative. It is likely that the qualitative, effective behavior types of the material in either the solid or liquid regime will be preserved.

More precisely, the effect of the shear rate heterogeneity is to smooth the apparent rheological trends. This is particularly true for the solid–liquid transition, which occurs progressively throughout all the gap as the applied stress increases beyond the critical level since the regions farther from the axis become liquid under higher torques ($C_c \propto \tau_c r^2$). This implies that under creep tests, just beyond the yield stress, the material is expected to flow (at a low shear rate) within a very thin layer along the inner cylinder, which means that the solid–liquid transition in that case corresponds to a strong localization of shear and the validity of the continuum assumption becomes questionable (see Section 3.3.4). A similar problem occurs with dynamic tests; as the strain or stress amplitude increases, the fluid becomes liquid in a progressively increasing thickness, so that the value of the apparent elastic modulus, which results from viscous effects in the liquid region and elastic effects in the solid region, does not drop suddenly to zero at the critical stress. As long as a liquid region and a solid region coexist, G' differs from zero. Thus, in practice, the $G'-\tau_0$ or γ_0 curve (Figure 3.5) appears to tend toward zero slightly more progressively beyond a critical stress value. This explains why it has often been assumed, although there is no substantive physical rationale for this assumption, that the solid–liquid transition corresponds to the intersection between the G' and G'' curves, a procedure that provides an arbitrary critical stress or strain value. In fact, it is more likely that the critical stress is reached along the inner cylinder when the elastic modulus begins to significantly decrease, which corresponds to the development of a thin liquid region.

These problems are even more critical when one studies thixotropy within a wide-gap Couette flow, where both the shear rate distribution and the fluid restructuring or destructuring are heterogeneous in space, so that relevant interpretations of data from macroscopic measurements in terms of structure evolutions are not obtained easily. In that case, in order to carry out consistent analyses of these flows in rheological terms, it is necessary to use exploration techniques capable of providing information concerning the local values of the velocity (see Chapter 4) or the internal structure.

3.2 WALL SLIP

3.2.1 Physical Origin

In the wall slip phenomenon there is an apparent discontinuity in velocity at the approach of a solid surface. An apparent slip may be assumed to occur when, on some scale of observation, the velocity profile intersects the solid surface at a finite velocity level (U_s) and with a finite slope (see Figure 3.9a). When this velocity level is close to zero, wall slip can be considered negligible. In the former case, on some smaller scale of observation, a slip layer appears in which the velocity varies more or less continuously from zero to U_s, the *slip velocity*, (see Figure 3.9b).

The question of the physical nature of wall slip remains somewhat open, and the answer might depend on material type. At this stage it seems useful to separate the materials into two types: those composed of a single type of elements and those composed of several types of elements. In the first category we find simple liquids for which wall slip is as a first approximation negligible because the perturbation due to the solid–fluid interface does not induce effects on a range much wider than that of the molecular interactions [2]. For more complex fluids such as entangled polymers, it has been suggested [3] that the range of

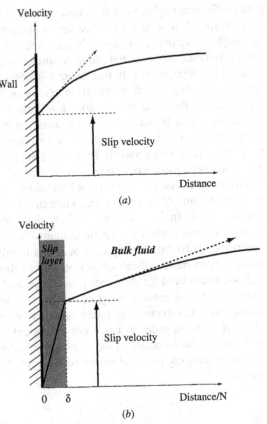

Figure 3.9   Typical aspects of the velocity profile in a paste flowing along a solid surface and exhibiting wall slip. On our scale of observation, the velocity profile intersects the ordinate axis at a finite value (the slip velocity) with a finite slope (a). A smaller observation scale might reveal a thin sheared layer of another material type between the homogeneous paste (bulk fluid) and the solid wall (b).

interactions with solid surfaces can be much larger than for simple, molecular liquids. Liliane Léger et al. [5], using near-field laser velocimetry, have shown that the apparent wall slip in polymer melts is effectively associated with disentanglements between surface chains and bulk chains, which leads to specific friction law along the interface. Such trends, coupled with polymer elasticity, lead to complex flow regimes, in particular in capillary [4–8] or sliding-plate flows [9], depending on the flow rate and the nature of the polymer and the solid surface. In this context the surface treatment may also strongly affect the flow characteristics of polymer melts. Since fluid–solid interactions are assumed to play a critical role in extrusion, a usual operation in material processing, there is a vast literature on the effects of wall slip, elasticity, flow regime on sharkskin, rupture, "undulations," and so on, on the shape of extrudates.

The last category of fluids, of specific interest in this book, are those consisting of more than two element species. Generally one of these species is a liquid matrix in which some elements are immersed. This in particular includes pastes and slurries. With such materials wall slip has been directly observed by different means: visualization of the periphery of a sample in a cone–plate system [10,11], NMR measurements [12], or laser anemometry [13]. For these heterogeneous fluids, wall slip was considered as resulting from a decrease in the concentration of elements on approach of the wall [14]. With such systems there is indeed a natural depletion of element concentration on approach of the fluid–solid interface at rest. If the elements are homogeneously distributed in the liquid, no elements can be centered at a distance from the wall shorter than their characteristic length. It follows that the mean concentration of elements decreases on approach of the wall (see Figure 3.10), an effect directly observed in specific cases [15]. In general the viscosity of the less concentrated material is lower than that of the rest of the fluid. As a consequence, for a similar shear stress, the shear rate in the depleted region (of lower element concentration) is greater than the shear rate slightly farther from the wall ($\dot{\gamma}_s$), in the homogeneous region. This is at the origin of the apparent discontinuity in the slope of the velocity field on a small scale of observation, which leads to an apparent velocity discontinuity on a larger scale (Figure 3.9). In the following the region of low element concentration and higher shear rate will be referred to as the *slip*

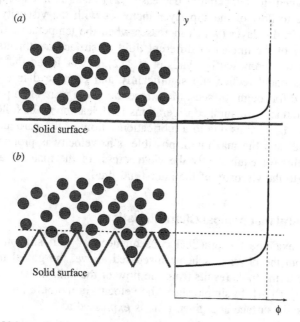

Figure 3.10 Volume fraction of suspended elements as a function of the distance (y) from the farthest point along the solid surface: smooth solid surface (a); rough solid surface (b).

layer (of thickness δ, not necessarily constant), while the remaining material, assumed to be approximately homogeneous and equal to the initial material, will be referred to as the *bulk fluid*.

Note that this natural element depletion might be enhanced by the flow, which could induce a migration from the high-shear region (here the slip layer) toward the low-shear region (here the bulk material) (see Section 3.6), which in turn tends to increase the shear rate in the slip layer and to decrease that of the bulk material. This implies that wall slip has a dynamic character resulting from the time necessary for this phenomenon to develop and stabilize. In particular, it has been shown that all occurs as if wall slip began with some sort of rupture of the links with the wall independently of the nature of the surface [16]. When small stress values are applied to a material in a Couette geometry with smooth surfaces, the resulting rotation velocity is negligible up to a critical deformation beyond which the materials starts to flow by slippage. This effect is reminiscent of the stick–slip effects suggested for polymer materials. Some theoretical approaches have also been proposed for predicting the particle migration under flow [17–20], but these theories have not been validated on a large panel of materials.

Finally, there is actually no general theory enabling one to predict the exact conditions under which wall slip occurs and to which extent it will occur as a function of flow, material, and tool characteristics. This clearly remains a wide-open field. Let us review some of the major trends observed. Wall slip tends to increase when the temperature decreases [21]. This is not unexpected since the resistance to flow of the slip layer increases with the viscosity of the (liquid) material in this layer (which increases when the temperature decreases). A detailed study of the impact of different fluid or surface treatments on the slip of silica gels was proposed by Walls et al. [22]. It was found that hydrophobic solid surfaces could reduce slip significantly and produce data comparable to those obtained for rough surfaces. More recently, there were attempts to quantify wall slip from microstructural considerations [23], leading to the following result. Wall slip may be seen as due to a lubricational flow of the particulate elements along the wall and the maximum, possible, slip velocity is proportional to the elastic modulus (or equivalently the yield stress) of the material and inversely proportional to the viscosity of the interstitial fluid.

3.2.2 General Mechanical Characteristics

Let us now examine the practical consequences of wall slip on rheometrical analysis. From the physical scheme discussed above, the global motion of the fluid relative to the tools results from the flow of the bulk material and the shear of the slip layer [24]. In simple shear the velocity in the bulk fluid at a distance y from the solid surface at a given time is expressed as

$$V(y) = \int_0^\delta \dot{\gamma}\, d\zeta + \int_\delta^y \dot{\gamma}\, d\zeta \qquad (3.7)$$

Let us assume that the slip layer is composed of a homogeneous material of constitutive equation given in the form (1.51) with a stress tensor Σ_s. If normal stress components do not play a significant role, taking into account time variations, equation (3.7) may be rewritten with the help of (1.54):

$$V(y,t)_{\Sigma_0} = \int_0^{\delta(t)} \chi_s(\zeta,t)_{\Sigma_0} d\zeta + \int_{\delta(t)}^y \chi(\zeta,t)_{\Sigma_0} d\zeta \qquad (3.8)$$

In the vicinity of the solid surface the shear stress (τ) in the bulk does not vary significantly. In the following text we will also neglect the time variations of the rheological behavior of the material in both regions. Omitting discussion of the state of reference Σ_0 and assuming that the thickness of the slip layer now depends solely on the local shear stress, we deduce the following from (3.8):

$$V(y) = \chi_s(\tau)\delta(\tau) + \chi(\tau)y \qquad (3.9)$$

The slip velocity may then be defined as follows:

$$U_s = U_s(\tau) = \chi_s(\tau)\delta(\tau) \qquad (3.10)$$

If wall slip occurs, its effects on the flow characteristics at a distance H will be negligible if $U_s \ll V(H)$:

$$\delta \ll \frac{\chi(\tau)}{\chi_s(\tau)} H \qquad (3.11)$$

This implies that *slip is significant only if the slip layer is sufficiently thick or the viscosity of the slip layer* $(\tau/\chi_s(\tau))$ *is sufficiently low.* However, it is worth noting that for pastes and slurries slip effects are in theory readily significant at stresses lower than the yield stress since $\chi(\tau < \tau_c) = 0$ in steady state.

The slip velocity (3.10) depends on the rheological properties of the material in the slip layer, but our current knowledge on this subject is still rather limited. All experiments have shown a power-law dependence of the slip velocity as a function of the shear stress $U_s \propto \tau^n$, but the exponents found depend on materials. For several different suspension types [16,21,25,26], it was found to be very close to 1, suggesting a Newtonian behavior of the slip layer, but with other materials such as foams, emulsions, or polymer solutions [9,23,27,28], n was found to assume values between 1 and 4, suggesting a general power-law behavior

$$\chi_s(\tau) = \alpha\tau^n \qquad (3.12)$$

in which α is a material parameter.

Let us consider the flow of a fluid layer between two parallel, smooth plates separated by a distance H. If wall slip occurs along each solid wall, the relative velocity of the plates follows from (3.9) and (3.10)

$$V = 2U_s + \chi(\tau)H \qquad (3.13)$$

which, in terms of shear rate, may be rewritten as:

$$\dot{\gamma}_{app} = \frac{2U_S}{H} + \dot{\gamma}_{eff} \tag{3.14}$$

where $\dot{\gamma}_{eff}$ is the (effective) shear rate in the bulk fluid and $\dot{\gamma}_{app} = U/H$ the apparent shear rate. Neglecting wall slip leads to underestimating the effective apparent viscosity (η_{eff}) of the fluid since the "apparent" viscosity (η_{app}) directly deduced from the macroscopic measurements is $\eta_{app} = \tau/\dot{\gamma}_{app} < \tau/\dot{\gamma}_{eff} = \eta_{eff}$. If a series of tests with different separating distances are carried out under the same stress, we have $\dot{\gamma}_{app} \rightarrow \dot{\gamma}_{eff}$ when $1/H \rightarrow 0$, which simply means, in other words, that wall slip becomes negligible when the characteristic size of the sample becomes sufficiently large. Also, the measurement of the apparent shear rate under two similar experiments carried out under the same shear stress but with different separating distances (H_1 and H_2) provides a set of two equations of the type (3.14) from which we can deduce the effective shear rate and the slip velocity. In the following text we review the corresponding procedures for the different viscometric flow types.

3.2.3 Couette Flow

Under flow conditions as described in Section 1.4.4 and the wall slip assumptions of Section 3.2.2, equation (3.7) is valid at any depth within the gap and the wall slip velocity along each solid surface is uniform. The rotation velocity of the inner cylinder may thus be expressed as

$$\Omega = \int_{r_1}^{r_1+\delta_1} \frac{\dot{\gamma}}{r} dr + \int_{r_1+\delta_1}^{r_2-\delta_2} \frac{\dot{\gamma}}{r} dr + \int_{r_2-\delta_2}^{r_2} \frac{\dot{\gamma}}{r} dr \tag{3.15}$$

Using (3.10) and the change of variable as in equation (1.79), we get

$$\Omega(M, t) = \Omega_s + \Omega_F \tag{3.16}$$

in which

$$\Omega_s = \frac{U_s(\tau_1)}{r_1} + \frac{U_s(\tau_2)}{r_2}; \qquad \Omega_F = \frac{1}{2} \int_{\tau_2}^{\tau_1} \frac{\chi(\tau)}{\tau} d\tau \tag{3.17}$$

are respectively the wall slip and the bulk fluid flow contributions to the rotation velocity.

Let us consider an experiment carried out with another Couette system with inner and outer radii r_1^* and r_2^* such that $r_2^*/r_1^* = r_2/r_1$, and under an applied torque M^* such that $M^*/M = (r_1^*/r_1)^2$ and leading to a rotation velocity of the inner cylinder Ω^*. Under such conditions, the stress distributions in the two geometries are similar, and in particular $\tau_1^* = \tau_1$ and $\tau_2^* = \tau_2$. From (3.17) we deduce that $\Omega_s^* = r_1/\Omega_s r_1^*$ and $\Omega_F^* = \Omega_F$ and from (3.16) we get

$$\Omega_F = \frac{r_1^* \Omega^* - r_1 \Omega}{r_1^* - r_1}; \qquad \Omega_s = \frac{\Omega - \Omega^*}{1 - r_1/r_1^*} \tag{3.18}$$

Two cases are of special interest. When the gap is small compared to the cylinder radius (i.e., $r_2/r_1 \approx 1$), we have $\tau_1 \approx \tau_2$ so that $U_s(\tau_1) \approx U_s(\tau_2)$ and $\Omega_s = 2U_s/r_1$. When the gap is infinitely large (i.e., $r_1/r_2 \ll 1$), equation (3.17) leads to $\Omega_s = U_s/r_1$.

3.2.4 Parallel Disks

For parallel disks [29], under flow conditions as described in Section 1.4.2 but now with wall slip, the velocity at the distance r along the top rotating disk is

$$\Omega r = H\chi(\tau) + 2U_S(\tau) \tag{3.19}$$

Using the apparent shear rate at distance r

$$\dot{\gamma}_{app} = \frac{\Omega r}{H} \tag{3.20}$$

equation (3.19) may be rewritten in the form of a relation between the apparent shear rate and the stress at the distance r. In particular, for $r = R$, we have

$$\dot{\gamma}_{app}(R) = \dot{\gamma}(\tau_R) + \frac{2U_S(\tau_R)}{H} \tag{3.21}$$

Since equation (3.21) shows that there exists a relation between the shear stress and the apparent shear rate, we can proceed to an analysis similar to that of Section 1.4.2 leading to an expression analogous to (1.69), now relating τ_R and $\dot{\gamma}_{app}(R)$.

Let us consider a set of two experiments with the same value of τ_R but with two different gap heights (H and H^*). In that case the effective shear rates in the two experiments should be similar, so that equation (3.21) provides a set of two equations (for the two gaps), the solution of which is

$$\dot{\gamma}(\tau_R) = \frac{H\dot{\gamma}_{app} - H^*\dot{\gamma}_{app}^*}{H - H^*}; \qquad U_S(\tau_R) = \frac{\dot{\gamma}_{app} - \dot{\gamma}_{app}^*}{2\left(\dfrac{1}{H} - \dfrac{1}{H^*}\right)} \tag{3.22}$$

Here the problem has been solved by taking advantage of the fact that the shear stress and shear rate at the periphery can be related to macroscopic quantities, but let us recall that this analysis does not provide any information concerning the stress and shear rate distribution along r.

3.2.5 Cone–Plate Flow

We could expect to have a more favorable situation with a cone–plate geometry, since in the absence of slip the stress distribution is almost homogeneous (see Section 1.4.3). If we assume that the shear stress remains uniform in the presence of wall slip, the effective shear rate and the slip velocity, which basically depend on the stress, should be uniform. Under these conditions the relative rotation velocity of the solid surfaces would be

$$\omega(r) = \frac{v_\varphi}{r} = \frac{2U_S}{r} + \dot{\gamma}\theta_0 \tag{3.23}$$

However, this expression is in general not compatible with the absence of variation of the rotation velocity with r, as imposed by the geometry and the solid nature of these surfaces: $\omega(r) = \Omega$. This means that our initial assumption is not valid. In practice, as soon as wall slip occurs, the shear stress distribution in a cone–plate geometry becomes heterogeneous [30,31] and it becomes difficult to separate wall slip effects from bulk flow.

3.2.6 Capillary Flows

The flow rate in a capillary in the presence of wall slip may be computed using the results of Section 1.4.5:

$$Q = 2\pi \int_{R-\delta}^{R} r v_z dr + 2\pi \int_{0}^{R-\delta} r v_z dr \tag{3.24}$$

Using a similar change of variable and the wall stress expression $\tau_p = RA/2$, this equation may be rewritten as follows:

$$\frac{Q}{\pi R^3} = \frac{U_S}{R} + \frac{1}{\tau_p^3} \int_{0}^{\tau_p} \chi(\tau)\tau^2 d\tau \tag{3.25}$$

From (3.25) it appears that

$$U_S = \left[\frac{d(Q/\pi R^3)}{d(1/R)}\right]_{\tau_w} \tag{3.26}$$

which implies that the slip velocity and, in a second step from (3.25), the bulk flow rate, may be determined from experiments with different capillary diameters (at least two) keeping the same wall stress value. An ingenious practical means has been suggested for determining, with a single apparatus, the slip velocity and the bulk flow rate on the basis of the theory given above [32]. It consists of two conduits of different sizes but with the same ratio R/L set in parallel at the exit of a reservoir containing the material and in which the pressure is controlled. Thus, under a given pressure level p, the pressure differences between the reservoir and the exit in each conduit are similar ($\Delta p = p - p_0$) and the stresses at the wall ($\tau_p = R\Delta p/2L$) in both conduits are similar so that the slip velocity and the bulk flow rate may be determined from (3.25) by measuring the corresponding discharge rates through each conduit: $U_s = (Q_2/\pi R_2^3 - Q_1/\pi R_1^3)/(1/R_2 - 1/R_1)$.

3.2.7 Wall Slip, Yielding, and Fracture

These theoretical considerations assume that slip is an effect that simply super-imposes any fluid flow along smooth solid surfaces. For yield stress fluids, when stresses smaller than the yield stress are applied, since we do not expect any bulk flow, only viscoelastic effects (see Section 2.2.1) or wall slip can induce motion. When stresses exceeding the yield stress are applied, a possible slip effect would take place in addition to bulk flow. In particular, in steady state, using (3.12) and (3.14), the resulting apparent behavior should take the form

$$\tau < \tau_c \Rightarrow \dot{\gamma}_{app} = \frac{2\alpha\delta}{H}\tau^n; \tau > \tau_c \Rightarrow \dot{\gamma}_{app} = \frac{2\alpha\delta}{H}\tau^n + \chi(\tau) \tag{3.27}$$

in which α is the parameter appearing in (3.12) and χ is an increasing function such that $\chi(\tau_c) = 0$. The corresponding apparent flow curve (see Figure 3.11) is thus composed of two main parts: a branch resembling that of a power-law model followed by a branch resembling the flow curve of a yield stress fluid model beyond some finite shear rate. The position of the transition between these two branches depends on the value of H (the definition of which depends on the type of geometry used); for sufficiently large H values, the first branch is situated in a region of very low shear rates so that the apparent yielding behavior of the

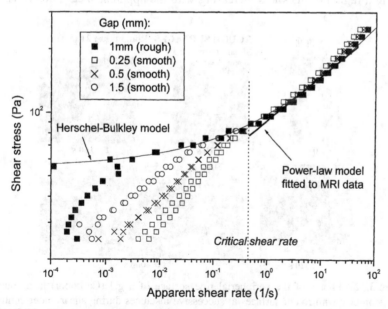

Figure 3.11 Flow curve of a model emulsion as determined from stress ramps with rough or smooth surfaces with parallel-plate geometry with different gaps. Fit of a Herschel–Bulkley model to the data with a rough surface ($\tau_c = 55$ Pa; $K = $ Pa $\cdot$ s$^{1/3}$; $n = \frac{1}{3}$). The truncated power-law model as deduced from MRI data has been represented by a straight, dark line. (From Ref. 28, Figure 7.)

material is not significantly affected by wall slip; for very low H values the stress plateau disappears and the flow curve may appear as basically composed of two parts of different slopes (see Figure 3.11). Note that the aspect of the flow curve for intermediate cases is somewhat similar to the apparent flow curve obtained from tests under increasing stress ramp (see Section 3.1.3), although in the latter case the first branch results mainly from fluid flow in the solid regime under low stress levels. Consequently, if wall slip occurs during an increasing stress ramp, these two effects overlap and it may become difficult to distinguish them.

In fact, the existing direct observations of the strain field suggest a slightly more complex scheme at a local scale. Effectively, they confirm [10,11,21,26] that below a critical shear stress, slip can occur whereas no bulk flow takes place. However, they also show that with rough surfaces fracture can occur (see Figure 3.12) within the material at low apparent shear rate, instead of the expected slow homogeneous bulk flow. Although such results have long been considered as artifacts of viscometric tests [33,34] that should be suppressed or avoided at best, more recent works (see Section 3.3) have shown that they more probably originate in the intrinsic rheological properties of pastes; pasty materials cannot flow steadily at a rate below a critical value, so the shear tends to localize in the regions of larger stresses when an apparent shear rate smaller than the critical shear rate is imposed by the tools (see Section 3.3). This finally suggests that the apparent start of bulk flow at the critical stress corresponds to a localized shear flow in a region of thickness increasing with the apparent shear rate. Thus, with

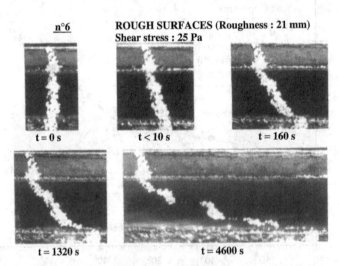

Figure 3.12 Photos of the peripherial free surface of a gel (Carbopol) in a cone and plate geometry with rough surfaces at successive instances during shear under controlled stress (here 25 Pa). The yield stress of the material was approximately 22 Pa. Placement of a vertical line of powder on the surface of the sample makes it possible to observe the strain field in time; the shear becomes heterogeneous beyond a critical deformation. Homogeneous shear was observed under higher stresses. (Reprinted from Ref. 11, Fig. 11(6). Copyright (1990), with permission from Elsevier.)

pasty materials, the transition from the pure wall slip regime to the homogeneous bulk flow regime is likely complex; beyond the critical stress wall slip effects *and* localized shear flow may coexist within the material [35] and wall slip effects become negligible as compared to bulk flow for greater applied stress [28].

3.2.8 How to Avoid Wall Slip

It is generally assumed that wall slip is negligible when the roughness of the wall becomes much greater than the typical element size. Indeed, in that case the volume fraction of suspended elements is, at least initially, homogeneous (see Figure 3.10) beyond the straight envelope of the wall roughness. This envelope plays an important role in the flow since, as a first approximation, it may be considered that between it and the solid surface the fluid does not move relatively to the solid so that this envelope forms an equivalent rigid boundary for the rest of the fluid. Finally, with sufficiently rough surfaces, at the approach of this equivalent rigid boundary there is no depletion in the volume fraction of elements. Note that it is difficult to quantify the critical value of roughness at which this condition applies; it depends on the shape of the roughness and the size and shape distribution of the elements. Roughly speaking, for a material with a wide element size distribution, a roughness equal to the maximum element size should in general make it possible to avoid wall slip, but for a material with uniform elements a roughness equal to several times the element size is necessary.

Instead of the inner cylinder in a Couette geometry, an extremely "rough" tool is often used for coarse suspensions [36] or in order to avoid wall slip for other pasty materials; this is a vane with either four, six, or eight blades. In that case the equivalent roughness of the solid surface is of the order of the blade half-width and thus is generally much greater than the diameter of paste elements. Moreover, it is reasonable to assume that the stress distribution in the fluid contained between the vane envelope and the outer cylinder is very close to that with a cylinder of the same diameter [36]. As a consequence, the viscometric theory for this geometry should be analogous to that with a usual Couette system (Section 1.4.4). Under these conditions it is assumed that the material in the interior of the envelope moves as a rigid body with the vane, but MRI observations of the flow field have shown that this hypothesis is somewhat erroneous [37].

Note that even with sufficiently rough walls the shear may seem to remain localized in a thin layer close to the solid surface. This effect, often referred to as *slippage*, should not be confused with wall slip. It more likely results from the inability of pasty materials to flow at a shear rate below a critical value (see Section 3.2.7). Note, however, that even when such an apparent slippage occurs, the stress needed to apply is in general much greater than that necessary for the same apparent motion with smooth surfaces, and this stress is often close to the material yield stress since it has to induce the relative motion of two almost rigid bodies containing more or less identical fluids. However, in this "discrete regime" (see also Section 3.3.4), we can also expect more complex effects with specific types of materials such as fluids exhibiting either a significant thixotropic character (in that case the corresponding stress level depends on the procedure), a certain crystal structuration leading to a friction force lower than the yield stress, or, on the contrary, a local flocculation leading to a larger force (see Section 3.3.5).

3.3 SHEAR LOCALIZATION

Since the shear stress slowly increases with shear rate at low flow rates, the flow curve of a pasty material can rarely be determined precisely from the usual experiments; a small heterogeneity in the stress distribution may lead to a large heterogeneity of the resulting shear rate. For example, if the stress distribution is not perfectly homogeneous within the gap of a rheometer, we can expect that under a given torque some regions will flow while others will remain at rest because the yield stress has not been overcome in the latter. At first glance this seems to provide some explanation for the common observation that, at the periphery of a pasty material sheared at low velocities in parallel disks or cone–plate geometries, the shear tends to localize in a thin layer while the rest of the material moves in mass with one of the tools.

Several other features have been observed at low, apparent shear rates with pastes: plateau or decreasing flow curve, hysteresis in increasing–decreasing stress ramps, and shear banding in cone–plate geometry. These phenomena tend to suggest that flow instabilities occur that might constitute serious deviations from the ideal scheme presented in Chapter 2 for soft jammed systems. In fact, we will see that they constitute the different rheometrical consequences of the peculiar, intrinsic, thixotropic character of jammed systems. In the following text we first describe the *plateau effect* and *viscosity bifurcation effect* as observed in rheometry and then review the physical or mechanical models proposed for representing or understanding these effect. Afterward we show that this plateau effect is basically associated with shear banding effects. Finally, we review the rheometrical implications of such phenomena. Since somewhat similar phenomena occur with a specific type of material, namely micellar solutions of surfactant molecules forming onions or wormlike micelles, we will use this well-documented example as a reference for the discussion, even if its physical origin is likely slightly different.

3.3.1 Plateau in Flow Curve and Viscosity Bifurcation

The flow curve of various micellar solutions [38–42] was shown to exhibit a stress plateau in the shear stress–shear rate curve at a particular stress value; thus, the stress appears to remain almost constant in a certain range of shear rates. The stress may in fact slightly increase in this region, but this increase appears negligible in comparison with its variations at shear rates outside this range (see .Figure 3.13). Under controlled stress experiments, when progressively increasing the stress level, one observes around a critical value (τ_c) a large increase of the resulting shear rate, which rapidly transforms from a small value ($\dot{\gamma}_1$) to a much

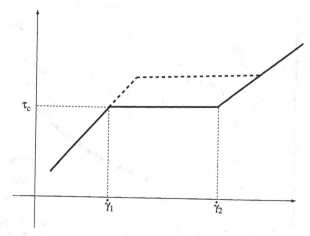

Figure 3.13 Schematic representation of the plateau phenomenon in flow curve as observed for micellar systems. The position of the plateau and the values of the critical shear rate may depend on flow history, leading to apparent flow curves situated between the continuous and the dashed lines.

larger value ($\dot{\gamma}_2$) associated with the end of the plateau. If the slow flow at shear rates below $\dot{\gamma}_1$ has not been detected, the fluid may seem to start to flow abruptly at τ_c. In several cases it has also been shown that the position of the plateau is not precisely defined; in a portion of the flow curve before the plateau and a portion after the plateau, the flows are extremely stable, but at the approach of the plateau different shear stress–shear rate data may be obtained depending on flow history [43,44] (see Figure 3.13). This in particular implies that flow curves obtained under an increasing or a decreasing ramp of stress may differ around the plateau.

The overall aspect of the flow curve of pastes under an increasing stress ramp and in the absence of wall slip (see Figure 3.4) is similar to a flow curve with a plateau as described above. However, the initial steep slope is generally associated with the viscoelastic solid regime for which the steady-state shear rate values are a priori equal to zero, and since the flow curve should describe the steady state constitutive equation in the liquid regime there is a priori no abrupt slope change in the rest of the curve. A plateau in flow curve can nevertheless be observed with various pasty and granular materials under increasing stress ramps [45,46] (Figure 3.14). This plateau depends strongly on flow history and in particular on the time of rest before starting the test, and there also may be a strong difference between flow curves obtained under increasing and decreasing stress or shear rate (see Figures 3.7 and 3.14b,c).

Within this frame a critical issue is whether the flows associated with the different parts of such a flow curve for pastes are stable. To clarify this point, it is necessary to carry out creep tests over sufficiently long times under different stress levels. It has appeared that even for materials having a priori a moderated

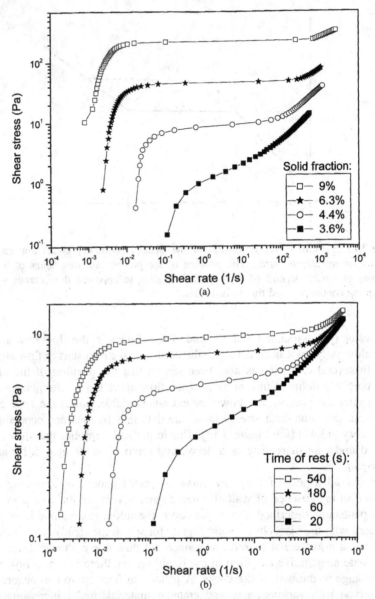

Figure 3.14 Apparent flow curve of pasty materials under increasing stress ramps with a cone–plate rheometer: (a) bentonite suspension at different volume fractions, preshear at 500 s^{-1} for 20 s followed by 20 s rest, stress ramp in 60 s; (b) bentonite suspension ($\phi = 3.6\%$) with the same preparation but different times of rest; (c) foam (Gilette Co.), preshear at 200 s^{-1} for 10 s, then 20 s rest and stress ramp from 80 to 250 Pa in 120 s (increase) + 120 s (decrease). (Reprinted with permission from Ref. 45, Figure 1b. Copyright (2002) by the American Physical Society.)

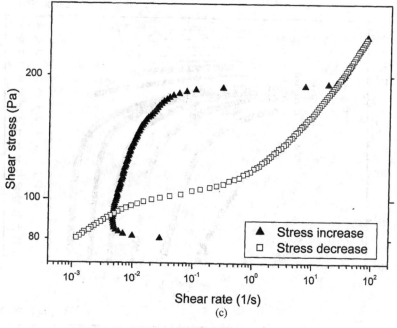

Figure 3.14 (*Continued*)

thixotropic character, the flow regimes in the plateau region are unstable; for stress levels beyond a critical value the shear rate rather rapidly reaches a constant value, but for stresses lower than this critical value the apparent shear rate continuously decreases in time so that the material tends to ultimately stop flowing [22,45,47] (Figures 3.15). For a shear stress slightly greater than the critical value, the shear rate reaches its lowest level $\dot{\gamma}_c$. For an ideal yield stress fluid, we expect $\dot{\gamma}_c = 0$ (see Chapter 2), but this is not the case in practice; creep tests show that there is a range of shear rates (i.e., below $\dot{\gamma}_c$) for which no stable flow can be obtained in steady state (see Figure 3.15), an effect referred to as *viscosity bifurcation* [47,48].

3.3.2 Modeling

Flow Curve with a Decreasing Part

Some characteristics of the plateau or viscosity bifurcation effect, such as dramatic increase in shear rate for a small increase of stress and decrease in shear rate as a function of time, suggest that this phenomenon could be related to an instability of the fluid flow. In a first stage it was thus considered as associated with a constitutive equation exhibiting a decreasing part in the flow curve in steady state [49–52], such that $d\tau/d\dot{\gamma} < 0$ in a certain range of shear rate. Two types of curve have been suggested: an S-shape curve, with an increasing part at low shear rates followed by a decrease and an increase in shear stress; and a flow

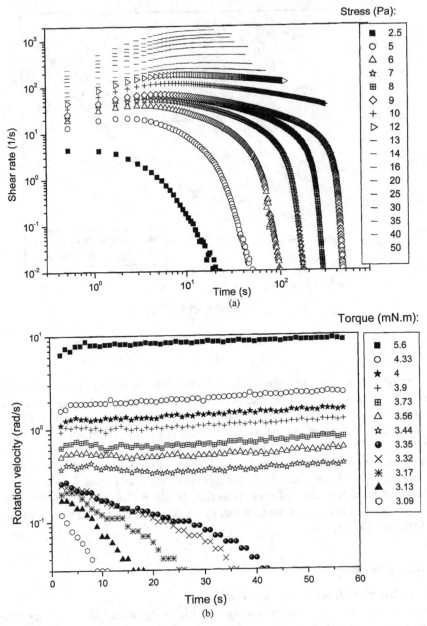

Figure 3.15 Creep tests with pasty materials in a cone–plate geometry after pre-shear and then at rest (2 s): (a) bentonite suspension ($\phi = 4.5\%$); (b) foam (Gilette Co.). (c) Analogous tests with a Couette geometry for a mustard with 5 min of preshear and 2 min rest. (From Ref. 70, Figure 2.)

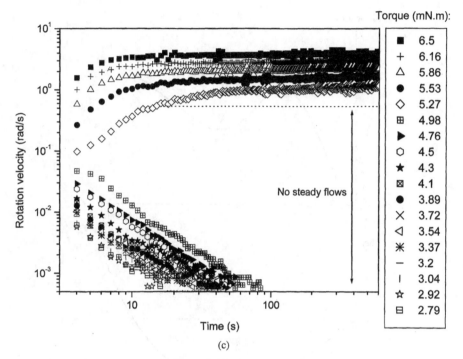

Torque (mN.m):

■	6.5
+	6.16
△	5.86
●	5.53
◇	5.27
⊞	4.98
▶	4.76
○	4.5
★	4.3
⊠	4.1
●	3.89
×	3.72
◁	3.54
✳	3.37
−	3.2
∣	3.04
☆	2.92
⊟	2.79

Figure 3.15   (*Continued*)

curve with only one extremum (a minimum) (see Figure 3.16). The interest of these models is that, as may be shown from the following, simple linear stability analysis, no steady-state flow in the decreasing part of the flow curve of a fluid exhibiting such a (theoretical) constitutive equation can be obtained.

Let us consider the simple shear flow of such a material between two parallel plates. Since the structures of the main momentum equations are similar (see Chapter 1) we would get similar results for other, more complex, geometries. We assume that a constant (apparent) shear rate $\dot{\gamma}$, situated within the decreasing region of the flow curve, is imposed on the material. Thus the velocity in the direction x under stable conditions is $u_0(y) = \dot{\gamma} y$, while the other velocity components remain equal to zero. We now assume that the velocity field is slightly perturbed, so that its expression becomes

$$u(y) = u_0(y) + u_1 \exp(iky + \omega t) \tag{3.28}$$

in which k and ω are two (real) parameters and u_1 is small compared to u_0. Under these conditions, assuming negligible gravity and normal stress effects, the momentum equation (1.16) in projection along the direction x reduces to

$$\rho \frac{\partial u(y)}{\partial t} = \frac{\partial \tau}{\partial y} \tag{3.29}$$

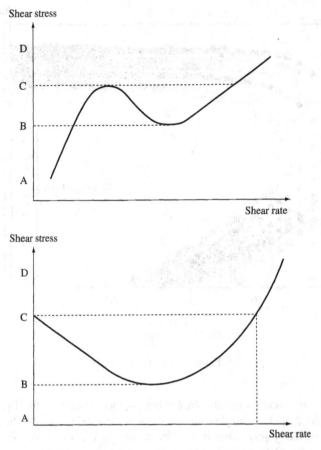

Figure 3.16 Schematic aspects of flows curves with an unstable flow region: (a) S-shape curve; (b) flow curve with a decreasing part.

Since we can write $\partial\tau/\partial y = (\partial\dot{\gamma}/\partial y)(\partial\tau/\partial\dot{\gamma}) = (\partial^2 u/\partial y^2)(\partial\tau/\partial\dot{\gamma})$, (3.29) may be rewritten

$$\rho\omega = -k^2\left(\frac{\partial\tau}{\partial\dot{\gamma}}\right) \tag{3.30}$$

This equation implies that in the region of decreasing stress ω is positive, which means that the amplitude of the perturbation, proportional to $\exp(\omega t)$, constantly increases in time. Since a real perturbation (of any form) can be decomposed as the sum of an infinity of sinusoidal perturbations of the form (3.28), each leading to flow instability, the flow is unstable. Thus, in practice, a "flow curve with a decreasing part" is fundamentally nonsense since steady state, homogeneous flows cannot be obtained in this part of the curve. When such a decreasing part is nevertheless observed in practice [53,54] it likely reflects a strong localization of shear in a thickness of the order of few diameters of material elements, and the apparent behavior in fact reflects the local behavior of this thin layer for which the validity of the continuum assumption is doubtful (see Section 3.3.4).

Let us now consider the response of a material with a (virtual) flow curve of this type. When a shear stress in the region AB or CD is applied (see Figure 3.16), the resulting shear rate is unique: a finite value for an S-shape curve and zero for a flow curve with a decreasing part. When a stress in the region BC is applied, for a flow curve with a decreasing part, stable flows may be obtained either at a zero shear rate (rest) or at a shear rate in the increasing part of the flow curve. For an S-shape curve a flow can be obtained in either the first or the second increasing part of the flow curve, but no stable flow can be obtained in the decreasing part. For both flow curve types the fluid will thus "choose" the part of flow curve in which it flows, but the simple description of its behavior in steady state does not provide enough elements for determining which of them applies here. We expect some sandwichlike structure of the fluid with different layers sheared alternately at the first or second shear rate values. At this point we can note that the coexistence of two distinct flow rates suggests that they correspond to two different internal structures of the fluid, leading to two very different viscosity levels under the same stress values. In this context the transition from one state to the other should be a dynamical process depending on flow history, so a more complete approach of the constitutive equation of pastes requires us to account for time effects.

Time-Dependent Fluid Behavior

Additional elements to the constitutive equation attempting to describe the evolution of the fluid structure in time have been proposed: the Johnson–Segalman model for polymers [50], the model of Spenley et al. [55], or that of Bautista et al. [56] for wormlike micelles, which hold that the basic viscoelasticity of the fluid is affected by processes of breaking and re-formation of micelles. These models predict that the flow curve should have an S shape. For pasty materials various thixotropic models have been developed, which in general hold that the viscoplastic characteristics of the fluid are affected by evolution of the structure described by a kinetic equation (see Section 2.2.4). However, only few such models [47,49,52,57] predict a flow curve with a decreasing part. The simplest of them rely on the conventional approach of thixotropy (see Section 2.2.4), in which the apparent viscosity is a function of a structure parameter varying as a function of flow history [equation (2.24)]. Using an apparent viscosity depending only on the structure parameter ($\eta = \eta(\lambda)$), the constitutive equation of the material in steady state ($d\lambda/dt = 0$) is

$$\tau = \dot{\gamma}\eta(k^{-1}(1/\theta\dot{\gamma})) \tag{3.31}$$

which, roughly speaking, decreases at low shear rates only when the viscosity function increases more rapidly with λ than with the function k. This is, for example, the case with $k = \lambda$ and $\eta = (1 + \lambda^n)\eta_0$ or $\eta = \eta_0 \exp a\lambda$ (with $n > 1$ and $a > 0$) [47,48]. Looking at the predictions of such a model in transient flows, we find that it predicts a flow instability at low shear rates as a result of the simple competition between structuration and restructuration processes; at low shear rates the rate of restructuring remains higher than the rate of destructuring, so that $d\lambda/dt$ remains larger than a finite value and the fluid viscosity tends to continuously increase, a phenomenon that is not compatible with a steady-state flow. On the contrary, at sufficiently high shear rates there can be a balance between restructuring and destructuring so as to obtain a steady flow at a (relatively) low viscosity. Under controlled stress such a model thus predicts the phenomena (viscosity bifurcation) observed with various pasty materials [47,48] (see Section 2.2.4).

3.3.3 Shear Banding

Various techniques (small-angle neutron scattering, small-angle X-ray scattering, birefringence, etc.) have been used to study the suggested structural transition in wormlike micelles exhibiting this peculiar rheological behavior. The observations in general led to the conclusion that the plateau is associated with a phase transition, such as the transition between isotropic and nematic phases in wormlike micelles [58,59], the coexistence of the lamellar phase and onions in other surfactant mixtures [60], or the coexistence of different orientational structures [42]. Since such measurements concerned the whole material or at least all the material in a volume across the gap, they provided global information and it could be conceived that the phase transition was progressive in time but approximately homogeneous in the bulk at any given time. However, from observations by flow birefringence on wormlike micelles, it was finally noticed that this transition occurred in space [40]; in a Couette system the second phase appears at the first critical shear rate ($\dot{\gamma}_1$) and progressively invades the gap as the apparent shear rate ($\dot{\gamma}$) increases, until it completely occupies the gap for $\dot{\gamma} = \dot{\gamma}_2$. We will refer to this effect as *shear banding*. It should not be confused with the heterogeneity of the shear rate distribution, possibly with some unsheared regions, which naturally develops in yield stress fluid flows. With similar techniques and analogous materials it was confirmed that the shear rates in the two phases strongly differ [61,64].

These observations have encouraged investigators to focus on the velocity field within flows of soft jammed systems using different techniques [MRI (see Figure 3.17), light scattering, simple microscopy]. All the results show that "shear banding" develops, in particular in the plateau region. The velocity profiles obtained for laponite suspensions [65], wormlike micelles [66,67], and various other pasty materials such as chocolate, clay suspensions (bentonite [37], laponite [68]), mayonnaise, silica suspensions [69], Carbopol gel [70], cement paste [71], concentrated emulsion [28], foam [72], and sewage sludges [16], generally exhibit two regions in which the shear rates are almost constant but the two corresponding, average values significantly differ. We note that there were hints of such an effect from the pioneering observations of Magnin and Piau [11] (see Section 3.2.7). Finally, a similar shear banding phenomenon was obtained from molecular dynamics simulations of a model glass system [73], leading to an

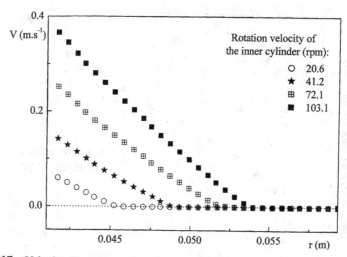

Figure 3.17   Velocity profiles as determined from MRI in a Couette geometry for a white cement paste under different rotation velocities of the inner cylinder. (Courtesy of S. Jarny.)

overall picture of flow characteristics very close to those described in the present section. These simulations, in which particles interact through Lennard-Jones potentials, tend to confirm that this is here a generic property of pastes, directly deriving from the general features of their internal structures, as described in Section 2.2.

In contrast with the predictions of flow characteristics in the case of simple yield stress models (see Section 4.2.1), the consequence of shear banding is a discontinuity in shear rate along the interface between the two regions. Moreover, in cone–plate geometries, the thickness of one of the bands increases linearly with the apparent shear rate while the shear rate in the sheared region does not change significantly [69,74,75] (see Figure 3.18). Note that the meaning of *shear banding* in this context slightly differs from that in soil or solid mechanics. Here the continuum assumption may remain valid within each band, in particular when their thickness is much greater than the size of fluid elements. In soil or solid mechanics shear banding generally takes the form of a strong localization of shear in a thickness typically of 5–15 particle diameters.

In contrast with micellar solutions, for pastes there is apparently no flow in one of the regions so that the shear rate seems to abruptly drop from a finite value ($\dot{\gamma}_c$) to zero through the interface between the two regions (see Figure 3.17). In Couette flow the sheared region is close to the inner cylinder, where the shear stress is larger. In that case, since the stress decreases with the distance from the axis, the shear rate may significantly vary in the sheared region. As a consequence, the discontinuity in shear rate at the interface between the sheared and the unsheared regions can be observed only at an appropriate scale that depends on the value of $\dot{\gamma}_c$ (see Figure 3.19). Note that for a Couette flow,

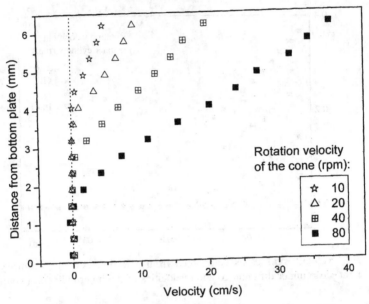

Figure 3.18 Tangential velocity profiles of a bentonite suspension within a cone–plate geometry under different rotation velocities of the upper cone as measured by MRI (from Ref. 69).

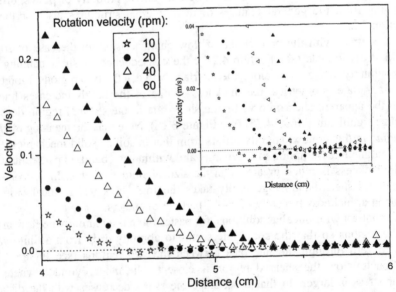

Figure 3.19 Velocity profiles in a Couette geometry for a foam (Gilette Co.). The insert illustrates the velocity on a smaller scale, clearly showing the slope break at the interface between the sheared and the unsheared regions. (Reprinted with permission from Ref. 72, Figure 1. Copyright (2005) by the EDP Sciences.)

although the relation (1.74) between the shear rate and the velocity is not simple, the slope discontinuity of the velocity is directly related to the critical shear rate. Indeed, since $v_\theta(r_c) = 0$, equation (1.74), in which we use $\omega = v_\theta/r$, yields:

$$\dot{\gamma}_c = \dot{\gamma}(r_c) = \left(\frac{\partial v}{\partial r}\right)_{r=r_c} \tag{3.32}$$

Assuming that we can neglect time effects, a simple means of representing these observations is to consider that the fluid in steady-state flow can be in a state with a behavior either (1) in simple shear of the form $\tau < \tau_c \Rightarrow \tau = f(\dot{\gamma})$ or (2) of the form $\tau > \tau_c \Rightarrow \tau = g(\dot{\gamma})$, in which f and g are two increasing, positive functions such that $f^{-1}(\tau_c) = \dot{\gamma}_1 < \dot{\gamma}_2 = g^{-1}(\tau_c)$. Although such a representation seems devoted primarily to the description of the apparent behavior of wormlike micelles in steady state, it can also be used for pasty materials by taking $\dot{\gamma}_1 = 0$ (which implies that f is not defined in that case). The apparent flow curve of such a fluid has a plateau between $\dot{\gamma}_1$ and $\dot{\gamma}_2$ similar to that described in Figure 3.13. Indeed, let us consider a situation for which some shear rate is imposed on the fluid while the shear stress is approximately homogeneous (e.g., in a cone–plate geometry). For an apparent shear rate $\dot{\gamma}$ in the range $[\dot{\gamma}_1; \dot{\gamma}_2]$, the stress is necessarily equal to τ_c; otherwise the shear rate would be given by one of the constitutive equations presented above and would be outside the range $[\dot{\gamma}_1; \dot{\gamma}_2]$. The shear rates of the material in states 1 and 2 are respectively $\dot{\gamma}_1$ and $\dot{\gamma}_2$, and the sample develops two parallel bands in states 1 and 2 of respective thicknesses h_1 and h_2. The relative velocity of the boundaries may then be written $V = h_1\dot{\gamma}_1 + h_2\dot{\gamma}_2 = H\dot{\gamma}$, from which we deduce that, in the absence of wall slip or other perturbing effects, the corresponding relative gap fractions of the material in states 1 and 2, respectively $\varepsilon_1 = h_1/H$ and $\varepsilon_2 = h_2/H$, should respect the following relation:

$$\dot{\gamma} = \varepsilon_1\dot{\gamma}_1 + \varepsilon_2\dot{\gamma}_2 \tag{3.33}$$

This is the so-called "lever rule."

The results described above have been obtained under the assumption that the stress is perfectly homogeneous within the gap. In reality there always remain some sources of heterogeneity, which will determine the spatial distribution of the regions 1 and 2. State 1 thus preferentially localizes in the regions of smaller stresses and state 2, in the other regions. In the case of pastes, the lever rule is expressed as

$$\dot{\gamma} = \frac{y_c}{H}\dot{\gamma}_c \tag{3.34}$$

in which y_c is the thickness of the sheared region. For example, when the sheared region corresponds to the largest y values, since we have $\dot{\gamma}_{\text{eff}} = du/dy = \dot{\gamma}_c$ in the sheared region, the velocity profile is given as $u(y) = 0$ for $y < y_c$ and $u(y) = \dot{\gamma}_c(y - y_c)$ for $y > y_c$.

With pasty materials the steady-state velocity profiles exhibiting shear bands seem to be stable over short time periods, but it was shown [69] that the extent of the unsheared region in steady state increases with the solid fraction for suspensions and with the time of rest before imposing shear for some thixotropic materials. This means that, more generally, the thickness of the rigid region increases with the "degree of jamming" of the structure, which implies that the critical shear rate increases as well. Also, for materials with a marked thixotropic character, it was observed that the thickness of the unsheared region increases or decreases in time respectively after a sudden velocity decrease or increase [37]. These observations seem to favor an explanation based on reversible time-dependent effects such as those of the various relatively sophisticated models mentioned in Section 3.3.2. These models generally assume that the material structure, although remaining in some similar disorder on a mesoscopic scale, may evolve in time so as to be somewhat jammed depending on flow conditions. In this frame the thixotropy model (see Section 3.3.2) provides a useful, generic explanation for these effects; around a critical stress there is a viscosity bifurcation that is in fact associated with a bifurcation of the jamming state. A slight heterogeneity in stress distribution may thus lead to the formation of two regions with very different flow properties.

The situation seems different for micellar solutions; the two coexisting regions appear to contain two materials of very different mesoscopic structures [67], and various studies have suggested that the phenomena were more complex than the simple, stable shear banding as described above. For example, it has been shown that the band stability is questionable, the localization of shear in the regions of larger stresses do not seem obvious for all systems [76], and the "lever rule" implies an extreme localization of shear at low velocities, which is not in agreement with birefringence observations. Generally speaking, the shear banding has been shown to be a dynamical process [77], with an evolution depending on the flow history. For wormlike micelles it was also observed [78], from study of the spatial distribution of the transmitted light intensity through the gap, that in one band the flow can be inhomogeneous on a small scale (say, $150\mu m$); the band consists of small subbands closely aligned in the flow direction. Finally the most complete study [67,79] on this subject showed that the shear banding is stable only when averaging the velocity profiles over sufficiently long times, but instantaneous velocity profiles (taken over ~ 1 s) appear to significantly fluctuate around this mean value. In particular, the position of the interface between the two regions significantly varies in time (Figure 3.20).

3.3.4 Implications for Paste Rheometry in the Continuum Regime

Generalities

Shear banding occurs as soon as the critical shear rate falls in (or is higher than) the range of apparent shear rates observed by the experimenter. The point is that shear banding is not an artefact or a perturbing effect that can be avoided such as wall slip or drying, it reflects the effective behavior of the material, which is uncapable of flowing steadily in a certain range of shear rates. Shear banding should thus be considered at best in order to deduce the effective behavior of the

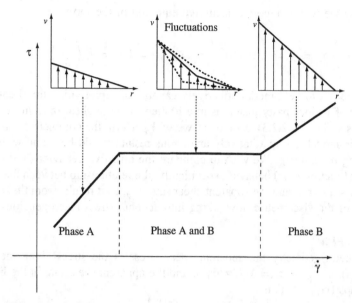

Figure 3.20 Schematic aspects of the velocity profiles for micellar solutions in Couette flows under different rotation velocities of the inner cylinder as observed by Salmon et al. [67] and corresponding state of the material (in two possible phases). The continuous lines correspond to the velocity profiles averaged over some time and the dotted lines, to the "instantaneous" velocity profiles.

material. In this frame two cases may be distinguished depending on whether the sheared layer can be considered as a continuum medium exhibiting an intrinsic constitutive equation. The first case (*continuum regime*) in general corresponds to flow regimes such that the size of the sheared layer is much larger than the maximum size of the suspended elements, while the second case (*discrete regime*) is generally encountered otherwise.

In the continuum regime the obvious impact of shear banding on conventional rheometry is that the apparent shear rate differs from the effective shear rate in the sheared region [see equation (3.34)] as long as there exists an unsheared region. The corresponding, apparent flow curve thus does not reflect the effective behavior of the material. For example, one can obtain apparent flows at any shear rate. The exact form of the apparent constitutive equation in fact depends on the stress distribution in the gap, which conditions the shear rate distribution. In some cases it is possible to have an idea of these effects. A quantitative approach may be proposed for a Couette geometry, since the stress distribution is well known in that case; a more qualitative approach may also be appropriate for a cone–plate geometry on the basis of some simple assumption concerning the stress distribution.

In the following text we attempt to compare the effective constitutive equation of the material in steady state with its apparent flow curve. For the sake of simplicity we will assume a constitutive equation of the form

$$\tau < \tau_c \Rightarrow \dot{\gamma} = 0; \tau > \tau_c \Rightarrow \tau = g(\dot{\gamma}) = k\dot{\gamma}^n = \tau_c \left(\frac{\dot{\gamma}}{\dot{\gamma}_c}\right)^n \qquad (3.35)$$

which appears to be a model effectively capable of representing the rheological behavior of various pasty materials in a limited range of shear rates in the liquid regime (see Section 2.2.3). Moreover, we will perform the complete calculations using this model, but qualitatively analogous results would be found with other increasing functions g. We wish to compare the effective behavior (3.35) to the apparent (macroscopic) behavior, specifically, the relationship between the apparent shear stress τ_{app} and the apparent shear rate $\dot{\gamma}_{app}$, as it results from the detailed solution of the viscometric flow taking into account stress heterogeneities.

Couette Flow

In a Couette geometry the apparent shear stress is the stress along the inner cylinder (i.e., $\tau_{app} = \tau_1 = M/2\pi h r_1^2$), and the apparent shear rate in the limit of small gaps ($r_1/r_2 \approx 1$) is

$$\dot{\gamma}_{app} \approx \frac{v_\theta(r_1)}{(r_2 - r_1)} \qquad (3.36)$$

Let us recall that the shear stress decreases with the distance r in the gap (see Section 2.5). For a fluid following (3.35), there is no motion as long as the applied torque is such that the apparent shear stress is less than τ_c. For a slightly larger, apparent shear stress [i.e., $\tau(r_2) < \tau_c < \tau_{app}$], there is an unsheared region ($v_\theta(r) = 0$) between $r_c = (C/2\pi h \tau_c)^{1/2}$, the distance at which τ_c is reached, and r_2. Using the stress distribution (1.76) and the relation between the velocity and the shear rate (1.75), we obtain by integration the velocity profile in the fluid in the sheared region (equation (4.24b)) from which we deduce the velocity along the inner cylinder:

$$v_\theta(r_1) = \frac{n}{2}\dot{\gamma}_c \left[\left(\frac{r_c}{r_1}\right)^{2/n} - 1\right] r_1 \qquad (3.37)$$

The local stress at a distance r may be expressed as $\tau(r) = \tau_c(r_c/r)^2$ (see Section 4.2.2), so that in particular

$$\tau_{app} = \tau_c \left(\frac{r_c}{r_1}\right)^2 \qquad (3.38)$$

Using (3.36) and eliminating r_c/r_i from (3.37) and (3.38), we deduce the relation between the apparent shear stress and the apparent shear rate in this regime:

$$\tau_2 < \tau_c < \tau_{app} < \tau_c \left(\frac{r_2}{r_1}\right)^2 \Rightarrow \tau_{app} = \tau_c \left[1 + K\frac{\dot{\gamma}_{app}}{\dot{\gamma}_c}\right]^n \qquad (3.39)$$

in which $K = 2(r_2 - r_1)/nr_1$. Note that this apparent behavior tends to a Bingham model in the limit of low $\dot{\gamma}_{app}$.

When $\tau_c < \tau_2$, all the fluid in the gap is sheared and the integration of the constitutive now gives

$$v_\theta(r_1) = \frac{n}{2}\left(\frac{\tau(r_1)}{k}\right)^{1/n}\left[1 - \left(\frac{r_1}{r_2}\right)^{2/n}\right]r_1 \qquad (3.40)$$

from which we deduce

$$\tau_{app} > \tau_c(r_2/r_1)^2 \Rightarrow \tau_{app} = \tau_c\left[\frac{K}{(1 - (r_1/r_2)^{2/n})}\frac{\dot{\gamma}_{app}}{\dot{\gamma}_c}\right]^n \qquad (3.41)$$

The transition between the regimes (3.39) and (3.41) occurs when the yield stress is reached along the outer cylinder, which means that $r_2 = r_c$ so that $\tau_{app} = \tau_c(r_e/r_i)^2$. The corresponding, critical, apparent shear rate is

$$\dot{\gamma}_X = \frac{\dot{\gamma}_t}{K}\left[\left(\frac{r_2}{r_1}\right)^{2/n} - 1\right] \qquad (3.42)$$

Thus the corresponding, complete, apparent flow curve, as given by (3.39)–(3.41), is that of a typical, simple yield stress fluid.

Cone–Plate Flow

Although we can seldom know the exact stress distribution in a cone–plate geometry, it is possible to estimate the qualitative effect of shear banding with the help of simple assumptions. Let us consider the flow of a fluid with a constitutive equation defined by (3.35) in a cone–plate geometry of small angle θ. For a "macroscopic" simple shear we can again assume (as in Section 1.4.3) that the (slightly conical) layers of fluid situated at different angles (x) between the cone and the plate glide over one another. However, here we take into account some possible slight variations of the radius (R) of these fluid layers as a function of the angle x: $R = R(x)$. We also assume for the sake of simplicity that the applied torque (M) is completely transmitted through the successive conical sample layers at different angles, and that the shear stress along each of these layers is approximately constant. Under these conditions we can deduce an approximate expression for the shear stress in the layer situated at the angle x:

$$\tau(x) = \frac{3M}{2\pi R(x)^3} \qquad (3.43)$$

According to this equation, a variation in layer radius of 1 mm for a cone radius of 1 cm, as is typically observed in practice, would induce a difference of more than 30% between the maximum and the minimum stress values. Thus the imperfection of the shape of the sample periphery is certainly the major source of stress heterogeneity, far more than those due to the cone angle (see Section 1.4.3).

Here we have no information concerning the function $R(x)$, but for the sake of simplicity we assume that $R(x)$, although it remains close to R_0, is an increasing function of x given by

$$R = R_0 + ax^p \tag{3.44}$$

where a and p are two positive parameters. Since this shape results from a slight perturbation of the peripherial free surface, we have $a\theta^p \ll R_0$. Under these conditions, from (3.43) and (3.44), the stress at the angle x can be expressed as

$$\tau(x) \approx \frac{3M}{2\pi R_0^3}\left(1 - \frac{3a}{R_0}x^p\right) \tag{3.45}$$

From (3.35) and (3.45) we conclude that no flow can occur as long as the apparent stress $\tau_{\text{app}} = 3M/2\pi R_0^3$, which is also the maximum shear stress in the fluid, is lower than τ_c. For an apparent stress slightly higher than τ_c the fluid is sheared in a region comprised between $x = 0$ and $x = x_c$ such that

$$\tau_c \approx \tau_{\text{app}}\left(1 - \frac{3a}{R_0}x_c^p\right) \tag{3.46}$$

It follows from (1.71) that the apparent shear rate is in fact the mean shear rate over the gap

$$\dot{\gamma}_{\text{app}} = \frac{1}{\theta}\int_0^{x_c} d\omega = \frac{1}{\theta}\int_0^{x_c}\dot{\gamma}(x)dx \tag{3.47}$$

so that, using (3.45), we find

$$\dot{\gamma}_{\text{app}} = \frac{1}{\theta}\int_0^{x_c}\dot{\gamma}_c\left(\frac{\tau(x)}{\tau_c}\right)^{1/n}dx \approx \frac{\dot{\gamma}_c}{\theta}\left(\frac{\tau_{\text{app}}}{\tau_c}\right)^{1/n}\left[x_c - \frac{3a}{(p+1)nR_0}x_c^{p+1}\right] \tag{3.48}$$

Introducing (3.46) in (3.48), we find

$$\dot{\gamma}_{\text{app}} \approx \dot{\gamma}_c\frac{x_c}{\theta}\left[1 + \frac{3ap}{(p+1)nR_0}x_c^p\right] \approx \dot{\gamma}_c\frac{x_c}{\theta} \tag{3.49}$$

From (3.46) and (3.49) it follows (using $K = 3a\theta^p/R_0$) that the apparent constitutive equation is

$$\tau(\theta) < \tau_c < \tau(0) \Rightarrow \frac{\tau_{\text{app}}}{\tau_c} \approx 1 + K\left(\frac{\dot{\gamma}_{\text{app}}}{\dot{\gamma}_c}\right)^p \tag{3.50}$$

which appears to be a Herschel–Bulkley model. However, we emphasize that some characteristics of this apparent Herschel–Bulkley behavior such as the power p and the coefficient K are dictated by the shape of the peripherial free surface, and not by the effective behavior of the fluid.

When the applied stress is sufficiently large [i.e., when $\tau(\theta) > \tau_c$], the apparent shear rate is again given by (3.48), in which we can now insert $x_c = \theta$ to obtain:

$$\tau(\theta) > \tau_c \Rightarrow \frac{\tau_{\text{app}}}{\tau_c} \approx \left(\frac{\dot{\gamma}_{\text{app}}}{\dot{\gamma}_c}\right)^n \left[1 + \frac{K}{(p+1)}\right] \tag{3.51}$$

Obviously in that case, since $K/(p+1) \ll 1$, the resulting apparent constitutive equation is close to the effective behavior (3.35) of the material beyond the critical stress. The transition between the regimes (3.50) and (3.51) occurs when $\tau(\theta) = \tau_c$, that is, at the critical, apparent shear rate

$$\dot{\gamma}_X = \dot{\gamma}_c \left[1 + \frac{pK}{(p+1)n}\right] \tag{3.52}$$

which is close to $\dot{\gamma}_c$.

Discussion

The complete apparent flow curve as given by (3.50) and (3.51), along with some qualitative views of the shear factor in the material, is shown in Figure 3.21. Note that in this figure, for the sake of clarity, the sheared thickness has been assumed around the center of the gap, which does not exactly correspond to the predictions of the theory stated above. As the apparent shear rate is increased from zero, the sheared thickness increases as predicted by (3.49). When $\dot{\gamma}_{\text{app}} > \dot{\gamma}_X$, the fluid is completely sheared in the gap and the apparent flow curve is close to the effective behavior, with a slight shift resulting from the slight stress inhomogeneity within the gap. Qualitatively analogous results would be obtained with a parallel disk geometry or a capillary geometry. The fundamental result is that in the continuum regime the presence of shear banding implies that the apparent flow curve of the material is that of a simple yield stress fluid even if the effective behavior corresponds to a fluid uncapable to flow steadily under a critical shear rate.

Detection or Control of Shear Banding

The results given above show that the usual rheometrical approaches are insufficient for detecting shear banding since the apparent flow curve is a priori interpreted as that of the homogeneous flow of a simple yield stress fluid. Shear banding can nevertheless be suspected when the following effects are observed:

1. The shearing observed at the periphery of the material seems to be heterogeneous.
2. A wide plateau region appears during increasing stress ramps.
3. Below an apparent shear rate the flow seems to be unstable; under controlled stress the shear rate tends to decrease significantly in time for a stress below a critical value (viscosity bifurcation effect).

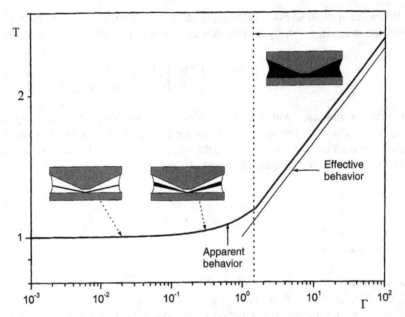

Figure 3.21   Apparent flow curve as obtained from rheometrical data with a cone and plate on a material with a constitutive equation of the type (3.35) according to the calculations given in Section 3.3.4 assuming some specific stress distribution within the sample. The effective flow curve is represented by the thin line and the apparent flow curve, by the thick line. The corresponding aspects of the distribution of shear within the material at different apparent shear rates are shown (unsheared regions in white sheared regions in dark).

In fact, a geometry in which the stress distribution is heterogeneous but controlled, such as the Couette geometry, makes it possible to go further in the analysis. Instead of using the rough approach leading to compute the average shear rate in the gap, we can use the systematic approach described in Section 1.4.4, which takes into account the shear rate heterogeneity. Let us recall that this approach consists in carrying out a series of experiments ($p = 0, 1, 2, \ldots$) under successive decreasing torque values in a ratio β, leading to successive shear stresses along the inner cylinder $\beta^p \tau_1$ and along the outer cylinder $\beta^{p+1} \tau_1$. Let us again assume that the fluid follows the constitutive equation (3.35). For the test corresponding to the value p, if $\beta^{p+1} \tau_1 > \tau_c$ the whole gap is sheared and we deduce from (3.40) the expression for the rotation velocity of the inner cylinder ($\Omega = v_\theta(r_1)/r_1$), which may usefully be written as

$$\Omega = \frac{n}{2} \left[\left(\frac{\beta^p \tau_1}{k} \right)^{1/n} - \left(\frac{\beta^{p+1} \tau_1}{k} \right)^{1/n} \right] \tag{3.53}$$

When $\beta^{p+1} \tau_1 < \tau_c$, there is an unsheared region in the gap and we deduce from (3.37) the expression for the rotation velocity as a function of the shear stress along the inner cylinder:

$$\Omega = \frac{n}{2}\left[\left(\frac{\beta^p \tau_1}{k}\right)^{1/n} - \left(\frac{\tau_c}{k}\right)^{1/n}\right] \tag{3.54}$$

Finally, when $\beta^p \tau_1 < \tau_c$, we obviously have $\Omega = 0$. Since for any value of τ_1 there exists a critical value p_c for which $\beta^{p_c+1}\tau_1 < \tau_c$, we may rewrite the sum (1.81) as

$$
\begin{aligned}
\dot{\gamma}(\tau_1) &= \left[\sum_{p=0}^{p_c-1}\left(2\tau\frac{\partial\Omega}{\partial\tau}\right)_{\beta^p\tau_1}\right] + \left(2\tau\frac{\partial\Omega}{\partial\tau}\right)_{\beta^{p_c}\tau_1} \\
&= \left[k^{-1/n}\sum_{p=0}^{p_c-1}((\beta^p\tau_1)^{1/n} - (\beta^{p+1}\tau_1)^{1/n})\right] + k^{-1/n}(\beta^{p_c}\tau_1)^{1/n} = \left(\frac{\tau_1}{k}\right)^{1/n}
\end{aligned}
\tag{3.55}
$$

This analysis of rheometrical data predicts a constitutive equation such that $\dot{\gamma}(\tau_1) = 0$ if $\tau_1 < \tau_c$ and $\dot{\gamma}(\tau_1) = (\tau_1/k)^{1/n}$ otherwise, which corresponds to the effective constitutive equation of the fluid. This result can be generalized to constitutive equation of the type (3.35) with other functions describing the behavior in the sheared region. Thus the analysis of Couette flows following the exact theoretical approach described in Section 1.4.4, which accounts for the shear rate heterogeneity, gives the effective constitutive equation even in the presence of shear banding. This is so as long as shear banding develops within the frame of the continuum assumption (so that the local constitutive equation is that of the fluid), and such an approach is possible as long as the heterogeneous stress distribution is controlled.

3.3.5   Paste Rheometry in the Discrete Regime

The preceding results imply that under (apparent) shear rates sufficiently lower than $\dot{\gamma}_c$ the thickness of the sheared region may be of the order of a few times the size of the elements of the fluid. In that case the continuum assumption is no longer valid, and thus the flow of the corresponding band does not correspond to a flow of the homogeneous paste (on a larger scale) and the rheological behavior of the sheared layer may be inconsistent with that of the sample. For example, the apparent behavior of a single layer of droplets sheared between two plates should a priori differ from the behavior of a much larger three-dimensional volume of an emulsion made of the same droplets in an interstitial liquid. Depending on material structure, different situations seem to be encountered for steady flows under controlled (apparent) shear rate.

For certain materials there is likely some specific arrangement (layering) of the suspended elements on the scale of the layer so that the shear stress needed to maintain a shear is lower than that resulting from a macroscopic flow. This

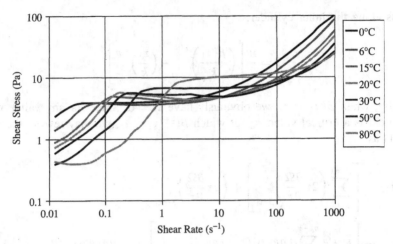

Figure 3.22 Flow curves for drilling fluids under different temperatures (with rough surfaces). The data have been obtained from a logarithmic shear rate decrease step by step (40 levels, 2 min per level) from 1000 to 0.01 s⁻¹. (From Ref. 80, Figure 1.)

leads to a drop in the flow curve at low shear rates (Figure 3.22). In that case the apparent flow curve might seem to increase at low shear rates before reaching the plateau. This effect is reminiscent of elastic effects in a stress ramp (see Section 3.1), although it is obtained in steady state here, and reminiscent of wall slip effects (see Section 3.2), although it is obtained with rough surfaces here.

For other materials the flow on a small scale requires a stress higher than that needed for the macroscopic flow of the homogeneous paste under some shear rate, so the apparent flow curve tends to decrease at low shear rates before reaching a plateau (see Figures 3.23 and 3.24). This is, for example, the case for a foam [72], an emulsion [80], or a laponite suspension [65]. Such a situation cannot reflect the effective rheological behavior of a unique, homogeneous material since it would lead to flow instability (see Section 3.3.2). For such materials under controlled stress no steady-state data can be obtained in the range of shear rates for which such a decreasing flow curve is obtained under controlled velocity. This effect is reminiscent of the decrease of the friction coefficient for solids at low velocities, and for pastes it indeed likely corresponds to some kind of "frictional" effects in a thin thickness of such particulate materials.

At last, for many pasty materials the apparent flow curve is that of a simple yield stress fluid; the shear stress needed to shear a very thin layer of material tends toward a plateau at low flow rates, which is in the continuity with the rest of the flow curve, and the level of this plateau does not significantly differ from that expected from a continuous approach using the effective yield stress (Section 3.3.4). One thus obtains the usual, apparent, simple yielding behavior represented by a Herschel–Bulkley model [81]. This might result from the fact that even under apparent shear rates for which the sheared thickness should in theory be of the order of the element size, there is some kind of mixing of the

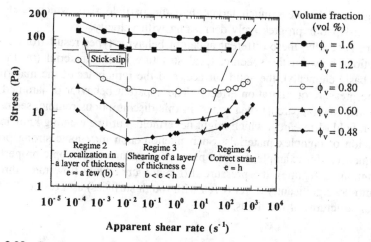

Figure 3.23 Steady-state flow curve for a laponite suspension at different solid volume fractions as obtained from cone–plate rheometry under controlled rotation velocity. The different regimes have been identified from direct observations of the velocity profile within this transparent material; h is the gap; b, the particle size; and e, the sheared thickness. (From Ref. 65, Figure 2, with kind permission from F. Pignon.)

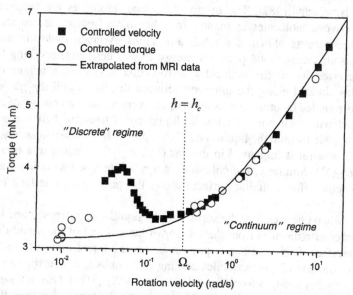

Figure 3.24 Foam "flow curve" obtained under different conditions from experiments in a Couette geometry: from controlled torque or controlled velocity tests. Each data point corresponds to the stress–shear rate pair obtained after 60 s of flow. The continuous line corresponds to the flow curve as deduced from MRI data assuming that the sheared layer stays in the continuous regime regardless of its thickness. (Reprinted with permission from Ref. 72, Figure 4. Copyright (2005) by the EDP Sciences.)

material in a thickness much larger than the particle size, which leads to an energy dissipation progressively decreasing with velocity.

More generally, these different apparent behavior types result from complex interactions between the specific, local structure of the material (on the scale of the basic elements), the solid surface, and the remainder of the material (via osmotic pressure or migration effects). The apparent behavior of jammed materials under low (apparent) shear rates, in this discontinuous regime, seems to be an open field of research, which might nevertheless utilize existing knowledge in lubrication of complex materials. Such phenomenon may have strong practical consequences; for example, for two pastes of similar macroscopic properties (in the continuum regime), the pressure difference under low flow rates through a conduit may significantly differ (even with rough surfaces) as a result of different internal structures of the fluids.

3.4   SURFACE TENSION EFFECTS

3.4.1   Surface Tension in Simple Liquids

Let us review some basic processes at the origin of surface tension effects. For more details in the case of simple liquids, the reader is referred to the text by Israelachvili [82]. The origin of surface tension is the cohesive force acting between molecules in matter. In a liquid the molecules are stacked to each other because of Van der Waals attractions. A given molecule within the liquid has an average total potential energy E_L (of cohesion) resulting from the forces exerted by all the molecules surrounding it (its close neighbors). The molecules situated along the interface between the liquid and the gas are only partly surrounded by other molecules and, on average, have an energy of cohesion associated with their contact with the liquid equal to about half the energy of those embedded within the liquid (i.e., $E_L/2$), plus half the cohesion energy of this molecule when it is immersed in the gas (i.e., $E_{LG}/2$), leading to a total energy $(E_L + E_{LG})/2$. Similarly, the cohesion energy of the corresponding fraction of gas molecules (for a similar surface) along the gas–liquid interface is $(E_G + E_{LG})/2$.

Now, when one increases the surface of the liquid–gas interface, one increases the number of molecules from the interior of the gas and of the liquid that come to the interface. Each elementary process involves a liquid molecule and a corresponding fraction of gas molecules coming into contact, so that the sum of their cohesion energy changes from $E_L + E_G$ to $(E_L + E_{LG})/2 + (E_G + E_{LG})/2$. The work required for this elementary process is the difference between these two energies: $\delta w = E_L/2 + E_G/2 - E_{LG}$. The total work for an increase dS of the surface of the interface is $dW = \delta w N\, dS$, where N is the number of molecules per unit surface. Surface tension effects occur when $E_{LG} \neq (E_L + E_G)/2$ so that $\delta w \neq 0$, which is generally the case. Using the *interfacial tension* between the gas and the liquid (i.e., $\gamma_{LG} = N\delta w$), also often referred to as the *surface tension* of the liquid, the surface energy may be rewritten

$$dW = \gamma_{LG} dS \qquad (3.56)$$

Similar definitions for the interfacial tension between a material A and a material B may be established. The surface tension for water in air at 20°C is 0.073 Pa·m. It varies from less than 10% around this value between 0 and 50°C. For other liquids it generally ranges from 0.02 to 0.08 Pa·m.

Let us now consider a droplet of liquid immersed in a gas. We assume that the pressure p_L in the liquid and p_G in the gas remain constant. Note that from a physical perspective these pressures are also related to the energy of cohesion; the work necessary to slightly increase the volume of the liquid or the gas under a given pressure is the work of the total cohesive forces between molecules over a (short) distance associated with the volume increase. Let us now assume that the droplet volume is changed by a small amount $d\Omega$, which leads to a change of the gas volume by a quantity $-d\Omega$ and a change of the surface of the gas–liquid interface by an amount dS. The total work to be supplied is equal to the sum of the work of change of the liquid volume $(-p_L d\Omega)$ and the gas volume $(p_G d\Omega)$ plus the surface energy associated with the interface change: $dW = -p_L d\Omega + p_G d\Omega + \gamma_{LG} dS$. We expect that there exists a position of equilibrium, for which the energy of the system (W) reaches a minimum, such that $dW = 0$. For a droplet assumed to remain spherical with a radius R, we have $d\Omega = 4\pi R^2 dR$ and $dS = 8\pi R\, dR$, and the equilibrium corresponds to $dW/dR = 0$, from which we deduce the following equilibrium condition for the pressure difference at the interface $(\Delta p = p_L - p_G)$:

$$\Delta p = \frac{2\gamma_{LG}}{R} \qquad (3.57)$$

More generally, for an interface of any form, it may be shown that the boundary condition along the interface of normal unit vector $\mathbf{n}$ between two materials (A and B) is

$$\Sigma_B \cdot \mathbf{n} - \Sigma_A \cdot \mathbf{n} = -\gamma \left(\frac{1}{R_1} + \frac{1}{R_2} \right) \mathbf{n} \qquad (3.58)$$

in which Σ_A and Σ_B are the stress tensors in each material and R_1 and R_2 are two radii of curvature of the interface in orthogonal planes containing $\mathbf{n}$. Note that the normal vector $\mathbf{n}$ here is oriented from A to B, and the sign of the radii of curvature must be taken into account; for example they are in general negative when A is a compact droplet.

Let us now consider the coexistence of the three phases (liquid, solid, and gas) along a line of contact. We assume that the solid surface is planar and that along the line of contact the angle between the gas–liquid and the liquid–solid interface is θ_E. When this angle is slightly changed along a length L of the line of contact, this induces a small displacement of this line by a distance Δx, where x is the abscissa on the solid surface in the direction perpendicular to the line of contact.

Around this region of the line of contact the liquid–solid interface increases by a surface amount $L \, \Delta x$, the solid–gas interface by a surface amount $-L \, \Delta x$, and the liquid–gas interface by a surface amount $L \, \Delta x \, \cos \theta_E$. Neglecting other possible changes in the rest of the liquid volume, at equilibrium the variation of interfacial energy resulting from this angle change must be equal to zero, which is also expressed as $\Delta W / \Delta x = 0$. Since $\Delta W = \gamma_{LS} L \, \Delta x - \gamma_{SG} L \, \Delta x + \gamma_{LG} L \, \Delta x \, \cos \theta_E$, in which γ_{SG} and γ_{SL} are the solid–gas and solid–liquid interfacial energies, we obtain the following condition for equilibrium of the interface between the three phases:

$$\gamma_{SG} = \gamma_{LG} \cos \theta_E + \gamma_{SL} \tag{3.59}$$

The relationships (3.57) and (3.59) provide practical means for measuring the surface tension of simple liquids.

3.4.2   Surface Tension for Non-Newtonian Liquids

The preceding description of the physical properties at the origin of surface tension basically accounts for simple Van der Waals attractive forces between molecules in a simple liquid. A relatively significant departure from this scheme may be feared for non-Newtonian fluids since their mechanical characteristics result from more complex local interactions between their elements (see Section 2.1). For simple liquids, a change of the material volume induces a configuration change, which in turn induces mainly a change in the total cohesive energy. A paste contains not only a liquid (the interstitial fluid), the molecules of which are stacked to each other, but also a lot of elements of different types and sizes dispersed in the liquid and that develop various interactions at different scales. In general, all these elements remain surrounded by liquid molecules, which in particular form the interface of the paste with the ambient gas. A change in the surface of this interface induces a change of the number of liquid molecules in contact with gas molecules, as in a simple liquid. This also implies some changes in the element configuration, which appears to be rather difficult to quantify. Here, for the sake of simplicity, we will assume that the latter effect is completely accounted for by the rheological behavior of the paste, which remains the same after any type of deformation.

Finally, we can separate the potential energies acting at different scales within the fluid into two classes:

- The cohesion energy between liquid molecules
- The other interactions between all elements dispersed in the liquid

In this context, assuming that there is always some liquid layer around the elements, a deformation of the material possibly leading to interface changes will require an energy strictly related to the change of the liquid–gas interface *and* an energy related to the change of element configuration. The total energy will thus be equal to the sum of an energy strictly related to surface tension effects of the interstitial liquid and mechanical effects strictly related to the macroscopic flow of the material assumed to remain homogeneous and constant. Under these conditions *the surface tension of pastes is equal to that of the interstitial liquid.* It is thus close to that of pure water in the absence of any additives but may be modified by the presence of hydrophilic or hydrophobic molecules. The different mechanical effects of surface tension described in the previous paragraph remain valid in this context. Note that the practical determination of the surface tension of pastes is problematic since both the pasty behavior of the material and surface tension effects may tend to fix the gas–paste interface in some specific position.

3.5 DRYING

In rheometrical tests with rotational geometries there is generally some boundary of the sample that is a free surface, that is, an interface between the fluid and the ambient gas. The liquid in the sample may then evaporate in the gas. Since the rate of evaporation increases with the relative velocity of the liquid and the gas, the sample may dry much faster than if it were left at rest in a container. For simple liquids, the consequence of such a phenomenon is simply a decrease in sample volume. But for pasty materials, a decrease of the liquid fraction may also affect their rheological properties. For these materials, it is even generally expected that drying tends to lead to the formation of a crust poor in liquid along the gas–sample interface. In fact, we will see that if the drying rate is not too large, the elements in the liquid have time to disperse as a result of their mutual interactions so that the bulk sample shrinks almost homogeneously. Here we review basic drying processes and their effects on pastes. Then we examine the consequences of these processes in rheometry. Finally, we see the different practical means for avoiding this problem.

3.5.1 Evaporation

Evaporation from a simple, pure liquid occurs because the molecules submitted to thermal agitation tend to randomly escape from the liquid if their kinetic energy happens to overcome some energy barrier due to Van der Waals (force) attractions. The corresponding rate of departure thus increases with the temperature of the liquid. In parallel, some escaped vapor molecules moving through the gas may rejoin the liquid after some collisions with other liquid molecules. The corresponding rate of return depends on the motion of vapor molecules on approach of the interface and thus on flow characteristics and temperature of the ambient gas. The net rate of evaporation from the liquid–gas interface is equal to the difference between the rates of departure and return. At equilibrium these two rates are equal and the rate of evaporation is equal to zero. This means that the gas is saturated in vapor molecules.

If the gas is not saturated in vapor molecules and is macroscopically at rest, the extracted molecules tend to diffuse through it so that the rate of evaporation

(J_e) is equal to the rate of diffusion of vapor through the ambient gas. The rate of this diffusion process is proportional to the gradient of *vapor density* n (i.e., the ratio of liquid vapor volume to total gas volume) in the direction (**x**) normal to the liquid–gas interface. This can be expressed in the form of the following flux

$$J_e = -D\frac{dn}{dx} \qquad (3.60)$$

where D is the coefficient of diffusion of the vapor in the gas, which results from collisions between the different molecules [83]. As a consequence, at leading order, D is proportional to the square root of the temperature. Typically, for water in air at a temperature of 20°C we have $D = 2.4 \times 10^{-5} \text{m}^2/\text{s}$. In parallel, n increases with pressure and decreases with temperature. However, in practice, the main variations of the rate of evaporation result from the gradient of vapor density, which significantly varies with the dynamic characteristics of the gas along the interface.

Two asymptotic situations may be considered depending on whether the gas is at rest or in rapid motion along the interface. If the gas is macroscopically at rest along the interface, the evaporation occurs principally in the form of a free diffusion process. Under these conditions, for a simple, unidirectional diffusion process we have in steady state $d^2n/dx^2 = 0$, which implies that dn/dx is constant. If the interface is at a distance H, much larger than the mean free path of vapor molecules, from a surface along which the vapor density may be assumed to remain constant and equal to n_0, and if there is no gas flow between the two surfaces, the rate of evaporation is

$$J_e = D\frac{(n_v - n_0)}{H} = \frac{K}{H} \qquad (3.61)$$

in which n_v is the saturation vapor density and $K = K(T, p)$ a function of the temperature and pressure.

If, on the contrary, the interface is in direct contact with a (vapor free) convection current, the gas flow brings with it vapor molecules just extracted from the surface, which damps the rate of return of the vapor molecules to the interface. Now the distribution of vapor molecules in the gas depends on the coupling between their diffusion rate and the flow characteristics. Since gas flows are generally turbulent, the velocity significantly varies only over a short distance, namely, the *boundary layer*, from the interface. Within this frame a rough, although useful, approximation consists in considering that dn/dx is constant in a region of thickness δ of the order of that of the boundary layer, while the vapor density is maintained constant at n_0 in the region of nearly uniform velocity [84]. Thus we find

$$J_e = \frac{K}{\delta} \qquad (3.62)$$

Between the point ($y = 0$) where a gas flux starts to flow along an object, up to a certain point y_c, the boundary layer is laminar and, at a distance $y < y_c$, δ is expressed as [85]

$$\delta = 5\sqrt{\frac{\mu_g y}{\rho_g V}} \tag{3.63}$$

in which μ_g, ρ_g, and V are respectively the viscosity, the density, and the velocity of the gas. When $y > y_c$, the boundary layer is turbulent and δ will depend on the geometry of the flow and the roughness of the surface.

This theoretical approach makes it possible to distinguish asymptotic cases and provides the basic, qualitative trends of evaporation. However, in practice, the equation given above cannot be easily used to quantitatively predict the rate of evaporation in particular because it is very difficult to precisely control the gas velocity field and the temperature along the interface.

3.5.2 Drying Regimes

General

When the evaporating liquid is initially within a porous material, we speak of *drying*. Since pastes are generally composed of a liquid capable of evaporating, and elements (bubbles, polymers, solid particles, oil droplets, etc.) mostly uncapable of evaporating, we can assume that these elements form a porous material through which the liquid can evaporate so that the paste dries. Now the rate of drying and its impact on the paste properties depend not only on the rate of evaporation of the liquid in the ambient gas but also on the capability of the liquid to move through the porous system toward some preferential liquid–gas interfaces and continuously provide new molecules for evaporation. The rate of flow through a porous material may be estimated from Darcy's law, which is simply an extension of the Poiseuille law for the flow through a solid structure of complex shape

$$J_p = \frac{k}{\mu_0} \frac{dp}{dx} \tag{3.64}$$

in which μ_0 is the viscosity of the interstitial liquid and k the permeability of the porous system. This permeability is difficult to estimate a priori. Its order of magnitude in a general case may be estimated from the Kozeny–Carman equation for packing of spheres of radius R: $k = R^2(1 - \phi)^3/45\phi^2$, by using the typical pore size instead of R and the porosity instead of $1 - \phi$ in this equation.

Drying of Solid Porous Materials

For a solid porous material different drying regimes may be identified [86] depending on the relative values of the evaporation rates either from layers close to the external surface [equation (3.62)] or in the interior of the porous material [equation (3.61)], and a rate of flow by capillary effects. The latter phenomenon results from some imbalance between the pressure differences due to surface tension effects along the gas–liquid interfaces close to or far from the free surface of the porous material. This typically occurs when the evaporation has reduced the radius of curvature of the liquid–gas interfaces close to the free surface but left unaffected the liquid–gas interface far from this free surface. The corresponding *capillary flow rate* tending to recover the equilibrium of surface tension effects (in the absence of gravity effects) may be roughly estimated from equation (3.64) using a pressure gradient resulting from surface tension effects throughout the liquid network in the material.

When the rate of evaporation from the external surface of the material is smaller than the capillary flow rate, the liquid has enough time to move toward the free surface before evaporating, and the rate of drying is governed by the rate of evaporation imposed by boundary conditions (*evaporative regime* or *constant drying rate regime*). When the rate of evaporation from the external surface becomes larger than the capillary rate of flow (which decreases as the liquid fraction decreases), the latter begins to govern the rate of drying (*capillary regime*). Things continue like this as long as the capillary flow rate, which goes on decreasing and tends to zero for a critical fraction of liquid for which the liquid network becomes discontinuous, remains lower than the rate of evaporation from the interior of the sample. Otherwise the interface between the liquid network and the gas begins to significantly recede inside the sample and the rate of drying is governed by the rate of evaporation from the interior of the sample (*receding regime*).

Drying of Pasty Materials

For pasty materials things are slightly different. When some liquid is withdrawn from the system, the elements (bubbles, droplets, particles, molecules) immersed in the remaining liquid tend to gather, thus leading to a more concentrated, and generally more viscous, material. Let us again use the decomposition (2.32) in which now the second term (Σ_p) is associated with the network of interactions (in general at a distance) between paste elements:

$$\Sigma = -p\mathbf{I} + 2\mu_0\mathbf{D} + \Sigma_p \tag{3.65}$$

Since the stress tensor of the paste (Σ) should not significantly change in time for a sample at rest equation (3.65) shows that there exists a strong coupling between the liquid motion and the stress tensor due to the network of interactions between elements. Drying processes involve slow flows for which the second term of the right-hand side (RHS) of (3.65) is negligible. Thus, for a sample at macroscopic rest, if the pressure does not change significantly from one point to another, Σ_p remains approximately constant throughout the material. In that case the sample should shrink more or less homogeneously [87]. More precisely, this situation occurs when the pressure gradient (dp/dx) resulting from the liquid flow through the porous system is much weaker than the possible variations of the stress tensor due to element interactions, which can be roughly estimated as τ_c/h, in which h is the sample length in the direction perpendicular to the free surface. On the contrary, if dp/dx is of the order of or larger than τ_c/h, the liquid flow tends to induce spatial variations of the particulate stress tensor, so that we expect some differential shrinkage, in general leading to a larger concentration of elements close to the free surface.

Let us now consider the complete paste drying process (see Figure 3.25). In the very first instant there is some liquid surrounding the elements so that the evaporation from the free surface of the sample is similar to that from a pure liquid and is given by J_e, as estimated from boundary conditions in Section 3.5.1. Then the liquid starts to moves towards the free surface through the sample a priori at a rate $J_p = J_e$. The sample shrinks like a sponge as a result of this liquid

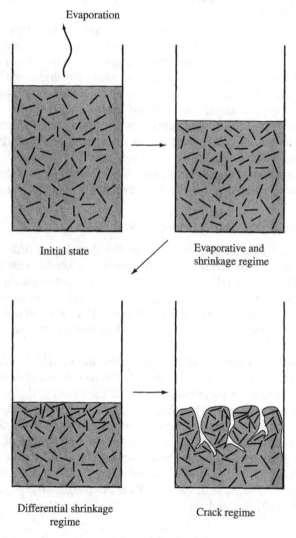

Figure 3.25 Schematic aspects of the cross section of a colloidal suspension under drying: different successive regimes (see text).

extraction, and the shrinkage remains homogeneous as long as

$$\frac{\mu_0}{k} J_e \ll \frac{\tau_c}{h} \tag{3.66}$$

In that case (*evaporative and shrinkage regime*) the rate of evaporation remains the same as that fixed by the boundary conditions (J_e) since a homogeneous shrinkage leads to a continuous supply of liquid molecules through the elements close to the free surface, and this paste region thus remains composed mainly of interstitial liquid. This regime corresponds to the "constant drying rate" regime typically observed with mineral suspensions [88], polymeric gels [89], or solid porous materials [86,90].

Now, when (3.66) is no longer valid, the shrinkage becomes heterogeneous and some dry crust tends to progressively form close to the free surface. This *differential shrinkage regime* occurs when the permeability of the material is or became (as a result of drying) too low, or when the drying rate is too high. In that case the strong concentration of elements developing close to the surface precludes a good migration of liquid molecules toward the free surface. This generally leads to a strong decrease in the drying rate, which is no longer imposed by the rate of evaporation but by the rate of flow through the (dense) porous system, which itself results from capillary effects such as in the capillary regime for solid porous materials (see text above). During this stage the material undergoes strong internal forces that may eventually lead to cracks.

3.5.3 Drying Rates during Rheometrical Tests

In rheometry drying processes occur while the fluid flows. Although the basic concepts discussed above remain valid, the pressure gradient in the fluid now also depends on flow characteristics. The question of the consequence of this effect on the effective liquid migration through the paste and on the transition between the two drying regimes for pastes remains open. Here we will only estimate the rate of drying in the evaporative regime, when the drying rate is imposed by boundary conditions.

For cone–plate or parallel disks geometries, the sample at the periphery is sheared at a rate $\dot{\gamma} = \Omega R / h$ in which Ω is the relative rotation velocity of the tools (here we assume that one tool is fixed), R the radius of the geometry, and h is the sample thickness at the periphery. The relative motion of the ambient gas and the paste is complex because the relative velocity varies from 0 to ΩR along h. As a first approximation we can assume that the free surface of the paste and the gas move relative to each other at a constant rate $\Omega R/2$. In the case of water evaporating in air, we have at 20°C: $n_v = 17$ g/m^3 = 17×10^{-6} m^3(liquid)/m^3; $\rho_g = 1.23$ kg/m^3, and $\mu_g = 1.8 \times 10^{-5}$ Pa·s. Assuming $n_0 = n_v/2$, taking $y \approx R$, and using (3.62), we find

$$J_e \approx \frac{n_v D}{10} \sqrt{\frac{\rho_g \Omega}{2 \mu_g}} \approx 7.5 \times 10^{-9} \sqrt{\Omega} \tag{3.67}$$

The fraction of sample volume lost by unit time by this process is $J_e 2\pi Rh/\pi R^2 h = 2J_e/R$ for the parallel disk geometry and twice this value for the cone–plate geometry. Finally, the fraction of sample volume lost after one hour of drying under such conditions for 1-cm-radius parallel disks is

$$F = 5.4 \times 10^{-3}\sqrt{\Omega} \tag{3.68}$$

For a Couette geometry we can proceed with analogous hypotheses, which lead to a fraction F equal to J_e/h in which h is the fluid depth. The largest rate of drying is obtained for the cone–plate geometry, and we find that after one hour under a rotation velocity $\Omega = 1$ rad/s typically about one percent of material has been lost by evaporation.

3.5.4 Effect of Drying on Rheometry

Two opposite effects on rheometrical tests occur as a result of the drying of pastes during experiments: a decrease of the material volume, which tends to decrease the apparent viscosity; and a decrease of the liquid fraction, which in general tends to increase the apparent viscosity of the material. Let us consider the impact of drying on the apparent yield stress ($\tau_{c,\mathrm{app}}$) as determined from rheometry, for a material composed of elements at an initial volume fraction ϕ and with an initial yield stress $\tau_c(\phi)$, in a parallel disk geometry. We assume that drying leads to a reduction of the sample radius from R to $R - e$ where $e \ll R$. This implies that the new concentration of elements in the liquid is

$$\phi^* = \phi\left(1 + \frac{2e}{R}\right) \tag{3.69}$$

Neglecting the shear rate heterogeneity (see Section 1.4.2), the critical torque for incipient motion is $M = \pi(R - e)^3 \tau_{c,\mathrm{eff}}/2$, in which $\tau_{c,\mathrm{eff}}$ is the effective yield stress. This yield stress is a function of the actual volume fraction of elements and may be written at leading order as $\tau_{c,\mathrm{eff}} = \tau_c(\phi^*) \approx \tau_c(\phi) + (2e/R)\phi(\partial\tau_c/\partial\phi)$. In parallel, the apparent yield stress may be expressed as a function of the torque as: $M = \pi R^3 \tau_{c,\mathrm{app}}/2$. It follows that at leading order

$$\tau_{c,\mathrm{app}} = \tau_c(\phi)\left[1 + \left(2\frac{\partial \ln \tau_c}{\partial \ln \phi} - 3\right)\frac{e}{R}\right] \tag{3.70}$$

In general, it has been found or predicted that the yield stress of pasty materials increases as a function of a power of ϕ, equal to or larger than 2 [34,91–93], which implies [from (3.70)] that the apparent yield stress increases with drying. In that case the relative error made on the estimation of the yield stress when neglecting drying effects is of the order of e/R.

3.5.5 Countermeasures

First, we can note that the total evaporation flux is proportional to the surface of the sample–gas interface. It follows that the effect of drying increases with the ratio of this surface to the sample volume. This means that a first possibility for minimizing drying effects is to increase the radius of the cone–plate or parallel disk geometries, and to increase the depth of the Couette geometry. Another possibility consists in increasing the confinement of the sample, which tends to increase the vapor density in the ambient gas, so that the evaporation rate decreases. The last and direct means to completely avoid evaporation consists in covering the free surface with oil. In general, an oil film rapidly spreads over the fluid so that the fluid–gas interface disappears, and this very thin layer does not significantly affect rheometrical measurements with a pasty material.

3.6 PHASE SEPARATION

As we have seen in Chapter 2, pasty or granular materials are heterogeneous at a mesoscopic scale; specifically, an elementary volume of material (Section 1.2.1) usually contains basic elements of different physical properties, such as liquid or gas molecules, and polymer chains, bubbles, droplets, or solid particles. In particular, here we will focus on their differences in size, shape, or density, which may induce relative motions between the elements. When this phenomenon occurs in one specific direction, it leads to a group motion of one phase through another phase (*phase separation*), which finally induces some heterogeneity of the material characteristics on a macroscopic scale. Here we will not consider other effects due to specific, local, physical properties with significant macroscopic consequences such as ordering occurring in liquid crystals or order–disorder transition in concentrated suspensions of uniform particle size.

Phase separation resulting from density heterogeneity may occur at rest simply because the gravity force differs from one element to the other (this is *sedimentation*), but phase separation resulting from size or density heterogeneities may be induced or fostered by flow. When coarse elements move through a fine material, that is, with typical elements much smaller in size than these coarse elements, as a result of flow, we speak of *migration*. When some particles move through an ensemble of slightly different particles but of sizes of similar order of magnitude, we speak of *segregation*. The general knowledge of these phenomena is rather limited and here we will simply review the basic concepts.

3.6.1 Sedimentation

Here we assume that the coarse elements found in the elementary volumes of the pastes are surrounded by a material that may be as a first approximation considered as a homogeneous paste. In this context our results (see Section 5.1.1) concerning the displacement of objects through a yield stress fluid at macroscopic rest apply. We may in particular estimate the critical conditions for which a coarse element would fall under the action of gravity through a paste; this is found from the balance between the gravity force ($\rho_P g \Omega$) minus the buoyancy force ($-\rho g \Omega$) and the drag force ($k_c \tau_c S$). For a sphere of radius R, we get

$$\tau_c = \frac{|\rho_P - \rho| g R}{3 k_c} \tag{3.71}$$

The direction of motion depends on the sign of $\rho_P - \rho$; the object falls when it is positive. Note that here the density (ρ) to be used in the buoyancy force is that of the complete fluid, including all types of elements, even the coarsest ones, because this force results from the vertical gradient (see Section 5.1.1), over the particle surface, of the pressure of the fluid surrounding the particle. Nevertheless, the result expressed by equation (3.71) relies on the assumption that one can define a stress tensor on each point of the surface of the solid particle, which means that the material around the particle may be seen as a continuum. As a consequence, this result at least does not apply for a particle in contact with other particles of similar size.

Equation (3.71) shows us that sedimentation in pasty materials at rest does not occur unless the particle size or the density difference are larger than critical values [which can be deduced from (3.71)]. This is so because the material behaves as a solid in its solid regime and thus can support a finite force without flowing. In contrast, sedimentation may occur under any condition within a flowing paste because as soon as a yield stress fluid is in its liquid regime, its apparent behavior is that of a simple fluid. More precisely, when a yield stress fluid flows, the stress distribution in the fluid is that of the equivalent simple fluid of similar, local apparent viscosity, as given by (2.20). *An object immersed in the flowing paste sees this viscous fluid around it and immediately starts to drop if its density differs from that of the rest of fluid.* This effect may be understood from a simple example. Let us consider a plate of large surface S embedded in a pasty material. When the paste is at rest the plate can remain in suspension, that is, does not settle, as long as the yield stress is sufficiently large. Now assume that the fluid is homogeneously sheared so that gliding fluid layers are parallel to the plate. Under these conditions, neglecting inertia and edge effects, the stress tensor is, by symmetry relative to the solid plane, identical along each plate side. Moreover, the total force exerted onto the plate as a result of the stress deviator is the sum of $(\mathbf{T} \cdot \mathbf{n})S$ along one side of the plate, in which $\mathbf{n}$ is the normal to the plate along this side, and $(\mathbf{T} \cdot (-\mathbf{n}))S$ along the other side. Thus the total force exerted onto the plate and resulting from viscous effects is equal to zero, and the plate will tend to settle if its density differs from that of the fluid.

3.6.2 Migration

Here we assume that gravity effects are negligible. Under these conditions a sphere embedded in a homogeneous flow (far from it) has no particular reason for migrating. Indeed, the stress tensor is symmetric relative to the sphere center so that the stress distribution around it is antisymmetric; any elementary force at one point of its surface is compensated by an opposite force at another point of its surface. For an object of arbitrary shape the problem has no more symmetries and the net force or torque exerted onto the object may be finite. However, they vary with the orientation of the object, and we can expect that for a set of such objects in various initial positions embedded in a macroscopically homogeneous flow, the net migration of the objects will be negligible because the induced motions in various directions balance. Thus migration in general occurs as a result of the large size, relative to the remaining element sizes, of an object embedded in a *heterogeneous flow*.

The sources of flow heterogeneities are mainly spatial variations of kinematics properties, such as a gradient of shear rate, or of fluid characteristics, such as a gradient of apparent viscosity. Migration is typically observed in Couette flows with suspensions of coarse particles in simple liquids [94–97]. In that case, since the shear stress distribution is heterogeneous [see equation (1.76)], the shear rate is already heterogeneous at the flow beginning when the material is homogeneous, but we can expect additional effects developing in time; as migration takes place, the distribution of particle concentration becomes heterogeneous, which leads to viscosity heterogeneities. Some theoretical approaches of these phenomena have been proposed [98–101], the results of which may be summarized as follows. The migrating flux, that is, the net number of particles of radius r moving through a unit surface per unit time, in a suspension with a solid volume fraction of particles ϕ and flowing in simple shear at a rate $\dot{\gamma}$, is roughly expressed in the form $\mathbf{N} \propto -r^2 \nabla (\mu \phi^2 \dot{\gamma})/\mu$ [102], in which μ is the viscosity of the suspension. This expression may be used to estimate the migration of coarse elements in pasty materials by using the apparent viscosity of the paste in the liquid regime. Roughly speaking, from the preceding expression for the net flux, the general result is a migration toward the regions of lowest shear rates. We note that migration was also considered to be the result of a gradient in suspension temperature, defined as the average kinetic energy of particle velocity fluctuations due to particle interactions [103]. Finally, we note that some differential migration of particles of different sizes may occur, leading to segregation effects.

Migration effects can also occur as a result of some difference between the particle density and that of the suspending medium [104–107]. This effect was observed and studied in capillary flows [104,107]. In a vertical capillary the particle migrates toward the walls if it is lighter than the fluid flowing upward (or denser than the fluid flowing downward), or migrates toward the central axis if it is denser than the fluid flowing upward (or lighter than fluid flowing downward). For a particle of the same density as the fluid, a migration toward a distance of the order of half the conduit radius occurs as a result of inertia effects.

Migration may also occur in squeeze flows with pastes when the rate of squeezing is not sufficiently rapid, allowing the interstitial liquid to flow through the porous material formed by the granular phase under the action of gradients of the interstitial pressure [108]. It has been proposed that such a phenomenon occurs for squeeze rates higher than a critical value increasing with the separating distance between the disks [109]

3.6.3 Segregation

Segregation effects have been observed in flowing or vibrated granular materials. A general result is that coarsest grains tend to migrate upward through a granular mass of smaller grains. This is so because as the granular mass is agitated, small grains tend to glide in small gaps formed below the large grains. However, there are also situations in which other phenomena seem to occur leading to other kinds of segregation [110–113].

3.6.4 Consequences in Rheometry

As phase separation develops, the studied fluid becomes heterogeneous on a macroscopic scale. Since the different possible effects are fundamentally time-dependent, their general consequence is an apparent thixotropy of the fluid under study; its apparent viscosity changes in time, becoming either larger or smaller than the initial one.

Sedimentation

Let us, for example, examine the consequence of this effect on Couette rheometry. We consider a pasty material containing coarse particles at a volume fraction ϕ and assume, for the sake of simplicity, that as a result of sedimentation, the solid fraction is $\phi - \Delta\phi^*$ (with $\Delta\phi^* \ll 1$) in the upper part of the cup ($h > z > \varepsilon$) while the solid fraction is $\phi + \Delta\phi$ in the lower part of the cup ($z < \varepsilon$). The macroscopic yield stress of the initial fluid may be expressed as a function of the volume fraction of coarse particles: $\tau_c (\phi)$. Using a Taylor development of this function, the yield stresses of the initial and final fluids are respectively $\tau_1 \approx \tau_c(\phi) - \Delta\phi^* \tau_c'(\phi) + (\Delta\phi^{*2}/2)\tau_c''(\phi)$ and $\tau_2 \approx \tau_c(\phi) + \Delta\phi\tau_c'(\phi) + (\Delta\phi^2/2)\tau_c''(\phi)$. Moreover, the conservation of particles implies that $\varepsilon\Delta\phi = (h - \varepsilon)\Delta\phi^*$. The apparent yield stress (τ_a) of this medium is associated with the torque necessary for solid–liquid transition in both regions: $M = 2\pi r_1^2[(h - \varepsilon)\tau_1 + \varepsilon\tau_2] = 2\pi r_1^2 \tau_a$, which yields

$$\tau_a = \tau_c + \frac{h\varepsilon}{h\varepsilon} \frac{\Delta\phi^2}{2}\tau_c''(\phi) \tag{3.72}$$

Thus the first-order terms disappear, which means that, because of some linearity of the problem, the possible effect of phase separation on the apparent yield stress is a priori of second order. Since the second derivative of the yield stress is in general positive (see Section 3.5.4), the effect of phase separation is an increase in the apparent yield stress. Note that when the volume fraction of coarse elements tends to the maximum packing fraction in some region, the yield stress of the material in this region tends to diverge, so that the effect of phase separation can be dramatic in that case.

Countermeasures of sedimentation in Couette rheometry consist in either recirculating the fluid vertically (in addition to the shear imposed by the tools) in order to maintain, on average, the solid fraction at its initial level [114,115], or shearing the material only in a specific region where it can be expected that the solid fraction remains constant [116].

Migration

In a Couette flow we have seen that the effect of migration is basically a decrease in the volume fraction of particles close to the inner cylinder. Let us now assume that we stop the flow and perform a test in view of determining the yield stress. Since the apparent solid–liquid transition of the whole material in fact corresponds to the solid–liquid transition of at least one (cylindrical) layer of material (see Figure 4.11), the apparent yield stress will be equal to the yield stress of the region close to the inner cylinder and thus will be smaller than that of the homogeneous suspension.

The problem is more complex for the apparent viscosity of the paste. Indeed, as the volume fraction of coarse elements decreases in the regions closest to the inner cylinder, thus tending to decrease the apparent viscosity, the volume fraction of the regions close to the outer cylinder increases, thus tending to increase the apparent viscosity of the sample. These two effects might more or less balance and, as for the effect of sedimentation on the yield stress, their total influence will generally be second-order in terms of volume fraction difference between the two regions and may become dramatic only for very high volume fractions.

3.7 CRACKING

In some cases viscometric flows of pastes evolve in cracking: the sample tendency to separate into two parts without contact. This phenomenon occurs particularly with polymer melts and polymer solutions, and with highly concentrated suspensions, and has been observed mainly with parallel disks and cone–plate geometries [117]. In those cases the phenomenon begins as a distortion of the peripherial surface, which quite rapidly propagates toward the interior of the sample, and in the final stage there remain two distinct regions of fluid moving rigidly with each tool without contact between them (see Figure 3.26). With concentrated suspensions there often remains, along the interface of one part with air, some particles that can touch the (irregular) interface of the other part during their relative rotation. This leads to the formation of concentric cylindrical tracks visible on the surface of each part after tool separation.

The origin of cracking remains unclear. For viscoelastic polymeric fluids, it has been argued that this could occur as a result of normal stress effects [118,119]. It is not clear whether this theory applies for concentrated suspensions since viscoelastic or normal stress effects during flows of such materials are often negligible (at least at moderate concentrations). In fact, it is worth noting that with high-yield-stress pastes it is likely that the critical shear rate beyond which stable flows exist is considerable (see Section 3.3), so that an extreme shear localization (discrete regime) should occur in a wider range of shear rate. This effect, coupled with the specific structure of concentrated suspensions, might be favorable for the development of cracks.

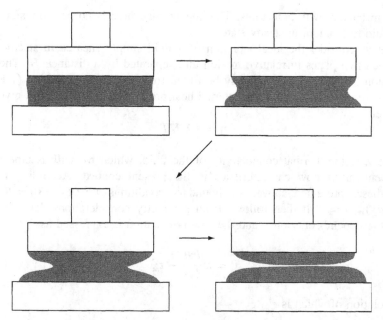

Figure 3.26   Development of a crack in a sample during a rheometrical test with a parallel disk geometry: successive aspects of a sample axial cross section in time (see text) (from left to right and from top to bottom).

3.8   TEMPERATURE EFFECTS

The usual treatment of rheometrical data assumes that the temperature is controlled. In practice, this is considered to have been achieved using a device that rapidly regulates the temperature of one tool in contact with the fluid. However, there remains some uncertainty regarding the value effectively reached by the fluid because energy dissipation due to shear may induce some increase in temperature, which in turns modifies the fluid viscosity. The complete treatment of the corresponding thermodynamic problems in various geometries and under different boundary conditions constitutes a vast domain. The interested reader may find detailed information in the books of Fredrickson [120] and Tanner [121]. Here we will focus on these effects from a basic point of view in an attempt to estimate by simple albeit approximate means the effect of viscous heating on rheometrical measurements. Two cases may be distinguished: rotational geometries, for which the fluid constantly recirculates; and capillary or free surface flows, for which the fluid is constantly supplied from upstream. The latter case is more complex than the former because the flows are fundamentally transient, and in particular the temperature may change in time and space. With rotational geometries we can assume that the temperature is imposed along one tool by an appropriate device, while the temperature of the other tool is imposed by the fluid in contact with both tools. This means that there is no heat transfer from the fluid to this tool in steady state.

Let us consider the ideal situation of a fluid simply sheared in steady state between two plates in relative motion and separated by a distance H. The distribution of fluid temperature may be found from the energy equation (1.18) in which we obviously assume no internal heat production and a heat flux given by Fourier's law

$$\mathbf{q} = -\lambda \, \nabla T \tag{3.73}$$

where λ is the thermal conductivity of the fluid, which we will assume to be independent of flow characteristics in the present context. According to the hypotheses presented above, the boundary conditions are $\theta(y = 0) = \theta_0$ and $(d\theta/dy)_{y=H} = 0$. It then follows from symmetry considerations that the temperature depends only on y and the energy equation is expressed as

$$0 = \lambda \frac{d^2\theta}{dy^2} + \tau \dot{\gamma} \tag{3.74}$$

the solution of which is

$$\theta(y) = \frac{\tau \dot{\gamma}}{\lambda} \left[Hy - \frac{y^2}{2} \right] + T_0 \tag{3.75}$$

The temperature difference between the two plates is

$$\Delta\theta = \frac{\tau \dot{\gamma} H^2}{2\lambda} \tag{3.76}$$

This result may be used to estimate the order of magnitude of the temperature increase due to viscous heating in a rotational geometry. For this purpose one has to use some mean value of the shear rate, shear stress, and the gap (as H) in (3.76). For a cone–plate flow, one may use for H half the maximum distance between the tools.

For most liquids the thermal conductivity is between 0.1 and 0.6 W m^{-1} K^{-1}, and we will assume that this is also the typical range for pastes since thermal transfers are basically ensured by the interstitial liquid in which various types of elements are immersed. The rheometer gap is typically between 0.1 and 10 mm depending on geometries. It follows that in most cases for pastes the temperature difference is very small. It may be large only with a very viscous paste under high shear rate. For example, if $H = 3.3$ mm; $\lambda = 0.5$ W m^{-1} K^{-1}; $\dot{\gamma} = 100$ s^{-1}; $\tau = 1000$ Pa, we get $\Delta\theta \approx 1$ K.

Let us now consider the effect of temperature on the apparent viscosity of pastes in the liquid regime. Since the yield stress results from the strength of the network of interactions between the elements in the interstitial liquid, and since the potentials of these interactions do not vary much with temperature in the usual range of study, it is likely that the impact of temperature changes on the value of the yield stress is minor. However, this conclusion is valid only for a material in the pasty regime, as long as hydrodynamic interactions of the interstitial liquid remain negligible. For example, it was observed that the flow curve of clay–water mixtures of relatively high yield stresses, under various temperatures in the range 5–30°C and in a limited range of shear rates, were similar as a first approximation [122]. On the contrary, the impact of temperature on the rheological behavior of pastes in the hydrodynamic regime is likely similar to that on simple liquids, since hydrodynamic interactions predominate in that case.

3.9 INERTIA EFFECTS AND TURBULENCE

From a general perspective inertia effects result from the (excessively) high kinetic energy of the material. Some inertia effects originate in the (*macroscopic*) acceleration terms in the momentum equation [left-hand side (LHS) of (1.15) or (1.16)]. Another type of inertia effect originates from the local flow instability when kinetic terms become sufficiently large in comparison to viscous terms; this is *turbulence* (see Section 1.2.3). Some of these effects lead to flow instability, but they must be distinguished from various types of flow instabilities occurring under specific conditions, such as shear banding (Section 3.3), fingering in confined flows (Section 5.2), and roll waves in free surface flows (Section 5.3), which are governed by the relative importance of different sources of energy dissipation.

3.9.1 "Macroscopic" Inertia Effects

Two subcases can be distinguished: inertia terms originating from transient effects (i.e., the first term of acceleration in Eulerian description: $\rho \partial \mathbf{v}/\partial t$); and those originating from the nonuniformity of the flow (i.e., the second term of acceleration: $\rho \nabla \mathbf{v} \cdot \mathbf{v}$). In order to estimate the importance of these effects, we must compare them to viscous forces. The first term of the LHS of (1.16) may be roughly written in the form $\rho \dot{\gamma} d / T$, where we assume that the material velocity transforms from zero to $v \approx \dot{\gamma} d$ in a characteristic time T (where d is the typical sheared thickness of material). For a paste, the corresponding viscous term in the momentum equation (stress divergence term) is of the order of τ_c/d. The inertia effects due to the transient character of the flow are negligible relative to viscous effects when the dimensionless number

$$I_t = \frac{\rho \dot{\gamma} d^2}{T \tau_c} \tag{3.77}$$

is much smaller than 1.

For rotating geometries a typical estimate of the second macroscopic inertia term of (1.16) is $\rho \dot{\gamma}^2 d^2 / r$, in which r is the typical radial distance in the fluid. The inertia effects due to the nonuniformity of the flow are negligible when

$$I_r = \frac{\rho \dot{\gamma}^2 d^3}{r \tau_c} \tag{3.78}$$

is much smaller than 1. For free surface or conduit flows the second inertia term is linked to spatial variations of the fluid cross section (S) (e.g., along a specific direction x), and inertia effects due to the nonuniformity of the flow are negligible when

$$I_x = \frac{\rho v Q d}{S^2 \tau_c} \frac{\Delta S}{\Delta x} \tag{3.79}$$

is much smaller than 1, in which Q is the flow discharge.

3.9.2 Turbulence

For Newtonian fluids this phenomenon is rather well characterized in various situations [123,124] and generally occurs when the ratio of the characteristic kinetic energy of the fluid (ρv^2) by unit volume to the characteristic viscous stress ($\mu \dot{\gamma} \approx \mu v / d$), specifically, the Reynolds number

$$\mathrm{Re} = \frac{\rho v d}{\mu} \tag{3.80}$$

is much larger than 1 (e.g., some turbulent effects may start to be observed for $\mathrm{Re} \approx 10$).

For other types of fluids, and especially for yield stress fluids, the situation is much less clear; the few existing theoretical works [125–130] for different flow geometries have not been fully validated by experiments. Moreover, they require knowledge of the exact form of the constitutive equation. To ensure a simple, general though approximate criterion, we can use the simple interpretation given above for Newtonian fluids and consider that turbulence in yield stress fluids occurs when the ratio of kinetic to viscous energies

$$\mathrm{Re}_y = \frac{\rho v^2}{\tau_c} \tag{3.81}$$

is much larger than 1. Here we assumed that the shear stress in the fluid is of the order of its yield stress, which means that the flow is not too rapid (pasty regime). When this assumption is no longer valid, the rate-dependent term in the constitutive equation (2.20) plays a major role in the shear stress (hydrodynamic regime) (see Section 2.2). In that case, since the apparent rheological behavior of the material is now Newtonian, it is sufficient for examining the occurrence of turbulence to use the criterion (3.80), in which μ would be replaced by the apparent viscosity of the fluid at high shear rates. To sum up one can consider that turbulence occurs in pasty flows either if $Re_y \gg 1$ in the pasty regime or if $Re \gg 1$ in the hydrodynamic regime. Note that it nevertheless remains conceptually difficult to fathom that turbulence could occur in the pasty regime since the elements in this regime are precisely embedded in a continuous network of soft interactions, but no clear view of this problem has been proposed.

The Taylor–Couette instability, which takes the form of rolls of the same order of magnitude as that of the gap, may also occur in Couette flow because of the predominance of inertia over viscous effects. For pasty materials only approximate and not validated criteria can be provided [131]. From a general point of view it is expected to occur at Reynolds numbers (Re or Re_y) slightly smaller than the critical one beyond which turbulence becomes significant.

REFERENCES

1. R. G. Larson, *The Structure and Rheology of Complex Fluids*, Oxford Univ. Press, Oxford, 1999
2. H. Hervet and L. Léger, *C. R. Phys.* **4**, 241 (2003).
3. P. G. de Gennes, *C. R. Acad. Sci. Paris B* **288**, 219 (1979).
4. A. V. Ramamurthy, *J. Rheol.* **30**, 337 (1986).
5. L. Léger, H. Hervet, T. Charitat, and V. Koutsos, *Adv. Colloid Interface Sci.* **94**, 39 (2001).
6. J. M. Piau, N. El Kissi, and B. Tremblay, *J. Non-Newtonian Fluid Mech.* **34**, 145 (1990).
7. N. El Kissi and J. M. Piau, *J. Non-Newtonian Fluid Mech.* **37**, 55 (1990).
8. N. El Kissi, L. Léger, J.-M. Piau, and A. Mezghani, *J. Non-Newtonian Fluid Mech.* **52**, 249 (1994).
9. S. G. Hatzikiriakos and J. M. Dealy, *J. Rheol.* **35**, 497 (1991).
10. A. Magnin and J. M. Piau, *J. Non-Newtonian Fluid Mech.* **23**, 91 (1987).
11. A. Magnin and J. M. Piau, *J. Non-Newtonian Fluid Mech.* **36**, 85 (1990).
12. S. G. Harding, O. Wik, A. Helander, N. O. Ahnfelt, and L. Kenne, *Carbohydrate Polymers* **47**, 109 (2002).
13. H. Müller-Mohnssen, D. Weiss, and A. Tipe, *J. Rheol.* **34**, 223 (1990).
14. H. A. Barnes, *J. Non-Newtonian Fluid Mech.* **56**, 221 (1995).
15. P. J. A. Hartman Kok, S. G. Kazarian, C. J. Lawrence, and B. J. Briscoe, *J. Rheol.* **46**, 481 (2002).
16. H. Tabuteau, J. C. Baudez, F. Bertrand, and P. Coussot, *Rheol. Acta* **43**, 168 (2004).
17. G. V. Vinogradov, G. B. Froishteter, K. K. Trilisky, and E. L. Smorodinsky, *Rheol. Acta* **14**, 202 (1975).
18. R. A. Mashelkar and A. Dutta, *Chem. Eng. Sci.* **37**, 969 (1982).
19. A. Delime and M. Moan, *Rheol. Acta* **30**, 131 (1991).
20. U. S. Agarwal, A. Dutta, and R. A. Mashelkar, *Chem. Eng. Sci.* **49**, 1693 (1994).
21. B. K. Aral and D. M. Kalyon, *J. Rheol.* **38**, 957 (1994).
22. H. J. Walls, S. B. Caines, A. M. Sanchez, and S. A. Khan, *J. Rheol.* **47**, 847 (2003).
23. S. P. Meeker, R. T. Bonnecaze, and M. Cloitre, *Phys. Rev. Lett.* **92**, 198302 (2004).
24. M. Mooney, *J. Rheol.* **2**, 210 (1931).
25. U. Yilmazer and D. M. Kalyon, *J. Rheol.* **33**, 1197 (1989).

26. D. M. Kalyon, P. Yaras, B. Aral, and U. Yilmazer, *J. Rheol.* **37**, 35 (1993).
27. Y. Cohen and A. B. Metzner, *J. Rheol.* **29**, 67 (1985).
28. V. Bertola, F. Bertrand, H. Tabuteau, D. Bonn, and P. Coussot, *J. Rheol.* **47**, 1211 (2003).
29. A. Yoshimura and R. K. Prud'homme, *J. Rheol.* **32**, 53 (1988).
30. P. O. Brunn, *Rheol. Acta* **37**, 196 (1998).
31. J. M. Dealy, *Rheol. Acta* **37**, 195 (1998).
32. W. Gleissle and E. Windhab, *Exper. Fluids* **3**, 177 (1985).
33. J. F. Hutton, *Rheol. Acta* **8**, 54 (1969).
34. P. Coussot, *Mudflow Rheology and Dynamics*, Balkema, North-Holland, Amsterdam, 1997.
35. J. B. Salmon, L. Bécu, S. Manneville, and A. Colin, *Eur. Phys. J. E* **10**, 209 (2003).
36. Q. D. Nguyen and D. V. Boger, *J. Rheol.* **27**, 321 (1983).
37. J. S. Raynaud, P. Moucheront, J. C. Baudez, F. Bertrand, J. P. Guilbaud, and P. Coussot, *J. Rheol.* **46**, 709 (2002).
38. H. Rehage and H. Hoffmann, *Mol. Phys.* **74**, 933 (1991).
39. D. Roux, F. Nallet, and O. Diat, *Europhys. Lett.* **24**, 53 (1993).
40. E. Cappelaere, J. F. Berret, J. P. Decruppe, R. Cressely, and P. Lindner, *Phys. Rev. E* **56**, 1869 (1997).
41. S. Hernandez-Acosta, A. Gonzalez-Alvarez, O. Manero, A. F. Mendez-Sanchez, J. Perez-Gonzalez, and L. de Vargas, *J. Non-Newtonian Fluid Mech.* **85**, 229 (1999).
42. P. Holmqvist, C. Daniel, I. W. Hamley, W. Mingvanish, and C. Booth, *Colloids Surfaces A, Physicochem. Eng. Aspects* **196**, 39 (2002).
43. D. Bonn, J. Meunier, O. Greffier, A. Al-Kahwaji, and H. Kellay, *Phys. Rev. E* **58**, 2115 (1998).
44. J. B. Salmon, A. Colin, and D. Roux, *Phys. Rev. E* **66**, 031505 (2002).
45. F. Da Cruz, F. Chevoir, D. Bonn, and P. Coussot, *Phys. Rev. E* **66**, 051305 (2002).
46. L. Heymann, S. Peukert, and N. Aksel, *Rheol. Acta* **41**, 307 (2002).
47. P. Coussot, Q. D. Nguyen, H. T. Huynh, and D. Bonn, *Phys. Rev. Lett.* **88**, 175501 (2002).
48. P. Coussot, Q. D. Nguyen, H. T. Huynh, and D. Bonn, *J. Rheol.* **46**, 573 (2002).
49. J. R. A. Pearson, *J. Rheol.* **38**, 309 (1994).
50. S. M. Fielding and P. D. Olmsted, *Eur. Phys. J.* **11**, 65 (2003).
51. M. E. Cates, T. C. B. McLeish, and G. Marrucci, *Europhys. Lett.* **21**, 451 (1993).
52. P. Coussot, A. I. Leonov, and J. M. Piau, *J. Non-Newtonian Fluid Mech.* **46**, 179 (1993).
53. R. Mas and A. Magnin, *J. Rheol.* **38**, 889 (1994).
54. S. Ducerf and J. M. Piau, *Les Cahiers de Rhéologie* (French Group of Rheology) **XIII**, 120–129 (1994).
55. N. A. Spenley, M. E. Cates, and T. C. B. McLeish, *Phys. Rev. Lett.* **71**, 939 (1993).
56. F. Bautista, J. F. A. Soltero, J. H. Pérez-Lopez, J. E. Puig, and O. Manero, *J. Non-Newtonian Fluid Mech.* **94**, 57 (2000).
57. D. Quemada, *Rhéologie* **6**, 1 (2004).
58. V. Schmitt, F. Lequeux, A. Pousse, and D. Roux, *Langmuir* **10**, 955 (1994).
59. J. F. Berret, D. C. Roux, G. Porte, and P. Lindner, *Europhys. Lett.* **25**, 521 (1994).
60. P. Partal, A. J. Kowalski, D. Machin, N. Kiratzis, M. G. Berni, and C. J. Lawrence, *Langmuir* **17**, 1331 (2001).
61. E. Cappelaere, R. Cressely, and J. P. Decruppe, *Colloids Surfaces A, Physicochem. Eng. Aspects* **104**, 353 (1995).

62. J. P. Decruppe, R. Cressely, R. Makhloufi, and E. Cappelaere, *Colloid Polymer Sci.* **273**, 346 (1995).
63. R. Makhloufi, J. P. Decruppe, A. Aït-Ali, and R. Cressely, *Europhys. Lett.* **32**, 253 (1995).
64. J. F. Berret, G. Porte, and J. P. Decruppe, *Phys. Rev. E* **55**, 1668 (1997).
65. F. Pignon, A. Magnin, and J. M. Piau, *J. Rheol.* **40**, 573 (1996).
66. R. W. Mair and P. T. Callaghan, *J. Rheol.* **41**, 901 (1997).
67. J. B. Salmon, S. Manneville, and A. Colin, *Phys. Rev. E* **68**, 051503 (2003).
68. D. Bonn, P. Coussot, H. T. Huynh, F. Bertrand, and G. Debregeas, *Europhys. Lett.* **59**, 786 (2002).
69. P. Coussot, J. S. Raynaud, F. Bertrand, P. Moucheront, J. P. Guilbaud, H. T. Huynh, S. Jarny, and D. Lesueur, *Phys. Rev. Lett.* **88**, 218301 (2002).
70. J. C. Baudez, S. Rodts, X. Chateau, and P. Coussot, *J. Rheol.* **48**, 69 (2004).
71. S. Jarny, N. Roussel, S. Rodts, R. Le Roy, and P. Coussot, submitted to *Concrete Cement Res.* (2005).
72. S. Rodts, J. C. Baudez, and P. Coussot, *Europhys. Lett.* **69**, 636 (2005).
73. F. Varnik, L. Bocquet, J. L. Barrat, and L. Berthier, *Phys. Rev. Lett.* **90**, 095702 (2003).
74. M. M. Britton and P. T. Callaghan, *Phys. Rev. Lett.* **78**, 4930 (1997).
75. J. B. Salmon, A. Colin, S. Manneville, and F. Molino, *Phys. Rev. Lett.* **90**, 228303 (2003).
76. E. Fischer and P. T. Callaghan, *Europhys. Lett.* **50**, 803 (2000).
77. J. P. Decruppe, S. Lerouge, and J. F. Berret, *Phys. Rev. E* **63**, 022501 (2001).
78. S. Lerouge, J. P. Decruppe, and J. F. Berret, *Langmuir* **16**, 6464 (2000).
79. S. Manneville, J. B. Salmon, and A. Colin, *Eur. Phys. J.* **13**, 197 (2004).
80. B. Herzhaft, L. Rousseau, L. Neau, M. Moan, and F. Bossard, *Soc. Petrol. Eng.* 77818 (2002).
81. P. Coussot, *Phys. Rev. Lett.* **74**, 3971 (1995).
82. J. Israelachvili, *Intermolecular and Surface Forces*, 5th ed., Academic Press, London, 1995.
83. D. Tabor, *Gases, Liquids and Solids, and Other States of Matter*, Cambridge Univ. Press, Cambridge, UK, 1991.
84. K. Hisatake, S. Tanaka, and Y. Aizawa, *J. Appl. Phys.* **73**, 7395 (1993).
85. B. R. Munson, D. F. Young, and T. H. Okiishi, *Fundamentals of Fluid Mechanics*, Wiley, New York, 1994.
86. P. Coussot, *Eur. Phys. J. B* **15**, 557 (2000).
87. G. W. Scherer, *Cement Concrete Res.* **29**, 1149 (1999).
88. H. H. Macey, *Trans. Br. Ceram. Soc.* **41**, 73 (1942).
89. G. W. Scherer, *J. Non-Cryst. Solids* **100**, 77 (1988).
90. E. U. Schlünder, *Chem. Eng. Sci.* **43**, 2685 (1988).
91. M. Dorget, *Rheological Properties of Silica/Silicone Composits*, Ph.D. thesis, INPG, Grenoble, 1995.
92. W. B. Russel, D. A. Saville, and W. R. Schowalter, *Colloidal Dispersions*, Cambridge Univ. Press, Cambridge, UK, 1989.
93. B. A. Firth and R. J. Hunter, *J. Colloid Interface Sci.* **57**, 249 (1976).
94. F. Gadala-Maria and A. Acrivos, *J. Rheol.* **24**, 799 (1980).
95. A. L. Graham, S. A. Altobelli, E. Fukushima, L. A. Mondy, and T. E. Stephens, *J. Rheol.* **35**, 191 (1991).
96. S. A. Altobelli, R. C. Givler, and E. Fukushima, *J. Rheol.* **35**, 721 (1991).

97. A. W. Chow, S. W. Sinton, J. H. Iwamiya, and T. S. Stephens, *Phys. Fluids* **6**, 2561 (1994).

98. A. Shauly, A. Wachs, and A. Nir, *J. Rheol.* **42**, 1329 (1998).

99. D. Leighton and A. Acrivos, *J. Fluid Mech.* **181**, 415 (1987).

100. R. J. Phillips, R. C. Armstrong, R. A. Brown, A. L. Graham, and J. R. Abbott, *Phys. Fluids* **A4**, 30 (1992).

101. G. P. Krishnan, S. Beimfohr, and D. T. Leighton, *J. Fluid Mech.* **321**, 371 (1996).

102. P. Coussot, *Rheophysics of Pastes and Suspensions*, EDP Sciences, Paris, 1999 (in French).

103. N. C. Shapley, R. A. Brown, and R. C. Armstrong, *J. Rheol.* **48**, 255 (2004).

104. R. G. Cox and S. G. Mason, *Ann. Rev. Fluid Mech.* **3**, 291 (1971).

105. P. Yaras, D. M. Kalyon, and U. Yilmazer, *Rheol. Acta* **33**, 48 (1994).

106. G. Segré and A. Silberberg, *J. Fluid Mech.* **14**, 115 (1962).

107. A. Karnis, H. L. Goldsmith, and S. G. Mason, *Nature* **200**, 159 (1963).

108. A. Poitou and G. Racineux, *J. Rheol.* **45**, 609 (2001).

109. J. Collomb, F. Chaari, and M. Chaouche, *J. Rheol.* **48**, 405 (2004).

110. D. J. Stephens and J. Bridgewater, *Powder Technol.* **21**, 17 (1978).

111. S. B. Savage and C. K. K. Lun, *J. Fluid Mech.* **189**, 311 (1988).

112. W. S. Foo and J. Bridgewater, *Powder Technol.* **36**, 271 (1983).

113. J. A. Drahun and J. Bridgewater, *Powder Technol.* **36**, 39 (1983).

114. T. J. Reeves, *Coal Prepar.*, **8**, 1 (1990).

115. B. Clarke, *Trans. Inst. Chem. Eng.* **45**, T251 (1967).

116. B. Klein, J. S. Laskowski, and S. J. Partridge, *J. Rheol.* **39**, 827 (1995).

117. J. F. Hutton, *Rheol. Acta* **14**, 979 (1975).

118. R. I. Tanner and M. Keentok, *J. Rheol.* **27**, 47 (1983).

119. M. Keentok and S.-C. Xue, *Rheol. Acta* **38**, 321 (1999).

120. A. G. Fredrickson, *Principles and Applications of Rheology*, Prentice-Hall, Englewood, Cliffs, NJ, 1964.

121. R. I. Tanner, *Engineering Rheology*, Clarendon Press, Oxford, UK, 1985.

122. P. Coussot and J. M. Piau, *Rheol. Acta* **33**, 175 (1994).

123. J. O. Hinze, *Turbulence: An Introduction to its Mechanism and Theory*, Mc Graw-Hill, New York, 1975.

124. M. Lesieur, *Turbulence in Fluids: Stochastic and Numerical Modelling*, 2nd ed., Kluwer, Boston, 1990.

125. R. W. Hanks, *Am. Inst. Chem. Eng. J.* **9**, 306 (1963).

126. R. W. Hanks and B. H. Dadia, *Am. Inst. Chem. Eng. J.* **17**, 554 (1971).

127. N. I. Heywood and J. F. Richardson, *J. Rheol.* **22**, 599 (1978).

128. A. B. Metzner, *J. Rheol.* **24**, 115 (1980).

129. R. J. Mannheimer, *J. Rheol.* **35**, 113 (1991).

130. I. A. Frigaard, S. D. Howison, and I. J. Sobey, *J. Fluid Mech.* **263**, 133 (1994).

131. P. Coussot and J. M. Piau, *J. Rheol.* **39**, 105 (1995).

CHAPTER 4

LOCAL RHEOMETRY

Typical rheometers measure macroscopic characteristics of the flow such as the force or the torque exerted along some boundary of the fluid and the relative velocity of boundaries. The corresponding data are interpreted in terms of the constitutive equation of the fluid using rigorous or approximate approaches (see Section 1.4). However, interpretations of the peculiar macroscopic trends often observed with pastes suggest that perturbating effects occur in the fluid or that the fluid exhibits a strongly nonlinear behavior (see Section 3.3). This justifies the more recent development of techniques aimed at directly measuring the velocity profile within flowing pasty and granular materials. Indeed, the basic interest in such approaches is that as soon as the stress distribution is controlled, one can relate the local stress to the measured local velocity and obtain the constitutive equation of the material on a local scale.

In Section 4.1 we review the techniques enabling determination of the velocity field within a flowing material. In fact, we discuss only magnetic resonance imaging (MRI), which appears as by far the most practical technique for nontransparent materials and may also provide various additional information concerning the material structure. In Section 4.2 we examine how velocity profiles determined under given flow conditions may be interpreted in rheological terms. In Section 4.3 we turn to a particular technique that consists in interpreting conventional rheometrical data in terms of velocity profiles, which provides an alternate method for "local rheometry" without expensive equipment.

Rheometry of Pastes, Suspensions, and Granular Materials: Applications in Industry and Environment
By Philippe Coussot   Copyright © 2005 John Wiley & Sons, Inc.

4.1 TECHNIQUES FOR MEASURING THE VELOCITY FIELD IN FLUIDS

Different methods exist for measuring the velocity profiles in a flowing fluid: particle imaging velocimetry (PIV) [1], dynamic light diffusion (DLD) [2,3], laser Doppler anemometry (LDA) [4], X or gamma ray [5], and ultrasounds [6]. These techniques are based on the analysis of signal attenuation or scattering, often with the help of autocorrelation functions. On the contrary, MRI is based on the direct, local excitation of material particles within the sample under study. Although some of these techniques have been used marginally for studying complex fluids, it is mainly MRI that has been used and developed with rheological objectives [7–19], and in the following text we review the principles of this technique, its potentialities, and problems encountered with this technique.

4.1.1 Principles of NMR

Free Precession
Atomic nuclei are characterized by different states of energy. When they are set in a magnetic field, their equilibrium distribution in the corresponding energy levels is displaced. The relevant description of the mutual interactions between nuclei requires a treatment in the field of quantum mechanics. However, it is possible to view the basic processes by employing a classical mechanics treatment. In that frame the nuclei are considered as charged particles exhibiting a magnetic moment (spin) $\mathbf{M}$ (this is in fact related to an energy vector characterizing the quantum energy repartition of an ensemble of particles). Within the frame of this classical mechanics analogy the distribution of charges is related to the distribution of masses in the particles so that the particle spin is simply proportional to its kinetic moment $\mathbf{K}$: $\mathbf{M} = \gamma \mathbf{K}$, in which γ is the gyromagnetic ratio of the particle. Electromagnetism laws tell us that when immersed in a magnetic field ($\mathbf{B}$), this spin is submitted to a torque $\mathbf{C} = \mathbf{M} \times \mathbf{B}$. In parallel, the kinetic momentum theorem, that is, the balance of the torque and the rate of change of angular momentum for a solid, writes $d\mathbf{K}/dt = \mathbf{C}$, which, taking into account the preceding equations, finally gives

$$\frac{d\mathbf{M}}{dt} = \gamma \mathbf{M} \times \mathbf{B} \tag{4.1}$$

The solution of equation (4.1) when $\mathbf{B}$ is a constant magnetic field of amplitude B_0 corresponds to a *precession* of the spin around this magnetic field at a rate $\omega_0 = \gamma B_0$; the *Larmor frequency*, that is, the spin, of initial value $\mathbf{M}(0) = \mathbf{M}_0$, rotates around the field direction, keeping its initial angle with this direction (see Figure 4.1). In complex coordinates this may also be expressed as

$$M(t) = M_0 \exp(i\omega_0 t) \tag{4.2}$$

Since there are also energy exchanges due to mutual interactions between material components, the spin progressively tends to align along the direction of the magnetic field. In practice, this may be described by simple relaxation processes involving separately the spin components parallel and perpendicular to the field direction, namely, $\mathbf{M}_{\|}$ and $\mathbf{M}_{\perp}$. The corresponding relaxation times are respectively T_1 and T_2, and simple relaxation equations may thus be used in each case.

Resonance

When some perturbating magnetic field ($\mathbf{B}_1$, orthogonal to $\mathbf{B}_0$ and such that $B_1 \ll B_0$) oscillating in time at frequency ω_1 along its direction is added to the main field, the precession process may be altered. In complex coordinates the oscillating magnetic field $\mathbf{B}_1$ rotates at rate ω_1. If this rate is equal to the rate of the spin precession in the presence of the total field ($\mathbf{B} = \mathbf{B}_0 + \mathbf{B}_1$) (i.e., $\omega_1 = \gamma B \approx \gamma B_0$), this additional field will significantly affect the spin evolution by periodically slightly perturbing the spin dynamics in the same place. The spin direction may thus completely change; an effect of resonance will occur. A detailed mathematical solution of this process may be found from equation (4.1), but here we will simply note that, after a short duration under the action of the additional field $\mathbf{B}_1$ (i.e., a pulse), $\mathbf{M}_{\|}$ can be equal to zero. At the end of this pulse (when releasing $\mathbf{B}_1$) the spin continues to precess around $\mathbf{B}_0$ and generally rapidly relaxes toward its equilibrium position (see Figure 4.1). Because it is easier to record the perpendicular component of the spin, the typical NMR signal corresponds to the signal emitted by the material during this relaxation.

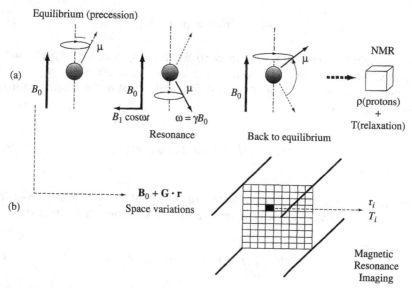

Figure 4.1 Principle of NMR and MRI: (a) free spin precession under the action of a large magnetic field $\mathbf{B}_0$, resonance under the action of a perturbating, radiomagnetic field $\mathbf{B}_1$, relaxation process after releasing $\mathbf{B}_1$; (b) local resonance in presence of a spatially varied field $\mathbf{B}$.

In fact, immediately after the end of the pulse the main magnetic field may be varied, which makes it possible to distinguish different physical phenomena (see text below).

4.1.2 Principles of MRI

The various possibilities of NMR spectroscopy and imaging are described in the usual textbooks (see, e.g., Ref. 20 for spectroscopy and Ref. 21 for imaging). Here we will consider only the effect of spatial gradients (represented by the tensor $\mathbf{G}$) of the main magnetic field on the signal in the form

$$\mathbf{B}(\mathbf{r}) = \mathbf{B}_0 + \mathbf{G} \cdot \mathbf{r} \tag{4.3}$$

which provides $B(r) = B_0 + \mathbf{G}_z \cdot \mathbf{r}$ in the direction of the main field z, where $\mathbf{G}_z = \mathbf{z} \cdot \mathbf{G}$. Now the spin precesses at a Larmor frequency $\omega(\mathbf{r}) = \gamma B$, which depends on its position $\mathbf{r}$ within the material. The first advantage of such gradients is to enable one to induce the resonance of spins situated in specific regions, that is, within the material layers for which $\omega_1 = \gamma B$. This makes it possible to proceed to selective excitation (selective pulse) in specific material layers.

The gradients have in fact additional interesting effects on the free precession process, which now occurs at a frequency depending on the position of the spin in the sample. Indeed, neglecting relaxation effects, we find from (4.2) that each elementary part of the material provides an elementary signal:

$$dS \propto \rho(\mathbf{r})dv \exp[i\omega(\mathbf{r})t] \tag{4.4}$$

This elementary signal is proportional to the local spin density ($\rho(\mathbf{r})$) and to a term varying with time and position that results from the *phase shift* $i\omega(\mathbf{r})t$. Omitting the constant of proportionality, the total signal in the frame rotating at ω_0 is expressed as

$$S(t) = \iiint \rho(\mathbf{r}) \exp[i\gamma\mathbf{G}_z \cdot \mathbf{r}t]d\mathbf{r} \tag{4.5}$$

This function is in fact a Fourier transform ($S(\mathbf{k})$) of the vector $\mathbf{k} = \gamma\mathbf{G}_z t/2\pi$, which can be inverted to give the spin density at any point in the material:

$$\rho(\mathbf{r}) = \iiint S(\mathbf{k}) \exp[-i\gamma\mathbf{G}_z \cdot \mathbf{r}t]d\mathbf{k} \tag{4.6}$$

Taking advantage of this equation, the basic objective of NMR imaging is to cover at best, within the shortest time, the $\mathbf{k}$ space by imposing successive magnetic field gradients of different amplitudes and directions, in order to obtain the most exact density distribution from (4.6). For this purpose a series of constant gradients of different amplitudes is generally applied to the material initially excited in the same manner. This means that the precision of the image in the real space, namely, the resolution, depends on how the **k** space has been covered (see Figure 4.2), and thus generally increases with the number of gradients imposed, which implies that the resolution improves with the duration of the experiment.

4.1.3 Principles of MRI Velocimetry

MRI velocimetry appears as a very practiceful means under various situations [22]. One basic technique (spin tagging) for MRI velocity imaging consists in using spin resonance in the presence of gradients as a marking system within the sample; one excites (with a pulse) the protons in a specific layer of material and creates an image as described above a short time after this excitation (before complete signal relaxation). Under appropriate conditions this image provides a view of the new distribution of the excited spins in space, and thus of the deformation of the layer as a result of the flow field (see Figure 4.3). There are two main problems with this technique. Since the time of flow needed to observe a significant deformation of the marked layer depends on the fluid velocity, it is difficult to obtain a good resolution in the low-velocity regions. Generally cumbersome image analysis techniques are required for a quantitative determination of the velocity profile from such data.

A more subtle technique consists in utilizing the fact that the phase shift of a moving material element varies in time and space as it goes through regions of different magnetic fields. Let us examine the effect of imposing a time-dependent gradient of the main magnetic field to a moving spin. Over a short duration t this motion may be expressed in terms of the current position

$$\mathbf{r}(t) = \mathbf{r}_0 + t\mathbf{v}_0 + \frac{t^2}{2}\mathbf{a}_0 + \mathbf{o}(t^2) \tag{4.7}$$

in which $\mathbf{r}_0 = \mathbf{r}(t=0)$, $\mathbf{v}_0 = \mathbf{v}(t=0)$, and $\mathbf{a}_0 = \mathbf{a}(t=0)$. This expression is valid within a small volume of material, in which the velocities of the different elements do not differ significantly during a typical duration of the elementary

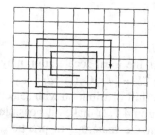 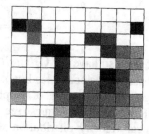

Figure 4.2 Principle of MRI: covering the **k** space (left) provides an image (right) in real space.

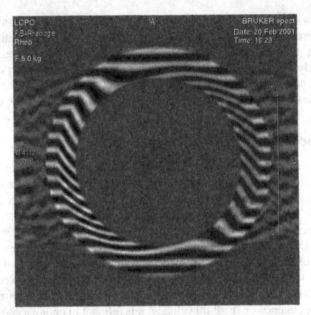

Figure 4.3 Apparent deformation in a bentonite suspension flowing in a Couette rheometer: view of the position of (excited) lines initially parallel to a (central) diameter after a short period of flow using a basic MRI velocimetry technique. [Courtesy of F. Bertrand (LMSGC, Champs sur Marne).]

step of a NMR sequence (say, several tenths of a microsecond). The total resulting phase shift of a fluid element after a time Δt is found from (4.4) by assuming that this element undergoes successive small phase shifts. In complex coordinates, since the signal is proportional to an exponential of the phase shift, the term associated with the successive phase shifts is the exponential of the sum of the phase shifts:

$$dS \propto \rho(\mathbf{r})dv \exp\left[i \int_{\Delta t} \mathbf{G}_z(t) \cdot \mathbf{r}(t)dt\right] \tag{4.8}$$

If $\int_{\Delta t} \mathbf{G}_z(t)dt = 0$ and $\mathbf{p} = \int_{\Delta t} \mathbf{G}_z(t)t\, dt \neq 0$, the phase shift at leading order is.

$$\delta = \mathbf{p} \cdot \mathbf{v}_0 \tag{4.9}$$

Note that if we also have $\int_{\Delta t} \mathbf{G}_z(t)t\, dt = 0$, the local acceleration may be found from the phase shift since at leading order $2\delta = [\int_{\Delta t} \mathbf{G}_z(t)t^2 dt] \cdot a_0$.

Coupling such observations of the phase shift distribution with selective pulses and/or NMR imaging techniques (see Section 4.1.2) makes it possible to determine the local velocity distribution within the material. In practice, different, more or less sophisticated sequences may be used [21,22] that account for various constraints of NMR imaging along with this specific type of gradient history.

4.1.4 Resolution and Difficulties

As for any NMR imaging technique, the absolute resolution and the quality of data from MRI velocimetry are difficult to define a priori. The richness of this technique lies in its complexity. The signal intensity first depends on material characteristics such as its relaxation times and liquid density. Basically the most favorable situation would thus a priori be that of a material with a large concentration of a simple liquid with a large time of relaxation. In practice, this is not so simple because in most sequences a sufficiently rapid relaxation of the signal is required between the successive pulses followed by the applications of different field gradients. Moreover, the overall quality or velocity resolution of the signal depends on

- The total experimental time, since the accumulation of data from the same procedure makes it possible to decrease the signal-to-noise ratio
- The size of the sample, since for a given time of experiment the ratio of the resolution to the sample size is almost constant
- The maximum possible intensity of gradients for the apparatus used, since the phase shift resolution is proportional to the gradient amplitude in (4.9)

These limitations concern any type of imaging experiment and are discussed in detail in several books or papers on the analysis of the interest of various procedures (sequences) with regard to particular situations. In fact, because of the infinite number of possibilities and the complexity of the possible effects, it is difficult to know a priori the most appropriate sequence for a given case. NMR imaging is in itself a scientific field. In the following paragraphs we examine in more detail some specific problems related to velocity measurements with this technique.

First, note that the NMR signal results from the excitation of spins of the various material types contained in an elementary volume of fluid. Usually the signal resulting from gas is negligible because of its very small density of protons, and the signal resulting from solid elements has such a short relaxation time that it is not recorded with the usual MRI setup. As a consequence, it is mainly the liquid phase of the fluid that is imaged. Under these conditions, neglecting the effect of possible, different relaxation times of different liquid species, a measure of the velocity of a given elementary fluid volume is relevant only if the different phases have the same "group velocity" within this volume. Otherwise it provides information only about the motion of the liquid phase. In practice with pasty materials, the relative velocity of two different phases is generally negligible. This may be appreciated from the Stokes number (St), which estimates the ratio of particle inertia to viscous drag. We may use the expression for this number from (2.26), in which the liquid viscosity should be replaced by the apparent viscosity of the fluid around an element. It appears that St is generally much smaller than 1 for typical elements leading to soft interactions (see Section 3.1). This means that the motion of fluid elements is governed by the fluid surrounding them so that there is no significant relative motion between the elements and the rest of the fluid. The situation may be different for granular pastes (see Section 3.6).

Another condition for the relevance of the velocity distribution as determined by NMR is that the medium be continuous at the scale of observation of this technique. Indeed, since, unless the velocity is uniform, the different parts of an elementary, imaged volume (associated with one data point) have different velocities, a MRI velocity profile is an approximation of the effective velocity profile. The velocity distribution as deduced from MRI may significantly depart from the effective velocity profile if, within some elementary imaged volume, there is a significant variation of the velocity as compared to the velocity difference between the boundaries.

4.2   RHEOLOGICAL INTERPRETATION OF VELOCITY PROFILES

The question we address here is how a velocity profile, as determined from any kind of technique, in the form of a series of velocity data at different distances from a solid boundary, can be interpreted in terms of the rheological behavior of the material. Fundamentally this requires some knowledge concerning the local stress in the material, since the constitutive equation relates the stress to the history of the local shear rate (see Section 1.3). This is obviously one of the five basic types of viscometric flows presented in Chapter 1, where we can hope to find a situation for which the stress distribution would be sufficiently simple and controlled. However, the cone–plate geometry is not appropriate in this purpose. Indeed, the shear stress distribution is almost homogeneous and the possible, slight heterogeneities deriving from imperfections of the shape of the periphery (see Section 3.3.4), and that would yield some shear rate variations in the gap, cannot be accurately predicted. The parallel disk geometry is also inappropriate because the radial stress distribution is a priori unknown, that is, it depends on the shear rate and on the material behavior. The other geometries (inclined plane, Couette, capillary) are appropriate for a rheological interpretation of velocity profiles; under the assumptions of Section 1.4, as soon as the material thickness is sufficient, the stress variations in the material, respectively given by equations (1.55), (1.76), and (1.86), are significant but independent of material behavior.

In the following text we will focus on the two geometries (Couette and capillary) for which it is most realistic to determine the velocity profiles from existing techniques. We start by reviewing (in Section 4.2.1) the typical trends of velocity profiles in Couette flow for the main types of behavior generally considered in the field of pastes or granular materials, namely, the power-law model (which in particular includes the Newtonian model), the Herschel–Bulkley model (which includes the Bingham model), and the truncated power-law model. Then we examine the different ways for interpreting velocity profiles in rheological terms, either from a single velocity profile (Section 4.2.2) or from a set of velocity profiles under different boundary velocities with the same material (Section 4.2.3).

It is worth noting that here we focus only on steady-state flows because the knowledge and the frame of interpretation of general properties of time-dependent fluid properties, from both experimental and theoretical perspectives, are as yet rather limited. As yet there seems to exist only one attempt to interpret the rheological behavior of thixotropic fluids from the velocity profiles in transient conditions [23].

4.2.1 Aspects of Velocity Profiles in Couette and Capillary Flow for Typical Rheological Behavior

Simple Power-Law Model

The constitutive equation of a simple power-law model in simple shear is given as

$$\tau = k\dot{\gamma}^n \tag{4.10}$$

in which k and n are two material parameters. For a Couette flow, after integration of (1.74) and (1.76) using (4.10), we obtain

$$v_\theta(r) = \frac{n}{2}\left(\frac{M}{2\pi hk}\right)^{1/n} r(r^{-2/n} - r_2^{-2/n}) \tag{4.11}$$

Using the maximum tangential velocity in the fluid, specifically, along the inner cylinder $v_{max} = \Omega r_1$, we can rewrite the tangential velocity as

$$V(R) = \frac{v_\theta(r)}{v_{max}} = \frac{r}{r_1}\left(\frac{r^{-2/n} - r_2^{-2/n}}{r_1^{-2/n} - r_2^{-2/n}}\right) = \frac{R}{\Lambda}\left(\frac{R^{-2/n} - 1}{\Lambda^{-2/n} - 1}\right) \tag{4.12}$$

in which $R = r/r_2$ and $\Lambda = r_1/r_2$. Examples of corresponding velocity profiles for different values of n are given in Figure 4.4.

For a capillary flow through a conduit of radius r_0, by integration of (1.84) using (4.10), we find

$$v_z(r) = \frac{n}{n+1}\left(\frac{1}{2k}\frac{\Delta p}{L}\right)^{1/n}(r_0^{(n+1)/n} - r^{(n+1)/n}) \tag{4.13}$$

and using the maximum velocity ($v_{max} = v_z(0)$) and the scaled distance $R = r/r_0$, we can rewrite the axial velocity as follows:

$$V(R) = \frac{v_z(r)}{v_{max}} = (1 - R^{(n+1)/n}) \tag{4.14}$$

It follows that for a given value of n the dimensionless velocity V depends only on the distance r and on neither the rotation velocity of the inner cylinder or the applied torque for the Couette flow nor the flow rate or the pressure loss by unit length for the capillary flow. This implies that for such a power-law fluid

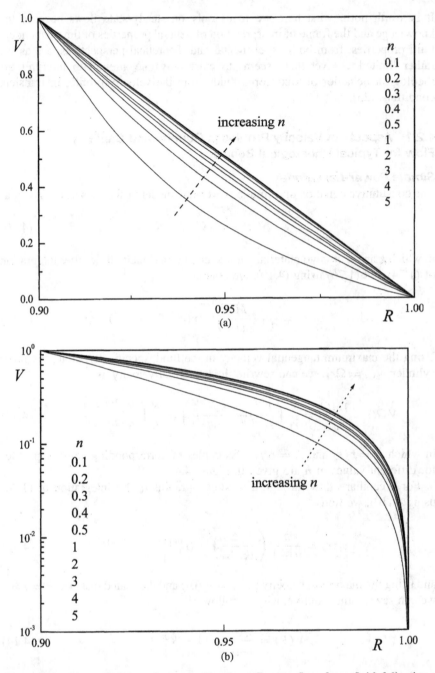

Figure 4.4 Dimensionless velocity profiles in a Couette flow for a fluid following a simple power-law model for different values of n on linear (a) and logarithmic (b) scales and for a inner:outer cylinder ratio of 0.9.

the different velocity profiles determined experimentally and scaled in this way should fall along a master curve.

Herschel–Bulkley Model

The constitutive equation in simple shear of a Herschel–Bulkley model is

$$\tau < \tau_c \Rightarrow \dot{\gamma} = 0; \ \tau > \tau_c \Rightarrow \tau = \tau_c + k\dot{\gamma}^n \tag{4.15}$$

For a Couette flow there are two cases depending on the relative values of r_2 and r_c, the critical distance at which the yield stress is reached:

$$r_c = \sqrt{\frac{M}{2\pi h \tau_c}} \tag{4.16}$$

When $r_c < r_2$, there is an unsheared region between the two cylinders and we have

$$r_c < r \leq r_2 \Rightarrow v_\theta(r) = 0 \tag{4.17a}$$

$$r_1 \leq r \leq r_c \Rightarrow v_\theta(r) = \left(\frac{M}{2\pi h k}\right)^{1/n} r \left(\int_r^{r_c} \frac{1}{x} \left(\frac{1}{x^2} - \frac{1}{r_c^2}\right)^{1/n} dx\right) \tag{4.17b}$$

When $r_c > r_2$ there is no unsheared region between the two cylinders and we have

$$v_\theta(r) = \left(\frac{M}{2\pi h k}\right)^{1/n} r \left(\int_r^{r_2} \frac{1}{x} \left(\frac{1}{x^2} - \frac{1}{r_2^2}\right)^{1/n} dx\right) \tag{4.18a}$$

There is no analytical solution for the integrals in (4.17b) and (4.18) except for entire values of n.

The velocity in the sheared region (4.17b) scaled by the maximum velocity is

$$V(R) = \frac{v_\theta(r)}{V_{max}} = \frac{R}{\Lambda^*} \left(\frac{\int_R^1 \frac{1}{x}(x^{-2} - 1)^{1/n} dx}{\int_{\Lambda^*}^1 \frac{1}{x}(x^{-2} - 1)^{1/n} dx}\right) \tag{4.18b}$$

in which here $R = r/r_c$ and $\Lambda^* = r_1/r_c$. Let us consider the different velocity profiles obtained for a given material under different rotation velocities of the inner cylinder. When scaled in this way these profiles are not similar since the dimensionless velocity depends not only on the dimensionless distance $R = r/r_c$ but also on the parameter Λ^*, which varies with the rotation velocity of the inner cylinder, or equivalently the torque, via r_c.

Using (4.16) we can also rewrite the expression for the velocity in the sheared region (4.17b) as

$$v_\theta(R) = r_c \left(\frac{\tau_c}{k}\right)^{1/n} R \left(\int_R^1 \frac{1}{x}(x^{-2} - 1)^{1/n} dx\right) \tag{4.19}$$

from which we see that the velocity profiles obtained for a material under different torques or rotation velocities should fall along a master curve when represented in the form of v_θ/r_c as a function of R. Examples of such velocity profiles for different values of n are given in Figure 4.5.

For a capillary flow the expression for the velocity depends on the relative position of the current radius and the critical radius at which the yield stress is reached

$$r_c = \frac{2\tau_c}{\Delta p/L} \tag{4.20}$$

and we deduce by integration of (4.15) and (1.86):

$$r \leq r_c \Rightarrow v_z(r) = \frac{n}{n+1} \left(\frac{1}{2k}\frac{\Delta p}{L}\right)^{1/n} (r_0 - r_c)^{1+1/n} \tag{4.21a}$$

$$r > r_c \Rightarrow v_z(r) = \frac{n}{n+1} \left(\frac{1}{2k}\frac{\Delta p}{L}\right)^{1/n} \left((r_0 - r_c)^{1+1/n} - (r - r_c)^{1+1/n}\right) \tag{4.21b}$$

The velocity in the sheared region may also be expressed as

$$v_{max} - v_z(R) = r_c \left(\frac{\tau_c}{k}\right)^{1/n} (R - 1)^{1+1/n} \tag{4.22}$$

For both capillary and Couette flows it follows from (4.17b) and (4.21b) that the variations of the velocity at the approach of the unsheared region as a function of the distance r are proportional to $|r - r_c|^{1+1/n}$. This means that for such fluids the velocity profile should retain a self-similar aspect at different scales of observations around the critical radius. We also deduce that $v_\theta(r) \to 0$ ($v_\theta(r) \to v_{max}$ for a capillary) when $r \to r_c$ and that the shear rate is continuous: $\dot{\gamma}(r) \to 0$ when $r \to r_c$. This result is in fact a straightforward consequence of the rheological behavior of the fluid, the same result that would be obtained for any yield stress model of the type (2.19) such that $f \to 0$ when $\dot{\gamma} \to 0$.

Truncated Power-Law Model
Let us recall the constitutive equation of a truncated power-law fluid in simple shear already presented [in equation (3.35)]:

$$\tau < \tau_c \Rightarrow \dot{\gamma} = 0; \quad \tau > \tau_c \Rightarrow \tau = k\dot{\gamma}^n = \tau_c \left(\frac{\dot{\gamma}}{\dot{\gamma}_c}\right)^n \tag{4.23}$$

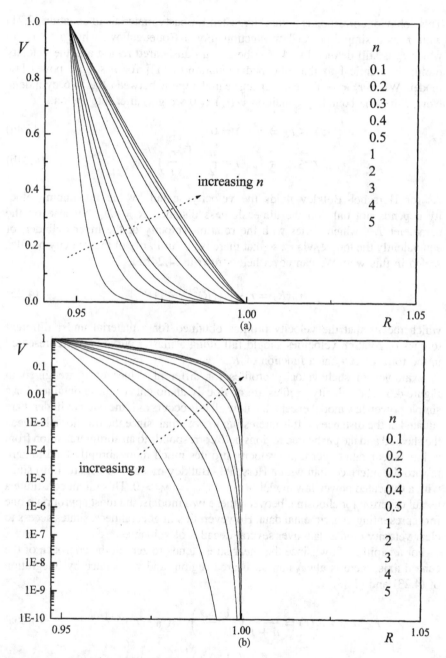

Figure 4.5 Dimensionless velocity profiles in a Couette flow for a fluid following a Herschel–Bulkley model for different values of n on linear (a) and logarithmic (b) scales and for a inner:outer cylinder ratio of 0.9.

This behavior corresponds to that suggested for pasty materials in equation (2.21) with f as a simple power-law function. For a Couette flow, when $r_c \geq r_2$, in which r_c is still defined by (4.16), there is no unsheared region and the velocity profile is identical to that obtained [equation (4.11)] for a simple power-law model. When $r_c < r_2$ there is an unsheared region between the two cylinders. Now, using the boundary condition $v_\theta(r_c) = 0$ we get, after integration:

$$r_c < r \leq r_2 \Rightarrow v_\theta(r) = 0 \tag{4.24a}$$

$$r_1 \leq r \leq r_c \Rightarrow v_\theta(r) = \frac{n}{2}\dot{\gamma}_c\left[\left(\frac{r_c}{r}\right)^{2/n} - 1\right] r \tag{4.24b}$$

As for Herschel–Bulkley fluids the velocity scaled by the maximum velocity depends not only on the dimensionless distance $R = r/r_c$ but also on the parameter Λ^*, which varies with the rotation velocity of the inner cylinder, or equivalently the torque, via r_c, so that there is no similarity of the velocity profiles scaled in this way. We can nevertheless rewrite (4.24b) as

$$v_\theta(R) = r_c\frac{n}{2}\dot{\gamma}_c[(R)^{2/n} - 1]R \tag{4.25}$$

which means that the velocity profiles obtained for a material under different torques or rotation velocities should fall along a master curve when represented in the form of v_θ/r_c as a function of R.

Examples of such velocity profiles for different values of n are given in Figure 4.6. The velocity profiles are obviously analogous to those obtained for a simple power-law model except that now all this occurs as if the outer cylinder were situated at the distance r_c. It is interesting to note that, since the transition between the sheared and the unsheared regions here corresponds to an abrupt transition from a finite shear rate to zero, the velocity profiles much more abruptly tend to zero toward the interface than with a Herschel–Bulkley model (see Figure 4.5b) since with a truncated power-law model $(\partial v_\theta/\partial r)_{r=r_c} = \dot{\gamma}_c > 0$. This might constitute a useful criterion for choosing, between these two models, the most appropriate one for representing experimental data. However, this in general necessitates access to clear velocity profile data over several decades of velocities.

For a capillary flow, since the shear stress tends to zero at the approach of the central axis, there is always an unsheared region, and we deduce by integration of (4.23) and (1.86)

$$r \leq r_c \Rightarrow v_z(r) = \frac{n}{n+1}\left(\frac{1}{2k}\frac{\Delta p}{L}\right)^{1/n}(r_0 - r_c)^{(n+1)/n} \tag{4.26a}$$

$$r > r_c \Rightarrow v_z(r) = \frac{n}{n+1}\left(\frac{1}{2k}\frac{\Delta p}{L}\right)^{1/n}(r_0^{(n+1)/n} - r^{(n+1)/n}) \tag{4.26b}$$

We may rewrite (4.26b) in a form similar to (4.25)

$$v_{max} - v_z(R) = r_c\dot{\gamma}_cR^{(n+1)/n} \tag{4.27}$$

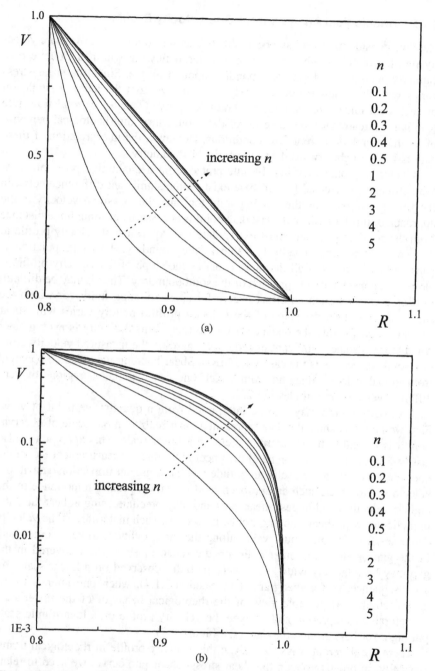

Figure 4.6 Dimensionless velocity profiles in a Couette flow for a fluid following a truncated power-law model for different values of n on linear (a) and logarithmic (b) scales and for a inner:outer cylinder ratio of 0.8.

4.2.2 Rheological Analysis from One Velocity Profile

Here we consider that we have been able to measure both the steady-state velocity profile (velocity–radius curve) of a material flowing within a Couette or a capillary geometry and the corresponding applied torque. Since the shear stress varies with the radius (or equivalently the distance from the axis), a single set of such data under one particular rotation velocity (Couette) or total flow rate (capillary) already provides a series of data concerning the rheological behavior of the fluid under different flow conditions. However, the interpretation of these data is less straightforward than conventional rheometry.

A velocity profile may first be interpreted from a qualitative point of view. Wall slip can be detected [17] if there exists some significant difference between the velocity imposed by the moving solid tool and the measured velocity at the approach of this tool. Since the last data point is necessarily at some finite distance from the solid surface, one must extrapolate the experimental velocity profile to deduce the (theoretical) velocity value at the boundary. This extrapolation is partly arbitrary but it is all the more exact as the slope of the velocity profile is closer to a constant at the approach of the solid boundary. Thus it may be difficult to decide whether such an effect occurs for shear thinning fluids for which the shear rate (and thus the slope of the velocity profile) rapidly varies with shear stress (or equivalently the distance from the tool). This difficulty is even greater when there is some significant uncertainty regarding the measured velocity close to a solid surface (as is the case with MRI). Shear banding may also be apparent from global velocity imaging from which one can distinguish regions of very different velocity amplitudes [24,25].

A velocity profile may also be interpreted from a quantitative point of view. First we note that since the data represent a clear reality on our scale of observation, it is logical to try to interpret them on a linear scale. This approach is also justified by the fact that in general the uncertainty due to measurement technique does not vary with the velocity amplitude and has a greater impact on data at low velocities. However, such an approach tends to attach greater importance to the highest velocities and largest shear rates and may preclude using at best the data, in particular with shear thinning fluids. Indeed, for such materials, as the velocity decreases from its maximum value along the inner cylinder to zero somewhere in the gap or along the outer cylinder, the range of shear rates covered in the gap may be relatively wide and is very partially observed on a linear scale. As a consequence, as for rheograms (see Section 3.1.4), when one intends to get an as complete as possible view of the rheological behavior of the material, it is preferable to represent and analyze the velocity profile on a logarithmic scale putting error bars around the data (see Figure 4.7).

There are different ways to analyze this velocity profile in rheological terms. Basically, in order to obtain the shear stress–shear rate curve, we need to relate the local slope of the velocity profile, which is related to the shear rate via (1.75) or (1.84), to the local shear stress, respectively given by (1.76) and (1.86). A direct approach thus consists to compute the local value of the shear rate from the local slope and correlate it with the local stress value to obtain point by point

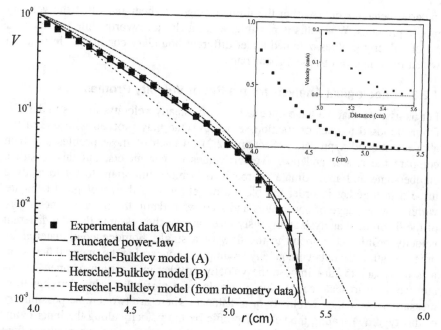

Figure 4.7   Velocity profile for a model emulsion in a Couette flow with error bars on logarithmic scale as obtained from MRI (shown on different linear scales in the inserts). Different models have been fitted to data on logarithmic scale: a Herschel–Bulkley model with different parameters and a truncated power-law model. [From Bertola et al., *J. Rheol.* **47**, 1211 (2003).]

the flow curve of the material. However, determination of the local shear rate from the velocity curve is cumbersome, subjective, and subject to multiple sources of errors; from a mathematical point of view, this is an ill-posed problem [26]. A simpler, straightforward, although still somewhat arbitrary, approach consists in fitting a model in the form of a function $f(r)$ to the velocity profile (Figure 4.7) from which one can deduce the shear rate: $\dot{\gamma} = r\partial(f(r)/r)/\partial r$ for a Couette flow and $\dot{\gamma} = \partial f(r)/\partial r$ for a capillary flow. For sufficiently simple expressions of f, it is possible to eliminate r between this expression and (1.76) (Couette flow) or (1.86) (capillary flow) to find analytical expressions for $\tau(\dot{\gamma})$. Otherwise it is at least possible to draw the flow curve by computing the shear stress and the shear rate for different values of r.

Another approach consists in assuming a specific rheological model ($\tau(\dot{\gamma})$) and comparing its predictions in terms of a velocity profile (see, e.g., Section 4.2.1) with experimental data, and fit the model parameters. An advantage of this approach is that it directly provides a simple rheological model in reasonable agreement with data, but it is not easy to find this model simply from the aspect of the velocity profile.

One must bear in mind that these different approaches based on the analysis of a single velocity profile in rheological terms lead to a model representing the material behavior only within the limited range of shear rates that the velocity profile provides. When this is possible, it is much more worthwhile to analyze a set of velocity profiles obtained under different boundary conditions, in order to widen the range of observed shear rates.

4.2.3 Rheological Analysis from a Set of Velocity Profiles

Let us assume that we have obtained a set of steady velocity profiles under different imposed torques or rotation velocities. One may proceed with one of the abovementioned approaches (Section 4.2.2) on each of these profiles and then compare the results and take average values of parameters, but this is still a cumbersome and approximate approach. It is more important to try to analyze these data together in order to obtain a direct view of the rheological behavior within a wide range of shear rates and somewhat damp the inherent uncertainty of the theoretical analysis. Obviously these approaches assume that the different velocity profiles correspond to the flow of a single, homogeneous material. In order to gather the data without any assumption concerning the material behavior, it is necessary to superimpose the velocity profiles in a certain manner. In this aim we need to scale each velocity profile by a particular value deriving from measurements, either a velocity or a distance. There are two simple possibilities of this type: (1) scaling the velocity profile by the velocity along the inner cylinder for a Couette flow, or the velocity along the central axis for a capillary flow and (2) scaling the distance by a possible critical radius at which the velocity drops to zero.

Scaling by the Maximum Velocity

Let us scale the velocity profiles in a Couette flow by the maximum velocity, reached along the inner cylinder. If the velocity profiles do not then superimpose along a single master curve, we can say nothing more about the constitutive equation of the material. If they fall along a single master curve (see e.g., Figure 4.8), we can analyze the data by utilizing this critical observation. In that case the velocity may be written $\Omega r_1 f(r)$, in which f is a function defined by the master curve, so that the shear rate at any distance r is now expressed as $\Omega r_1 r \partial (f/r)/\partial r = \Omega F(r)$, which defines F, a function of r and constant parameters. Since the shear stress is a function of the distance r via (1.76), we may also express it as $\Omega F(\sqrt{M/2\pi h \tau})$. Now, using the function $G = 1/2\pi h (F^{-1})^2$, we deduce a general relation between the shear rate and the shear stress at any point in the material when a torque M corresponding to the rotation velocity Ω of the inner cylinder is applied:

$$\frac{\tau}{M} = G\left(\frac{\dot{\gamma}}{\Omega}\right) \tag{4.28}$$

This equation is in particular valid for the torque M_1 for which the rotation velocity $\Omega = 1$ rad/s is reached: $\tau = M_1 G(\dot{\gamma})$. By eliminating the stress between this equation and that for any other torque M, we get

$$G(\alpha \dot{\gamma}) = \beta(\alpha) G(\dot{\gamma}) \tag{4.29}$$

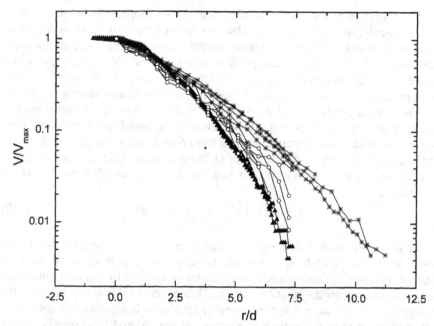

Figure 4.8 Dimensionless velocity profiles as determined by different techniques [Da Cruz, Ph.D. thesis, Marne la Vallée Univ., 2004; Mueth et al., *Nature* **406**, 385 (2000); Bocquet et al., *Nature* **396**, 735 (1998)] for dry granular materials under different velocities: V/V_{max} as a function of the ratio of the distance to the grain diameter. (Courtesy of F. Da Cruz.)

in which $\beta = M_1/M$ can be considered as a function of $\alpha = 1/\Omega$. From (4.29) we can successively show that $G(\alpha^p) = \beta(\alpha)^p G(1)$ for any entire p, then for any rational p/q, then for any real x. In particular, we deduce that $G(x) = G(\exp(\ln x)) = G(e)^{\ln x} G(1) = x^{\ln G(e)} G(1)$, which finally means that G is a simple power-law function. *Thus the behavior of a homogeneous material for which the velocity profiles scaled by the maximum velocity fall along a single master curve is of the power-law type* (4.10).

Following the same approach now using the (maximum) velocity along the central axis of the conduit, a similar result is obtained with the capillary flow. In that case the shear rate is directly proportional to the derivative of the fluid velocity [equation (1.74)] and the shear stress is proportional to the radius [equation (1.76)]. We easily obtain a relation of the type $\tau/(\Delta p/L) = G^*(\dot{\gamma}/V_{axis})$ analogous to (4.28), from which we can again deduce that the relation between the shear stress and the shear rate follows a power-law model.

Scaling by the Critical Distance of the Solid–Liquid Interface

Let us now assume that in Couette flow, beyond a critical distance r_c from the central axis, the velocity is equal to zero. Obviously, since the shear stress decreases with the distance from the axis, this reflects that the material is a yield stress fluid that cannot flow when the applied stress falls below a critical value, namely, the yield stress, τ_c. Here we assume that this yield stress, which dictates the position of the solid–liquid interface in the gap, remains constant under different boundary conditions and previous flow histories. This in particular means that the material remains homogeneous and that thixotropy effects possibly leading to different positions of the interface depending on flow history remain negligible.

Under such conditions, to each level of rotation velocity Ω corresponds a critical radius r_c associated with the location of the liquid–solid interface. Here we only consider experiments for which this critical radius remains smaller than the radius of the outer cylinder. From (1.76) we deduce that the corresponding applied torque is $M = 2\pi h r_c^2 \tau_c$, so that the stress at any distance r may be rewritten as

$$\tau(r) = \tau_c \left(\frac{r_c}{r}\right)^2 = \frac{\tau_c}{R^2} = \tau(R) \tag{4.30}$$

in which we again used $R = r/r_c$. In such a representation, because the critical stress at the interface $(R = 1)$ is fixed, the shape of stress distribution does not change with the imposed velocity; only the range covered by the stress increases as r_c increases with the rotation velocity. Moreover, in this representation, the distribution of the shear rate magnitude $(\dot{\gamma}(R))$ associated with this stress distribution via a unique constitutive equation of the sheared fluid should remain constant for different velocity levels. From (1.74) we derive that this shear rate distribution is also related to the velocity profile $v(R) = v_\theta/r_c$ via

$$\dot{\gamma}(R) = \left| R \frac{\partial(v/R)}{\partial R} \right| \tag{4.31}$$

Thus we deduce that the different velocity profiles expressed in the form $v(R)$, which define the unique shear rate expression $\dot{\gamma}(R)$, should be similar. In this context we expect that the representation of data in terms of v versus R provides a single master curve. We have seen that this is effectively the case for fluids following a Herschel–Bulkley or a truncated power-law model as proved by equations (4.19) and (4.25).

A similar approach can be carried out for capillary flows (Section 1.4.5); a yield stress fluid is not sheared below the critical radius $r_c = 2\tau_c/A$ so that the shear stress (1.86) may be written as $\tau(R) = \tau_c R$ and the shear rate as $\dot{\gamma}(R) = \partial v/\partial R$. We deduce that for a single material the velocity profiles $v(R)$ (within an additional constant) should fall along a master curve since they represent the link between the shear stress and the shear rate through a single constitutive equation. However, since the velocity in that case is maximum for the critical radius, it is necessary to use the dimensionless, reduced velocity $v^*(R) = (v_{max} - v_z)/r_c$, in order to directly obtain the master curve. We have seen that this is effectively the case for fluids following a Herschel–Bulkley or a truncated power-law model as proved by equations (4.22) and (4.27).

When it is possible to put all the data along a master curve by this way, this proves both the consistency of the assumption of a yielding behavior (with a

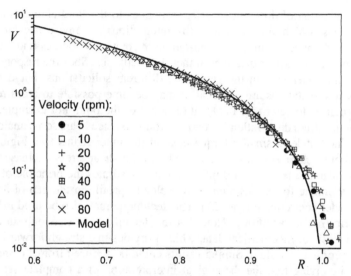

Figure 4.9 Dimensionless velocity profiles ($V = v/r_c$ as a function of $R = r/r_c$) for a bentonite suspension in Couette flow as determined by MRI under different rotation velocities of the inner cylinder. (Reprinted from Roussel et al., *J. Non-Newtonian Fluid Mech.* **117**, 85 (2004), with permission from Elsevier.)

unique yield stress under various flow conditions) and the existence of a single rheological behavior for the flowing region. In practice this has also the great advantage to provide a single curve including all data, along which one may fit a model for the velocity in order to deduce the constitutive equation by integration of (4.30) and (4.31). The validity of this approach has been shown for some materials (Figure 4.9). This requires one to first determine the values of r_c obtained under different torques or rotation velocities. For a fluid following a truncated power-law model, this determination remains relatively simple because the shear rate at the solid–liquid transition is finite. For a Herschel–Bulkley fluid, a direct, exact, determination of r_c from data is difficult because the shear rate progressively tends toward zero on approach of the liquid–solid interface (see Section 4.2.1). In practice, for such a fluid, the best way may consist in directly fitting the values of r_c so as to get a master curve for the velocity profiles $V(R)$. However, it seems that the sets of data available so far in the literature could be more appropriately represented by a model predicting a drop in shear rate at the liquid–solid interface, such as the truncated power-law model.

4.3   VELOCITY PROFILE RECONSTRUCTION FROM RHEOMETRY

As already mentioned, Couette flows have the great interest that the spatial heterogeneity of the shear stress distribution is controlled. This implies that under controlled torque, at a certain distance from the axis, the stress history experienced by a fluid element is known. A series of tests under different torque values lead to the application of different shear stress values to the fluid elements at different distances, which altogether lead (by integration) to some particular rotation velocity of the inner cylinder. The ensemble of rotation velocities under different torques provides a set of data that in fact indirectly describes the response of the different fluid elements in the gap under different solicitations, but if some of these local solicitations are identical, it may become possible to cross-reference the information to determine the local response of the fluid. For example, we can note that the stress distribution beyond a certain distance bears some analogy with that obtained under a smaller torque beyond the inner radius (see Figure 4.10); the maximum shear stresses are identical and the shear stress decreases to zero when the distance tends toward infinity. This constitutes the principle of the following technique for reconstructing the velocity profile in the gap of a Couette system under controlled torque [27]. This technique may also be used in the case of uniform free surface flows. Note that this technique consists only in an alternative analysis of rheometrical data, which provides in a straightforward way the velocity profile in the fluid. Similar results could be deduced from a cumbersome analysis involving first the rheological interpretation of a complete set of data, for example, via a model, and then computation of the flow predictions.

4.3.1 Local Flow Properties as a Function of Stress History

Kinematics
We consider the flow of a material between coaxial cylinders (only the inner cylinder rotates) as has been described in Section 1.4.4. Let us recall that as long

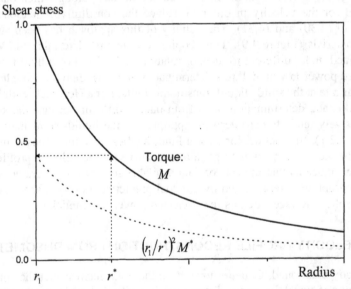

Figure 4.10 Shear stress as a function of the distance from the central axis in a Couette flow for two different torque values. The shear stress at a distance r^* under the torque M is equal to the shear stress along the inner cylinder under the torque $(r_1/r^*)^2 M$.

as the flow is stable, the cylindrical fluid layers located at different distances from the axis are in relative motion, and the shear stress distribution is given as

$$\tau(r, t) = \frac{M(t)}{2\pi h r^2} \tag{4.32}$$

in which $M(t)$ is the current torque applied to the inner cylinder. In order to describe the flow characteristics of the material, we can focus on the evolution of $\omega = \omega(r, t)$, the rotation velocity within the material at distance r from the axis and at time t in the frame of reference of the laboratory, which is related to the shear rate via (1.74).

Now we assume that the outer cylinder is at an infinite distance (i.e., $r_2/r_1 \rightarrow \infty$). Since, from (4.32), the shear stress tends to zero as the distance tends to infinity, the shear rate tends to zero for a viscous fluid and we have $\omega(r_2 \rightarrow \infty) = 0$. As a consequence, in the absence of wall slip, the rotation velocity $\Omega(t)$ of the inner cylinder at time t as given by (1.77) may be rewritten as

$$\Omega(t) = \int_{r_1}^{\infty} \frac{\dot{\gamma}}{r} dr \tag{4.33}$$

and more generally the rotation velocity at any distance r at the time t is

$$\omega(r, t) = \int_{r}^{\infty} \frac{\dot{\gamma}}{x} dx \tag{4.34}$$

The rotation angle reached a certain time t after the initial instant at distance r may be expressed as

$$\varphi(r, t) = \int_{0}^{t} \omega(r, t') dt' \tag{4.35}$$

Specific Form of the Constitutive Equation

The behavior in simple shear of an incompressible, homogeneous, jammed material can be expressed in the form of a relationship between the stress tensor and the history of the local shear (1.51). This relation may be inverted to get (1.52), which expresses the current shear rate as a function of the history of the shear stress and the normal stress differences. The shear stress history follows from the torque history via (4.32), which involves only the current torque $M(t)$, but we a priori ignore the variations of the normal stress differences. Intuitively, for this very simple flow, it seems that all flow characteristics should depend only on the unique boundary condition in terms of stress, namely, the torque applied. Here, following this idea, we will assume that the normal stress differences also depend only on $M(t)/r^2$. This might, in fact, constitute a general case in this particular context since any elementary solution of the type $\sigma_{rr} = \alpha M(t)/r^2$; $\sigma_{\theta\theta} = -\alpha M(t)/r^2$, in which α is a constant appears to be a solution of the momentum equation in the radial direction (1.16). Under these conditions the shear rate finally depends solely on the history of $M(t)/r^2$, which may be expressed as follows:

$$\dot{\gamma}(r, t) = \underset{0 < \vartheta < t}{\text{H}} (\Sigma_0, M(\vartheta)/r^2) = \dot{\gamma}_{\Sigma_0;M} (r, t) \tag{4.36}$$

4.3.2 Reconstruction of the Profiles of Rotation Velocity and Rotation Angle

Torque–Distance Equivalence Principle

Keeping the same geometry, the local, instantaneous shear stress at a given distance r when a torque M is applied is equal to the local, instantaneous shear stress at a given distance r^* when a torque M^* such that

$$M^* = \left(\frac{r^*}{r}\right)^2 M$$

is applied (see Figure 4.10). It follows from (4.36) that the instantaneous shear rate at a distance r under a torque history $M(\vartheta)$ is equal to the shear rate at a distance r^* when a torque history $M^*(\vartheta)$ such that $M^*(\vartheta) = (r^*/r)^2 M(\vartheta)$ at each time ϑ is applied:

$$\dot{\gamma}_{\Sigma_0;M} (r, t) = \dot{\gamma}_{\Sigma_0;M^*}(r^*, t) \tag{4.37}$$

This result may be used in (4.34), which, thanks to a change of variable ($\xi = (r^*/r)x$), transforms as

$$\omega_M(r, t) = \int_r^\infty \frac{\dot{\gamma}_{\Sigma_0;M}(x, t)}{x} dx = \int_{r^*}^\infty \frac{\dot{\gamma}_{\Sigma_0;M^*}(\xi, t)}{\xi} dr' = \omega_{M^*}(r^*, t) \tag{4.38}$$

in which $r^* = \alpha r$, where $\alpha = \sqrt{M^*(\vartheta)/M(\vartheta)}$ is a constant. This in particular implies that the rotation velocity at a distance r under the torque history $M(\vartheta)$ is equal to the rotation velocity of the inner cylinder $\Omega(r_1, t)$ under the torque history $M^*(\vartheta) = M(\vartheta)(r/r_1)^2$.

Reconstruction

It results that, at any time t, the distribution of $\{\omega_M(r, t); r > r_1\}$ corresponds to the distribution $\{\Omega_{M^*(t')}(r_1, t); M^*(\vartheta) < M(\vartheta)\}$ with $r = r_1(\sqrt{M^*(\vartheta)/M(\vartheta)})$, which means that the spatial distribution of rotation velocities in the material under a given torque history may be obtained from a set of rotation velocities of the inner cylinder under analogous torque histories of smaller amplitudes, starting from the same initial state.

Remarks

Since this result is valid for any given time period, an analogous result may be obtained for the angle of rotation ($\varphi(t)$) after a certain time of flow: the distribution of rotation angles within the gap under a given torque after a time t is equivalent to the ensemble of rotation angles of the inner cylinder after the same time under different, smaller torque histories, starting from the same initial state.

The correspondence of the velocity distribution within the gap with the set of rotation velocities obtained under various stress magnitudes in particular results from the fact that the upper boundary of the integrals in (4.33) or (4.34) remains constant; it is equal to infinity, which derives from the assumption of infinite outer radius. The correspondence between the velocity distribution and the rotation velocities under smaller torque histories cannot a priori be obtained in practice because the upper boundary is equal to the outer radius situated at a finite distance and our change of variable leads to a change in another finite value. Thus the reconstruction technique a priori provides an approximate velocity profile or rotation angle profile, which are closer to the effective profiles as the ratio r_1/r_2 is closer to zero. However, we will see that, because of the specific behavior characteristics of pasty materials, this technique can, under some conditions, provide the exact velocity or rotation angle profiles (see Section 4.3.4), so it is especially appropriate for that type of material.

4.3.3 Application to Simple Shear Stress Histories

Creep Tests
The reconstruction technique is simple to carry out for creep tests: $M(\vartheta) =$ constant. It consists in imposing on to the material in the same initial state a series of torques of different values below a given level M_0. This provides a set of creep curves (rotation angle–time curves) (see Figure 3.1) from which one can extract the profile of rotation velocity at each time via the torque–distance equivalence principle (Section 4.3.2). An example application of this technique along with a comparison with MRI data in steady state is shown in Figure 4.11.

Stress Ramp
Let us consider the case of a linear ramp of torque applied to the material, with the torque at the time t given as

$$M(t) = 2\pi h K t \tag{4.39}$$

in which K is a constant. We deduce from the abovementioned considerations that the distribution of $\{\omega_K(r, t); r > r_1\}$ is equal to the distribution $\{\Omega_{K^*}(r_1, t); K^* < K\}$ with $K^* = (r_1/r)^2 K$; that is, the spatial distribution at a given time during a torque ramp may be found from an ensemble of rotation velocities of the inner cylinder under torque ramps at smaller rates (of stress increase), starting from the same initial state.

Dynamic Tests
A similar approach may be used for dynamics tests under controlled torque. If the torque at the time t is given as

$$M = A \sin ft \tag{4.40}$$

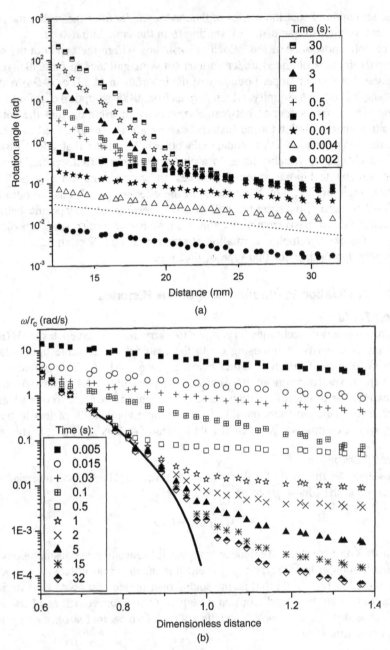

Figure 4.11 Reconstructed rotation angle (a) and rotation velocity (b) distribution at different times during creep tests for a hair gel in a Couette system (data of Figure 3.1). The continuous line for the rotation velocity corresponds to the steady-state profile as determined from MRI. (Reprinted with permission from Baudez and Coussot, *Phys. Rev. Lett.*, **93**, 128302 (2004), Figure 1a and 1b. Copyright (2004) by the American Physical Society.)

in which f is the frequency, we deduce that the distribution of $\{\omega_A(r, t); r > r_1\}$ is equal to the distribution $\{\Omega_{A^*}(r_1, t); A^* < A\}$ with $A^* = (r_1/r)^2 A$; thus, the spatial distribution at a given time during torque oscillations may be found from an ensemble of rotation velocities of the inner cylinder under torque oscillations at smaller amplitudes but with the same frequency, starting from the same initial state. From these data the effective, local rheological parameters under dynamic tests such as the elastic and viscous moduli, and the phase shift (see Section 3.1.2) may be determined.

Let us emphasize that these parameters can be determined only from analysis of the velocity or deformation profile in time and not directly from a set of the values of these variables under smaller torques or amplitudes. Indeed, the "apparent" values of these variables result from a generally complex analysis of the rotation velocity of the inner cylinder that results from the complete velocity profile. Thus the reconstruction technique directly applied to such variables, although it is more straightforward than the analysis of the velocity profile in time, provides the distribution of only the apparent variable through the gap. For example, a series of ramps of oscillations at different rates of increase in the stress amplitude (which means that we now use a parameter A varying in time as αt) provides the evolution of the distribution of the apparent phase shift in the material as a function of time (or equivalently the torque amplitude) for the faster ramp. An example of such a distribution is given in Figure 4.12.

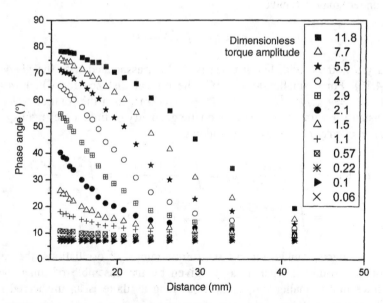

Figure 4.12 Reconstructed phase shift distribution under different maximum stress amplitudes during a ramp of oscillations under increasing torque amplitude with a hair gel in a Couette system. (Reprinted with permission from Baudez and Coussot, *Phys. Rev. Lett.*, **93**, 128302 (2004), Figure 3. Copyright (2004) by the American Physical Society.)

4.3.4 Specific Behavior Types

Viscoelastic Solid

Here we consider a viscoelastic material following the simple Kelvin–Voigt model. In that case, when a torque history $M(\vartheta) = M_0 f(\vartheta)$ is applied to the material, its deformation in time may be found from integration of the constitutive equation (2.14):

$$\gamma(t) = \left[\int^t \frac{M(\vartheta)}{2\pi \mu h r^2} \exp\left(\frac{G}{\mu}\vartheta\right) d\vartheta \right] \exp - \left(\frac{G}{\mu}t\right) = \frac{M_0}{r^2}\Gamma(t) \qquad (4.41)$$

in which Γ is a function of t and material parameters only. Thus the deformation and its time derivative (i.e., the shear rate) are simply proportional to $1/r^2$ and to the torque magnitude M_0. Thus, the rotation velocity is expressed as follows:

$$\omega(r, t) = \int_r^\infty \frac{\dot{\gamma}}{x} dx = \frac{M_0 \dot{\Gamma}(t)}{2r^2} \qquad (4.42)$$

We deduce that in the solid regime the distribution of rotation velocity of a paste is also proportional to $1/r^2$.

Yield Stress Fluids

Let us assume that there exists a distance r_c for which, at any time t, for the flow under study, we have

$$r_c < r \Rightarrow \dot{\gamma}(r, t) = 0 \qquad (4.43)$$

For a yield stress fluid this distance is, in fact, associated with the yield stress via (4.16). For a smaller torque M^*, the critical distance for which (4.43) is valid is $r_c^* = r_c\sqrt{M^*/M} < r_c$. If $M^* = M(r_1/r)^2$, then the rotation velocity at any distance r within the gap of a Couette geometry of finite outer radius r_2 such that $r_c < r_2$ may be expressed as follows:

$$\omega_M(r, t) = \int_r^{r_2} \frac{\dot{\gamma}_M(x, t)}{x} dx = \int_r^{r_c} \frac{\dot{\gamma}_M(x, t)}{x} dx = \int_{r_1}^{r_c^*} \frac{\dot{\gamma}_{M^*}(\xi, t)}{\xi} d\xi$$

$$= \int_{r_1}^{r_2} \frac{\dot{\gamma}_{M^*}(\xi, t)}{\xi} d\xi = \Omega_{M^*}(r_1, t) \qquad (4.44)$$

As a consequence, under creep tests, torque ramps, or oscillations, the velocity distribution within the gap is again given by the ensemble of inner cylinder velocities under smaller torque levels as long as there is an unsheared region within the gap.

The situation described by (4.43) is generally considered to occur for yield stress fluids when the applied stress along the outer cylinder is, at least in steady state, that is, for sufficiently large values of t, smaller than the yield stress of the fluid. However, the question of the validity of (4.43) for such materials under transient flows remains open since their yielding character was shown to be intimately related to their thixotropic character (see Section 3.3). For example, MRI observations of the velocity profiles of bentonite suspensions showed that after a sudden decrease in the rotation velocity of the inner cylinder, the material can continue to flow for a significant time in a region where the stress is lower than the material's yield stress in steady state [23]. The situation might nevertheless be more favorable for start flows; indeed, after a sufficient time of rest, for a sudden increase in rotation velocity of the inner cylinder, the position of the interface between the sheared and the unsheared regions appears to progressively increase before reaching an asymptotic value (r_c) [23], so that (4.43) remains valid if $r_c < r_2$. Besides these thixotropic effects, the possible elastic deformations of the material during the very first times of flow might concern the whole gap so that (4.43) would not be valid at the beginning of the flow.

4.3.5 Wall Slip Effect

The developments discussed above are valid only if the local motion is at any time strictly associated with a flow of the homogeneous material. However, different phenomena such as fracture, wall slip, and density heterogeneities may affect the reconstructed profile in an a priori unknown way. At least some of these phenomena should be detectable from a thorough analysis of the results. Let us, for example, consider the case of wall slip. We assume that it affects a layer of negligible thickness close to the inner cylinder and induces an additional rotation velocity of the inner cylinder Ω_S, which a priori depends on the local shear stress (τ_1) and on the time. Under these conditions equation (4.38) becomes

$$\omega_M(r, t) = \omega_{M^*}(r_1, t) - \Omega_S(\tau_1, t) \qquad (4.45)$$

in which τ_1 is the shear stress along the inner cylinder for a torque M^* (i.e., $\tau_1 = M^*/2\pi h r_1^2$). As a result, the distribution of $\{\omega_M(r, t); r > r_1\}$ within the material is now equal to the distribution $\{\Omega_{M^*}(r_1, t) - \Omega_S(M^*/2\pi h r_1^2, t); M^* < M\}$. Thus the reconstruction technique may be used in the presence of wall slip as long as the effect of wall slip can be quantified from independent measurements.

Conversely, if we a priori ignore wall slip even though it effectively occurs, and directly compute the distribution $\{\Omega_{M^*}(r_1, t); M^* < M\}$, we will in fact get the velocity distribution $\{\omega_M(r, t) + \Omega_S((r_1/r)^2 M, t); r > r_1\}$. For pasty materials the slip effect is dominant only at low stresses and seems to follow a power-law model at stresses lower than the fluid yield stress (see Section 4.2). It follows that in practice wall slip could be detected, at least qualitatively, from an apparent velocity profile composed of two regions: close to the wall a region of rapidly decreasing velocity profile, in which the shear thinning character of the bulk fluid predominates ($\omega_M \gg \Omega_S$), and a region exhibiting a much more gradual decrease in velocity farther from the inner cylinder, in which the slip velocity term dominates in (4.45) ($\omega_M \ll \Omega_S$) (see Figure 4.13).

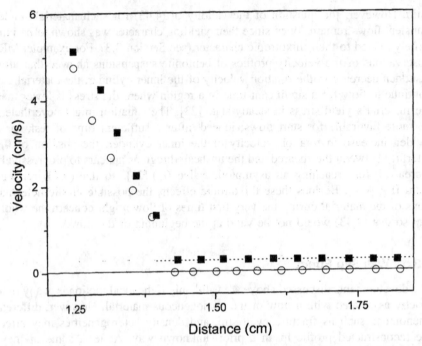

Figure 4.13 Reconstructed velocity profile for a mustard in steady state under a torque of 1.92 mN·m with rough (circles) and smooth (squares) surfaces. (From Ref. 27, Figure 4.)

4.3.6 Inclined Plane Flow

A similar approach can be applied for inclined plane flows. By integrating the shear rate between a particular height y and the free surface h, we get the difference between the maximum velocity (reached at the free surface) and the velocity within the material at y

$$U(h, t) - v_x(y, t) = \int_y^h \dot{\gamma} d\xi \qquad (4.46)$$

where the shear rate is a priori a function of the history of the shear stress and the normal stress differences. For the uniform flow of an infinitely wide and long fluid layer, the shear stress at the height ξ is given as $\rho g(h - \xi) \sin i$, while the normal stress differences are equal to zero (see Section 1.4.1). It follows that, as long as the flow remains uniform, the local, instantaneous shear rate writes:

$$\dot{\gamma}(\xi, t) = \underset{0 < \vartheta < t}{\mathrm{H}} (\Sigma_0, (h - \xi) \sin i(\vartheta)) = \dot{\gamma}_{h;i(\vartheta)}(\xi, t) \qquad (4.47)$$

In this expression we accounted only for possible changes of the slope in time because they can be controlled, and we assumed a constant fluid thickness.

Let us now consider another uniform flow of thickness h^* and slope history $i^*(\vartheta)$ such that, for a given height y, we have at any time ϑ:

$$h^* \sin i^*(\vartheta) = (h - y) \sin i(\vartheta) \tag{4.48}$$

Then in (4.47) we can modify the variables $\xi \leftrightarrow \zeta$ such that $(h - y)(h^* - \zeta) = h^*(h - \xi)$. In that case we have $(h - \xi) \sin i(\vartheta) = (h^* - \zeta) \sin i^*(\vartheta)$ for any time ϑ, which means that the shear stress histories in the flow at these corresponding depths are similar. As a consequence, from (4.47), we have $\dot{\gamma}_{h,i(\vartheta)}(\xi, t) = \dot{\gamma}_{h^*,i^*(\vartheta)}(\zeta, t)$, and we deduce

$$U(h, t) - v_x(y, t) = \int_y^h \dot{\gamma}_{h;i(\vartheta)}(\xi, t)d\xi = \frac{h - y}{h^*} \int_0^{h^*} \dot{\gamma}_{h^*;i^*(\vartheta)}(\zeta, t)d\zeta$$

$$= \frac{h - y}{h^*} U(h^*, t) \tag{4.49}$$

Thus the instantaneous distribution of fluid velocity within a uniform flow of thickness h under a slope history $i(\vartheta)$, specifically, $\{v_x(y, t); 0 < y < h\}$, may be simply found from the set of free surface velocities of uniform flows of thicknesses h^* and submitted to flow histories $i^*(\vartheta)$ following the condition (4.48). We can employ either of two simple practical approaches to achieve that objective (1) systematically vary the flow thickness h^* white keeping the same slope history $(i(\vartheta) = i^*(\vartheta))$ for all flows or (2) keep the same thickness $(h = h^*)$ in (4.48) and vary only the slope history.

REFERENCES

1. M. Raffel, C. Willert, and J. Kompenhans, *Particle Image Velocimetry—a Practical Guide*, Springer-Verlag, Berlin, 1998.
2. G. G. Fuller, *Optical Rheometry of Complex Fluids*, Oxford Univ. Press, Oxford, UK, 1995.
3. J.-B. Salmon, S. Manneville, A. Colin, and B. Pouligny, *Eur. Phys. J. Appl. Phys.* **22**, 143 (2003).
4. F. Durst, A. Melling, and J. H. Whitelaw, *Principles and Practice of Laser-Doppler Anemometry*, Academic Press, New York, 1976.
5. M. P. Luong, *Proc. 36th Congress of the French Group of Rheology*, Marne-la-Vallée, 2001, p. 174 (in French)
6. P. Corvisier, C. Nouar, R. Devienne, and M. Lebouché, *Exper. Fluids* **31**, 579 (2001).
7. S. W. Sinton and A. W. Chow, *J. Rheol.* **35**, 735 (1991).
8. J. R. Abbott, N. Tetlow, A. L. Graham, S. A. Altobelli, E. Fukushima, L. A. Mondy, and T. S. Stephens, *J. Rheol.* **35**, 773 (1991).
9. M. Nakagawa, S. A. Altobelli, A. Caprihan, E. Fukushima, and E. K. Jeong, *Exper. Fluids* **16**, 54 (1993).
10. C. J. Rofe, R. K. Lambert, and P. T. Callaghan, *J. Rheol.* **38**, 875 (1994).
11. A. W. Chow, S. W. Sinton, J. H. Iwamiya, and T. S. Stephens, *Phys. Fluids* **6**, 2561 (1994).

12. A. M. Corbett, R. J. Phillips, R. J. Kauten, and K. L. McCarthy, *J. Rheol.* **39**, 907 (1995).
13. S. J. Gibbs, K. L. James, L. D. Hall, D. E. Haycock, W. J. Frith, and S. Ablett, *J. Rheol.* **40**, 425 (1996).
14. D. F. Arola, G. A. Barrall, R. L. Powell, K. L. McCarthy, and M. J. McCarthy, *Chem. Eng. Sci.* **52**, 2049 (1997).
15. K. L. McCarthy and W. L. Kerr, *J. Food Eng.* **37**, 11 (1998).
16. A. D. Hanlon, S. J. Gibbs, L. D. Hall, D. E. Haycock, W. J. Frith, and S. Ablett, *Magn. Reson. Imag.* **16**, 953 (1998).
17. P. T. Callaghan, *Rep. Prog. Phys.* **62**, 599 (1999).
18. S. I. Han, O. Marseille, C. Gehlen, and B. Blümich, *J. Magn. Res.* **152**, 87 (2001).
19. J. Götz, W. Kreibich, M. Peciar, and H. Buggisch, *J. Non-Newtonian Fluid Mech.* **98**, 117 (2001).
20. R. R. Ernst, G. Bodenhausen, and A. Wokaun, *Principles of Nuclear Magnetic Resonance in One and Two Dimensions*, Oxford Science Publications, Oxford, 1987.
21. P. T. Callaghan, *Principles of Nuclear Magnetic Resonance Microscopy*, Clarendon Press, Oxford, UK, 1991.
22. E. Fukushima, *Ann. Rev. Fluid Mech.* **31**, 95 (1999).
23. J. S. Raynaud, P. Moucheront, J. C. Baudez, F. Bertrand, J. P. Guilbaud, and P. Coussot, *J. Rheol.* **46**, 709 (2002).
24. R. W. Mair and P. T. Callaghan, *J. Rheol.* **41**, 901 (1997).
25. M. M. Britton and P. T. Callaghan, *J. Rheol.* **41**, 1365 (1997).
26. Y. Yeow and J. W. Taylor, *J. Rheol.* **46**, 351 (2002).
27. J. C. Baudez, S. Rodts, X. Chateau, and P. Coussot, *J. Rheol.* **48**, 69 (2004).

CHAPTER 5

NONVISCOMETRIC FLOWS OF YIELD STRESS FLUIDS

Viscometric flow measurements can barely be obtained without the help of a sophisticated setup such as a rheometer. In practice it is easier to squeeze a paste between two solid surfaces, to pour it over a plane, or to displace an object through it. In each case one may obtain some information about the rheological behavior of the material from the relation between some force to which the fluid is submitted and a specific rate of motion. Another advantage of these flows is that various problems, such as migration, drying, wall slip, shear banding, and fracture, develop less frequently than during conventional rheometrical tests. In general, because we are dealing with complex nonviscometric flows, with no assumption concerning the constitutive equation of the fluids, the corresponding data can be used only to compare the apparent viscosities of different materials. Once a reasonable assumption regarding the rheological behavior of the fluid is reached, there exist several conditions for which simple, approximate flow solutions may be found and it becomes possible to relate some key rheological characteristics to straightforward measurement of macroscopic quantities. Here we review these conditions for Newtonian and yield stress fluids. In the latter case we assume that the fluid remains perfectly rigid in its solid regime, and follows, a constitutive equation such as (2.20) in its liquid regime. Moreover we mainly focus on slow on incipient motions, for which the exact form of the constitutive equation does not play any role; only the yielding criterion and the yield stress value are important. We review the displacement of an object through a fluid

Rheometry of Pastes, Suspensions, and Granular Materials: Applications in Industry and Environment
By Philippe Coussot   Copyright © 2005 John Wiley & Sons, Inc.

(Section 5.1), the squeezing of a material between two solid plates (Section 5.2), and spreading or coating flows (Section 5.3).

5.1 DISPLACEMENT OF AN OBJECT THROUGH A YIELD STRESS FLUID

Here we focus on the relative motion of an object and a volume of fluid, and estimate the (resisting) viscous force acting on the object as a result of its motion. We will express the relative, steady velocity of the object as $\mathbf{V} = V\mathbf{x}$. The fluid is globally at rest relative to the frame of reference; that is, its boundaries do not move. Moreover, the fluid boundaries are situated far from the object and thus have a negligible effect on the flow characteristics of the fluid around the object. For a sphere moving through a Newtonian fluid under such conditions, there exists a complete analytical solution [1,2]. For an object of complex shape, the problem is more complex and requires a numerical treatment. With a yield stress fluid, even for a sphere, no simple solution has been found yet and numerical simulations provide some key information. Here we first consider the case of a very long object in slow motion along its main axis through a yield stress fluid. Then we focus on the slow or incipient motion of a compact object through a Newtonian fluid and a yield stress fluid.

5.1.1 Drag and Buoyancy Forces

We consider a compact object of external surface S moving at a steady velocity ($\mathbf{V}$) through the fluid macroscopically at rest. As a result of the fluid viscosity, this object is submitted to a force ($\mathbf{F}$) equal to the opposite of the force that it exerts on the fluid. The latter results from the sum of the stresses exerted on each point of the solid–fluid interface S. Finally, using the decomposition (1.34), the net force on the object may be expressed as the sum of two components, respectively the *buoyancy* and the *drag force*

$$\mathbf{F} = \int_S \mathbf{\Sigma} \cdot \mathbf{n}\, ds = \int_S -p\mathbf{n}\, ds + \int_S \mathbf{T} \cdot \mathbf{n}\, ds \qquad (5.1)$$

in which $\mathbf{n}$ is the outer unit normal vector to the surface of the object. The first term of (5.1), resulting from pressure, is the buoyancy force $\mathbf{B}$, which, using an integral theorem, may be rewritten as $-\int_\Omega \nabla\, p\, dv$, in which Ω is the object volume. In the momentum equation it is often possible to assume that pressure and gravity effects separately balance (hydrostatic pressure distribution), specifically, $-\nabla p - \rho g\mathbf{z} = 0$, in which $\mathbf{z}$ is the vertical axis (ordinate; oriented upward). Under these conditions the expression for $\mathbf{B}$ becomes

$$\mathbf{B} = \rho g \Omega \mathbf{z} \qquad (5.2)$$

This corresponds to the well-known Archimedean principle, which tells us that an object immersed in a liquid undergoes, as a result of the vertical gradient of pressure, a net force corresponding (although in opposite direction) to the weight of the displaced liquid. Obviously when the particle density matches that of the fluid ($\rho_p = \rho$), the buoyancy force is balanced by the weight of the object ($-\rho_p g \Omega \mathbf{z}$), which thus remains in suspension within the fluid.

As long as the hydrostatic pressure assumption remains valid, the second term of (5.1) depends solely on viscous effects due to the relative motion between the solid surface and the fluid. This is the "drag force" $\mathbf{F}_D$, which corresponds to the force resisting motion and is thus oriented in the direction opposite that of the object motion: $\mathbf{F}_D = -F_D \mathbf{x}$, in which F_D is positive. Here we focus on this force in the case of laminar fluid flows. Let us apply the momentum balance to the fluid volume situated between S and S^*, the external surface of a larger volume, including the object. Under the assumption of hydrostatic pressure distribution gravity effects are balanced by pressure effects, and in the absence of inertia effects, we finally obtain

$$-\mathbf{F}_D + \int_{S^*} \mathbf{T} \cdot \mathbf{n}\, ds = 0 \qquad (5.3)$$

We deduce that for any type of surface S^* the average stress over the surface S^*, expressed as $1/S^* \int_{S^*} \mathbf{T} \cdot \mathbf{n}\, ds$, is equal to $\mathbf{F}_D/S^*$ and thus tends to zero when $S^* \to \infty$. The validity of (5.3) for any surface size and shape S^* implies that in any direction the local stress tensor $\mathbf{T}$ tends toward zero as the distance from the object tends toward infinity. The motion of the object through the fluid may thus be observed via the perturbation of the stress and strain fields that it induces in the fluid around the object, and which vanishes as the distance from the object increases.

5.1.2 Displacement of a Long Object along Its Axis

Generalities

Here we consider the motion of a long, cylindrical object with a circular basis (of radius R) and assume that this object moves in the direction $\mathbf{x}$ of its axis (Figure 5.1). Under these conditions, neglecting the fluid flow at the extremities of the cylinder (edge effects), the flow can be considered as uniform along the x axis, which means that the velocity field is independent of x. Moreover, for reasons of symmetry, all variables are independent of θ, the position around the axis. In the absence of any flow instability a simple strain field compatible with boundary conditions is such that only the longitudinal component of the velocity is different from zero and depends only on r: specifically, $v_x = f(r)$. A velocity field of this type means that the flow consists in the relative motion of concentric, cylindrical, material layers parallel to the central axis, as in a capillary flow (see Section 1.4.5). At a given point, the history of the relative deformation gradient is given entirely by the history of the relative velocity of the local cylindrical layers, namely, the shear rate, which is expressed as $\dot{\gamma} = f'(r)$. Thus we are dealing with a viscometric flow. From equation (1.51) we deduce that the stress tensor is a function of the distance from the axis, the time and a state of reference of the material [$\boldsymbol{\Sigma} = \boldsymbol{\Sigma}(r, t)_{\Sigma_0}$], but in the following discussion we will neglect

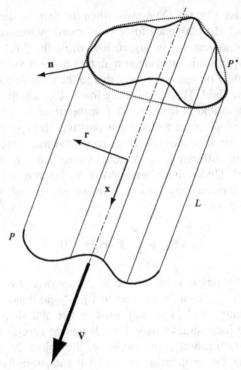

Figure 5.1 Displacement in the direction of its axis of a cylindrical object with a basis of any shape through a fluid.

the two latter dependences. From equation (5.3) applied to a cylindrical surface S^* of radius r and centered around the object axis, we obtain

$$2\pi r L \tau(r) = F_D \tag{5.4}$$

in which τ is the shear stress amplitude.

Newtonian Fluid

In that case the integration of (5.4) using the constitutive equation $\tau = \mu |\partial v_x / \partial r|$ and factoring in the boundary conditions $v_x(r_0) = 0$ (adherence along the cylindrical container of radius r_0) and $v_x(R) = V$ gives

$$F_D = \frac{2\pi L \mu V}{\ln(r_0/R)} \tag{5.5}$$

It is worth noting that the drag force depends on the container size and tends toward zero when this size tends toward infinity.

Yield Stress Fluid

We deduce from (5.4) that for a yield stress fluid there exists a critical distance, $r_c = F_D/2\pi L \tau_c$, beyond which the fluid remains unsheared since $\tau(r > r_c) < \tau_c$.

The critical force for incipient motion or stoppage corresponds to $r_c = R$, which yields

$$F_c = 2\pi L R \tau_c \tag{5.6}$$

Thus, in contrast with the Newtonian case, the drag force needed for any motion is independent of container size.

For a cylinder of basis P of any shape (in the plane perpendicular to the cylinder axis) (see Figure 5.1), the only velocity component differing from zero is still the longitudinal component v_x, but it now depends both on r and θ. The flow is a simple shear, but its direction varies in space and the stress distribution is a priori unknown. Far from the object the fluid remains unsheared, but in the limit of very slow motion the sheared thickness decreases as the sheared region comes closer to the object. In parallel, it is likely that the fluid in the convex regions of the object, close to several solid boundaries, will cease to flow before some other fluid regions. Consequently, it may considered as a first approximation that the ultimate fluid layer along which the fluid is sheared in the limit of very slow flows corresponds to the smallest, cylindrical, concave envelope of the object, of basis P^*. Since the shear stress along this surface is expressed as $-\tau_c \mathbf{x}$, the critical drag force on the object is

$$F_c = P^* L \tau_c \tag{5.7}$$

5.1.3 Displacement of a Compact Object of Any Shape

Newtonian Fluid

Here we are no longer dealing with a viscometric flow. The complete description [2] of the flow field from momentum equations shows that the perturbation of the flow field induced by the motion of sphere extends to an infinite distance r from the object since the fluid velocity approaches zero at large distances as r^{-1} [2]. The corresponding second invariant of the strain rate tensor (equivalent to the shear rate in simple shear) decreases as r^{-2}, which gives an idea of the variation in space of the stress perturbation. In simple shear thinning fluids (without yield stress) the viscosity decreases with this second invariant so that the velocity decreases much more rapidly but the perturbation theoretically still extends to an infinite distance. In parallel, in the absence of inertia effects, it may be shown that in the Newtonian case the fluid velocity and, as a consequence, the second invariant of the strain rate tensor, are proportional to the object velocity. For both Newtonian and simple shear thinning fluids, this invariant decreases to zero as the object velocity decreases to zero, which implies that the drag force tends toward zero.

More precisely, for an object moving through a Newtonian fluid of viscosity μ, it may be shown from simple dimensional considerations that, keeping constant the shape and orientation of the object, the drag force is proportional to V, μ, and L, a characteristic length of the object

$$\mathbf{F}_D = -k L \mu \mathbf{V} \tag{5.8}$$

in which k is a form factor that depends on the shape of the object and its orientation relative to the direction of motion [3]. For example, for a sphere of radius R, we have $kL = 6\pi R$, and for a disk of radius R, $kL/\pi R$ ranges from 3.4 to 5.2 as the disk axis varies from parallel to perpendicular to the direction of motion.

Yield Stress Fluid

Here we consider a yield stress fluid initially at rest. From preliminary considerations concerning the stress tensor decreasing to zero as the distance from the object tends toward infinity (see Section 5.1.1), it results that in any given direction, beyond some critical distance, the yielding criterion (2.18) is satisfied, so that the fluid should remain rigid despite object motion. Thus, unlike Newtonian or simple shear thinning fluids, a yield stress fluid through which an object moves remains unperturbed beyond a finite distance from the object. In fact, this is likely an approximation since the corresponding location of the predicted rigid region should displace and thus deform with the object motion. As noted by Lipscomb and Denn [4] for other flow types, this is in contradiction with the assumption of a rigid body, and only uniform flows can theoretically give rise to the formation of a truly rigid region.

On the basis of this rigid-region approximation, the set of the critical positions beyond which the yielding criterion is not fulfilled at a given time in the different directions forms a surface S_c surrounding the object (Figure 5.2) and, from equation (5.3), the amplitude of the drag force exerted onto the object moving in the direction of unit vector $\mathbf{x}$ may be written $F_D = -\int_{S_c} (\mathbf{T} \cdot \mathbf{n}) \cdot \mathbf{x} \, ds$. On approach of S_c, the velocity tends toward zero so that the amplitude of the strain rate tensor tend towards zero, which implies that, for a constitutive equation of the type (2.20), the deviator of the stress tensor is proportional to τ_c. Scaling the preceding integral by the object surface, it follows that the drag force may

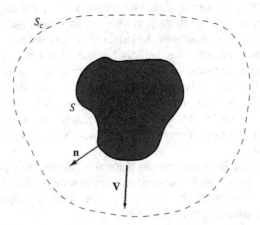

Figure 5.2 Critical surface surrounding a compact object and beyond which the fluid remains rigid.

simply be expressed as

$$F_D = k\tau_c S \tag{5.9}$$

in which k is an increasing, positive, dimensionless function of the extent of S_c relative to the surface of the object.

We obviously expect that, for a given object, the surface S_c increases as the object velocity increases, so that the drag force increases. On the contrary, when the velocity decreases to zero, S_c should approach S and k should decrease toward a critical, minimum value k_c. We deduce the expression for the critical drag force for incipient motion or stoppage of a given object:

$$F_c = k_c \tau_c S \tag{5.10}$$

There does not yet exist a general theoretical approach providing the value of k or k_c as a function of the shape and velocity of the object; thus, in the following paragraphs we review numerical and experimental literature data for estimating these parameters.

Let us first consider the incipient motion or stoppage of a sphere in a yield stress fluid. The simplest approach [5] consists in assuming that, immediately when flow starts or stops, the force exerted by the fluid corresponds to a shear stress vector (at the surface of the object) of amplitude equal to the yield stress and directed tangentially to the object surface in the direction the closest to $-\mathbf{x}$. After integration one finds $k_c = \pi/4$. This scheme is no doubt oversimplified since, for example, at the front and the back of the sphere (in the direction of motion) the velocity of the fluid is by symmetry directed along $\mathbf{x}$, which implies that the main shear flow around the central axis is by no means tangential to the object surface. Assuming that under static conditions a yield stress fluid can be considered as an ideal solid body with a yield criterion, a more complete analysis of the problem has been attempted with the help of the plasticity theory [6]. Within this frame the stress field in the fluid at rest is estimated or determined from maximum energy principles, and in particular, the surfaces along which rupture occurs in a solid body, or correspondingly where yield stress is reached for a yield stress fluid, are approached. However, the three-dimensional problem remains difficult. By borrowing ideas from the theory of two-dimensional slip lines in a rigid plastic, that is, approximating the yield surface as a kind of truncated toroid with its section centered on the surface of the sphere (see Figure 5.3), Ansley and Smith [7] have obtained $k_c = 7\pi/8$. Note that in that case there also appear two rigid regions at the front and at the tail of the sphere (see Figure 5.3). The simulations of Beris et al. [8], who combined variational inequalities with finite-element numerical approximations and effective minimization techniques to determine the velocity distribution and drag force around the sphere, agree reasonably well with the results of Ansley and Smith, and actually constitute a work of reference. In particular, Beris et al. found a shape of the critical surface close to that assumed by Ansley and Smith and a critical value:

$$k_c = 3.5 \tag{5.11}$$

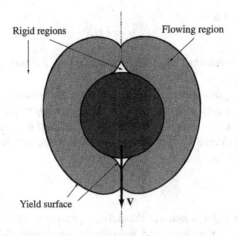

Figure 5.3   View of the critical surface surrounding the sphere and along which the yield-ing criterion is reached (envelop in the gray region), according to the Ansley–Smith [7] slip line theory.

It would be useful to extend the expression (5.10) to other object shapes, for example, by considering that, as for motion through a Newtonian fluid, the form factor is modulated by the shape and orientation of the object. However, there is actually no available formula of this type, and we can simply suggest that, for an object with an aspect ratio of the order of 1, the drag force as a first approximation may correspond to that on the smallest sphere including the object, as given by (5.10) with the value of k_c given by (5.11).

Experimental Problems
Various analytical expressions for the drag force [8–12] on a sphere moving through a yield stress fluid have also been proposed from experiments and sim-ulations. All the theoretical works, which assume a simple yielding behavior, predict a particle velocity progressively decreasing to zero as the applied force tends to the critical value for incipient motion. The experimental results presented in the literature are in agreement with this basic expectation, namely, that the object can move through the fluid only when the force applied to it becomes greater than a critical one, for example beyond a critical density or volume of the object. These data have been obtained from settling tests, in which a sphere is left to fall through a fluid under the action of gravity. However, as yet no clear quantitative agreement has been established between the numerical or ana-lytical predictions and experimental data, which remain somewhat approximate and contradictory [13]. For example, the existing data (in terms of critical drag coefficient) for the incipient motion [7,10,14–16] are in general significantly lower than the theoretical prediction (5.11). This might be due to either (1) the general uncertainty concerning the experimental determination of the incipient motion or (2) overestimation of the effective yield stress from rheometrical tests by extrapolation toward low shear rates of data obtained at high shear rates.

A further examination of the data in the light of the deeper insight of the behavior of pastes shows that the theory presented above might differ more significantly from reality. First, the data for a settling velocity smaller than 10 cm/s [17] are poorly reproducible. Other authors [10,18] pointed out that reproducible results could be obtained only after releasing 4–10 spheres in the fluid. Another critical observation is that all data in the literature [5,17–19] correspond to relatively large settling velocities; the velocity : particle diameter ratio, which is a characteristic shear rate for this motion, is never below 1 s^{-1} and is generally much larger. Finally, ultrasonic observations in bentonite suspensions show that velocity increases as the object goes deeper inside the fluid [20]; the final velocity is 5 times the initial value. Systematic tests with a laponite suspension also made it possible to find different regimes of fall [21] of a sphere in the fluid in different initial states of structure: a rapid fall at an almost constant rate, a slower fall at a progressively decreasing velocity, and a slow fall at a rapidly decreasing rate, finally leading to apparent stoppage.

Such problems do not occur in simple liquids. For example, by increasing the fluid viscosity or decreasing the solid–liquid density difference, one obtains a continuous set of fall velocities tending to zero. Relevant explanations for these problems cannot be found in the viscoplastic behavior of the fluids, either; as the solid density decreases or the yield stress increases toward critical values, the settling velocity should reach any value down to zero with an analogous reproducibility and without any significant time effects. The above mentioned problems thus necessarily result from some uncontrolled time and flow-dependent mechanical properties of the fluid, which suggests that the possible thixotropic behavior of the pasty materials should also be considered in predicting the characteristics of the slow fall of particles through these materials. For example, a simple model [21] using the basic features of the (thixotropic) rheological behavior of pasty materials made it possible to explain the regimes described above. The fall of an object in a paste thus appears to basically follow a bifurcation process; for a sufficiently large force applied onto the object, its rapid motion tends to sufficiently liquefy the fluid around it so that its subsequent motion is more rapid and so on until reaching a constant velocity. On the contrary, if the force applied to the object is not sufficient, the surrounding fluid has enough time to restructure, which slows down the motion and so on until complete stoppage of the object.

5.2 SQUEEZE OR STRETCH FLOWS

Let us assume the flow of a material situated between two parallel plates in relative motion (at a relative velocity $2\mathbf{V} = 2V\mathbf{z}$) along their common normal direction. For the sake of simplicity, we will consider mainly the case of a nearly cylindrical volume of fluid of radius R. When the plates come closer to each other, this is a squeeze flow; otherwise this is a stretch flow. Such flows are nonviscometric; the material flows in both the axial and radial directions in a complex way, and thus the strain field cannot a priori be described in the form of constant fluid layers gliding over one another. Nevertheless, under some specific conditions we can estimate the flow characteristics with a good precision via analytical approaches.

When the planes are very close to each other, that is, when the ratio of the separation distance ($2b$) to a typical length ($2R$) of the surface of contact (S) between the material and one of the planes is much smaller than 1, we can use the so-called lubrication approximation to solve flow equations (Section 5.2.1). On the contrary, when $b/R \gg 1$, the flow close to the planes has no effect, and we are dealing with a typical elongational flow (Section 5.2.2). In practice, gravity may also play a significant role when the weight of material ($2\rho g S b$) is of the order of or larger than a force of the order of that needed to deform the material ($2\tau_c S$), that is, when $\rho g b/\tau_c > 0.1$. This situation will be considered in Chapter 7, and here we will focus only on situations for which gravity effects can be neglected (i.e., $\rho g b/\tau_c \ll 1$). Finally, the interaction between the material and the plane can play a significant role. Two limiting regimes may be distinguished: adhesion or perfect slip at the wall. Adhesion has been implicitly assumed in the considerations above, but for a perfect slip, in the absence of gravity effects, all flows become elongational. Surface tension effects are not negligible when the force that they induce is of the order of or larger than the force necessary for squeezing or stretching the fluid [expression (5.25) or (5.29)]. Typically the radius of curvature in a plane r,z can be of the order of b. This implies that the total force associated with surface tension effects is equal to the pressure that they induce ($2\gamma_{LG}/b$) times the surface of liquid–gas interface ($\approx 2\pi Rb$), which gives $4\pi R\gamma_{LG}$.

Other simple squeeze flows can also be considered: squeeze flow between two plates forming a dihedral (Section 5.2.4) and squeeze flow of a band between two parallel plates (Section 5.2.5). Note that the calculations below assume that there is no flow instability; that is, the free surface remains symmetric by rotation around the axis, but fracture, shear banding, or fingering may occur during stretch or squeeze flow. We examine the conditions of occurrence of these effects in Section 5.2.6.

5.2.1 Lubricational Regime

Here we consider the case of perfect adhesion at the wall with $b/R \ll 1$. We describe the flow in the frame (O, r, θ, z) linked to the central plane (see Figure 5.4). Note that in this frame the fluid situated at $z = 0$ does not move in the direction z. In the absence of any flow instability, the free surface of the sample keeps its symmetry of revolution. It follows that the velocity and the stress components are independent of θ. There is also no reason why a motion would exist in one particular direction around the z axis, so that $v_\theta = 0$. For a fluid volume situated within the cylinder of radius r, the net fluid discharge in the longitudinal direction is expressed as $Q_z = \pi r^2 V$, where the rate at which the thickness of the material in the upper region evolves may also be written $V = \partial b/\partial t$. In parallel, the fluid discharge in the radial direction is

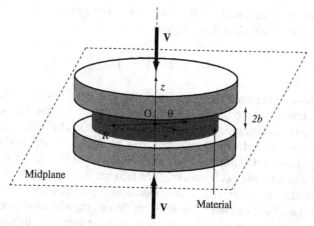

Figure 5.4 Schematic representation of a squeeze or stretch flow between two parallel plates.

$Q_r = 2\pi r \int_0^b v_r(r, z)dz$. The balance of these two quantities $(Q_r + Q_z = 0)$, which follows from the mass conservation, yields

$$U(r) = 1/b \int_0^b v_r(r, z)dz = -\frac{r}{2b}V \qquad (5.12)$$

in which U is the mean radial velocity at the distance r.

The lubrication assumption here consists in considering that in most of the fluid the gradients of velocity components in the directions r or θ are negligible compared to those in the direction z, along which the fluid thickness is much smaller than in the other directions. It follows that the only significant component of the strain rate tensor is

$$\dot{\gamma} = \dot{\gamma}(r, z) = \frac{\partial v_r}{\partial z} \qquad (5.13)$$

Under these conditions the squeeze flow might be seen as a kind of instantaneous, radial simple shear with characteristics evolving in time. It follows that for a fluid with a constitutive equation of the type (2.20) in the liquid regime, that is, with a shear-dependent term linear in **D**, the normal stress components of the stress deviator are equal to zero. In addition, we will assume that the pressure does not vary significantly in the gap thickness (i.e., $\partial p/\partial z = 0$).

Let us now apply the momentum equation in integral form to a small, cylindrical volume of height z above the midplane and contained between r and $r + dr$. Note that by reason of symmetry the radial component of the fluid velocity reaches an extremum in the midplane so that $\dot{\gamma}(r, 0) = 0$. Moreover, for the same reason, the shear stress along this plane is equal to zero. In the radial direction the force resulting from the shear stress along the upper surface (i.e., $2\pi r \, dr \tau_{rz}$), is balanced by the force due to the radial pressure drop [i.e., $2\pi rz(p(r) - p(r + dr))$], and we deduce

$$\tau_{rz}(r, z) = \left(\frac{\partial p}{\partial r}\right) z \tag{5.14}$$

In this expression the pressure gradient is negative for a squeeze flow because the pressure transmitted to the fluid by the plate tends to decrease from the center to the periphery. In contrast, the pressure gradient is positive for a stretch flow.

The shape of the velocity distribution can be found by integrating the constitutive equation containing only one variable [defined by (5.13)] in which the stress distribution (5.14) has been introduced. Since the pressure was assumed to be independent of z, $\partial p/\partial r$ can be considered as a parameter in this integration. In a second stage the kinematic boundary conditions may be used to compute the absolute value of the velocity. One then can find the pressure distribution, from which one deduces the net force exerted onto the plates. We will now follow this procedure for two specific cases (Newtonian and yield stress fluid).

Newtonian Fluid

For a Newtonian fluid, equation (2.2) would give $\tau_{rz} = \mu_0 \dot{\gamma}(r, z)$. From (5.13) and (5.14), after integration between b and z and taking into account the boundary condition $v_r(r, b) = 0$, we deduce the velocity distribution:

$$v_r(r, z) = \frac{1}{2\mu_0}\left(\frac{\partial p}{\partial r}\right)(z^2 - b^2) \tag{5.15}$$

A second integration provides the mean velocity at a given distance r:

$$U(r) = -\frac{b^2}{3\mu_0}\left(\frac{\partial p}{\partial r}\right) \tag{5.16}$$

From (5.12) and (5.16) we deduce $\partial p/\partial r = 3\mu_0 r V/2b^3$, which may be integrated between r and R to give the pressure distribution:

$$p(r) = \frac{3\mu_0(r^2 - R^2)V}{4b^3} \tag{5.17}$$

Equation (5.17), in which we neglected the atmospheric pressure and surface tension effects, [i.e., we used the boundary condition at the free surface $p(R) = 0$], may again be integrated to give the total force exerted onto the plates:

$$F = -\int_0^R 2\pi r p(r) dr = \frac{3\pi \mu_0 R^4 V}{8b^3} \tag{5.18}$$

If the fluid volume is constant (i.e., $\Omega_0 = 2\pi R^2 b$), this force may be expressed as a function of the separating distance:

$$F = \frac{3\mu_0 \Omega_0^2 V}{32\pi}\frac{1}{b^5} \tag{5.19}$$

In that case it should be sufficient to plot the F/b^5 curve and estimate the slope to deduce the fluid viscosity μ_0.

Yield Stress Fluid

Here, since $\sqrt{-D_{\mathrm{II}}} = |\dot{\gamma}|/2$, the constitutive equation (2.18)–(2.20) is

$$|\tau_{rz}| < \tau_c \Rightarrow \dot{\gamma} = 0 \qquad \text{and} \qquad \dot{\gamma} \neq 0 \Rightarrow \tau_{rz} = \varepsilon[\tau_c + F(|\dot{\gamma}|)] \qquad (5.20)$$

where we used $\varepsilon = |\dot{\gamma}|/\dot{\gamma}$, which is equal to -1 for a squeeze flow and to 1 for a stretch flow. From (5.14) and (5.20) we deduce that there are an unsheared region and a sheared region in the fluid: $\dot{\gamma} = 0$ for $|z| < z_0$; $\dot{\gamma} \neq 0$ for $|z| > z_0$, with

$$z_0 = \varepsilon \frac{\tau_c}{\partial p/\partial r} \qquad (5.21)$$

The velocity profile in the sheared region may be found from the integration of the second equation of (5.20) in which one uses the expression for the shear stress (5.14).

Note that, from a strict point of view, with a constitutive equation exactly defined as (2.18)–(2.20), such a strain field distribution is not realistic since it leads to a rigid region of shape evolving in time as the material deforms, which is incompatible with the concept of rigid-body characteristics of such a region [4]. However, as suggested by Lawrence and Cornfield [22], it would probably be overzealous to try to discuss the exact form of the strain field whereas thixotropic effects can play a role in various situations, the lubrication approximation is a simplifying assumption, and even the yield stress model is merely an approximation of reality. As a consequence, the result given in this above context must be considered as a reasonable approximation of reality. Although we can hardly expect that strictly rigid regions exist in the fluid, the surface $z = z_0$ likely separates a region in which the strain is extremely weak and a region in which the strain is larger.

An approach similar (successive integrations) to that used for a Newtonian fluid may be used to deduce an analytical expression for the force in the case of a yield stress fluid. This obviously requires an explicit expression of the shear stress as a function of the shear rate, that is, the specific form of the function F chosen in the constitutive equation (2.20). The reader is referred to six papers [23–28], which provide different expressions under limiting conditions or in more general situations, for Bingham and Herschel–Bulkley fluids, and with various assumptions concerning the fluid–solid interactions (no, partial, or complete wall slip). In contrast with viscometric flows, there is no means to interpret the flow data in rheological terms without assuming a priori a specific form for the constitutive equation. Here we focus only on a specific case, very slow flow or incipient motion, which may be interpreted in a straightforward way in terms of the clear rheological characteristics of the fluid, namely, its yield stress, since in that case the shear stress in the flowing regions tends toward this single value.

From above considerations we deduce that the whole fluid remains rigid when $z_0 > b$. In the limiting case, $z_0 = b$, which corresponds to incipient motion or flow stoppage, the integration of (5.21) between R and r assuming no surface tension effect and negligible atmospheric pressure provides the pressure distribution:

$$p(r) = \varepsilon \frac{\tau_c}{b}(r - R) \qquad (5.22)$$

The net normal force exerted on the plate in that case is found by integrating the pressure (5.22) over the surface of contact:

$$F_c = \varepsilon \frac{2\pi \tau_c R^3}{3b} \qquad (5.23)$$

Equation (5.23) thus provides an expression for the force applied in the case of extremely slow flows ($V \to 0$). Since the assumed constitutive equation is continuous, predicting a continuous transition from rest to slow flows around the yield stress, equation (5.23) also provides an expression for the minimum force needed to induce some motion for a given separation distance and a given radius. It is worth noting that a similar result may be obtained via an approach within the frame of plasticity since the natural strain field assumed under these boundary conditions consists of plane deformations equivalent to the velocity field deduced via the lubrication assumption [29].

For a given volume of material ($\Omega_0 = 2\pi R^2 b$) equation (5.23) provides the value of the critical separating distance for which the fluid stops (when squeezing) or starts to flow (when stretching) under a given force F_0:

$$b_c = \left(\frac{\tau_c^2 \Omega_0^3}{18\pi F_0^2} \right)^{1/5} \qquad (5.24)$$

When the radius of the fluid sample (R_0) remains constant because the fluid can freely get out of the space between the two disks that squeeze it, the ultimate separating distance under a given force F_0 is given by

$$b_0 = \varepsilon \frac{2\pi \tau_c R_0^3}{3F_0} \qquad (5.25)$$

This result assumes that the fluid out of the gap does not play any role, which is, for example, the case if it is immediately withdrawn.

5.2.2 Elongational Regime

Here we consider the cases of perfect slip at the wall *or* sample of long length compared to its radius ($b/R \gg 1$), and still negligible gravity effects ($\rho g b/\tau_c \ll 1$), although in that case this is seldom realistic. The consequence of these assumptions is that edge effects are negligible, and in the absence of any flow instability the material remains perfectly cylindrical during flow. Once again the flow is symmetric by rotation around the axis so that the variables do not depend on θ and $v_\theta = 0$. Moreover, the flow is similar in any fluid portion so that v_r does not depend on z. In addition, if the fluid velocity in the longitudinal direction ($\mathbf{z}$) depended on r, the cylindrical layers of materials at different distances would move relative to each other in some specific direction. This is incompatible with boundary conditions (uniform velocity along the solid plates) and would favor one specific direction. As a result, v_z does not depend on r and is simply proportional to the velocity of displacement of the boundary: $v_z = (z/b)V$. For a fluid volume situated within the cylinder of radius r and height z, the fluid discharge in the radial direction is expressed as $2\pi r z v_r$, while the fluid discharge in the longitudinal direction is $\pi r^2 v_z$. It follows from the mass conservation ($2\pi r z v_r + \pi r^2 v_z = 0$) that we can express the velocity field in the form

$$v_r = -\frac{d}{2}r; \quad v_z = dz \tag{5.26}$$

in which $d = V/b$ is the rate at which the fluid is strained. From (1.27) (applied to $\mathbf{D}$) and (5.26) we can compute $\sqrt{-D_{\mathrm{II}}}$, which is found to be equal to $\sqrt{3}|d|/2$. From the constitutive equation (2.20) the normal stress writes $\tau_{zz} = \varepsilon(2\tau_c/\sqrt{3} + F(d))$, in which $\varepsilon = |d|/d$. Now, neglecting pressure effects, the normal force exerted on the fluid when $d \to 0$ may be computed

$$F_c = \int_0^R 2\pi r \tau_{zz} dr = \varepsilon \frac{2}{\sqrt{3}} \pi R^2 \tau_c \tag{5.27}$$

which provides the critical force for incipient motion for a given radius R. The critical radius for incipient motion under the action of a given force F_0 may be deduced from (5.27):

$$R_c = \left(\frac{\varepsilon F_0}{\beta \pi \tau_c} \right)^{1/2} \tag{5.28}$$

The coefficient $\beta = 2/\sqrt{3}$ derives from the particular form of the yielding criterion in the constitutive equation, the Von Mises criterion.

5.2.3   Intermediate Cases

Here we assume that R is of the order of b and that there is a perfect adherence of the fluid along the boundaries. In such a case it is not possible to find an analytical expression for the critical force for slow flow or incipient motion. However, we can roughly estimate this force by considering that in such an intermediate case the flow results from the composition of a tendency to squeeze or stretch the fluid far from the boundaries and a tendency to shear the fluid along the solid boundaries; each of these phenomena becoming negligible for $R \gg b$ and $R \ll b$, respectively. Under these conditions, using the expressions in the two limiting cases (5.23) and (5.27), it is reasonable to assume that the critical force in any case is given by

$$F_c = \varepsilon \frac{2\pi R^2}{\sqrt{3}} \tau_c f\left(\frac{R}{b}\right) \tag{5.29}$$

in which f is a function such that $f \to 1$ when $R/b \to 0$ and $f \approx R/\sqrt{3}b$ when $R/b \to \infty$. This is, for example, the case with the function $f(R/b) = 1 + R/\sqrt{3}b$.

5.2.4 Squeeze or Stretch Flow between Two Planes Forming a Dihedral

Here we consider the case of a fluid contained between two plates linked along one of their sides and forming an angle (θ) varying in time (see Figure 5.5). The material thus undergoes a squeeze or stretch flow that has some analogy with that resulting from the motion of parallel plates or disks as described in Section 5.2.1. We assume that the angle remains small, and we consider the flow in the frame of reference (x,y,z), where y is perpendicular to the planes and z is the direction of the line of contact between the plates. The fluid thickness $2b \approx \theta x$ remains throughout much smaller than the width of the plane D (along z) and the distance (L) between the interface and the line of contact. Within the framework of the lubrication approximation only the component of the velocity along x, or more precisely its gradient $\dot{\gamma} = \partial v_x/\partial y$, induces significant stresses. If we neglect the fluid adhering to the planes behind the air–fluid interface (for a stretch flow) and the deformations of this interface, a simple mass balance ($2bL = $ constant) provides a relation between the velocity of displacement of this interface ($\dot{L} = dL/dt$) and the rate of change of the angle ($\dot{\theta} = d\theta/dt$): $\dot{L} = -L\dot{\theta}/2\theta$. For reason of symmetry the shear stress in the midplane is equal to zero.

Here we again assume that the pressure in the fluid depends only on x. Let us apply the momentum equation in integral form to a small fluid portion contained

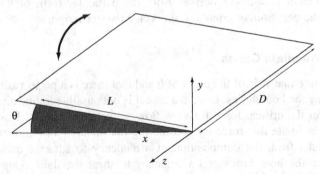

Figure 5.5 Schematic representation of a squeeze flow between two plates forming a dihedral.

between x and $x + dx$ and situated above the midplane. In the direction x the sum of the force resulting from the wall shear stress (i.e., $D\,dx\tau_p$) and that due to the pressure drop [i.e. $Dh(p(x) - p(x + dx))$] is equal to zero so that

$$\tau_p = \left(\frac{\partial p}{\partial x}\right)\frac{\theta x}{2} \tag{5.30}$$

The wall stress for incipient motion or slow flow is equal to $\varepsilon\tau_c$, in which $\varepsilon = |\dot{\gamma}|/\dot{\gamma}$ is equal to -1 for a squeeze flow and to 1 for a stretch flow. In that case the pressure distribution can be deduced from the integration of (5.30) between x and L:

$$p(x) = \varepsilon\frac{2\tau_c}{\theta}\ln\left(\frac{x}{L}\right) \tag{5.31}$$

in which we neglected the pressure at the air–fluid interface. The pressure distribution (5.31) is obviously not strictly valid since it predicts that the pressure would tend toward infinity when x tends toward zero, but this is simply a classical approximate result of the lubrication assumption for fluid thickness tending to zero. We then can compute the torque (which, as for it, is finite anyway) applied to the upper plane relative to the axis of rotation corresponding to the line of intersection between the planes. The critical torque associated with slow flow or incipient motion is as follows:

$$M_c = -\int_0^L Dxp(x)dx = \varepsilon\frac{\tau_c L^2 D}{2\theta} \tag{5.32}$$

Obviously again this expression may be used to get the critical angle at which the flow of a given volume of material $\Omega = \theta DL^2$ stops (when squeezing) or starts (when stretching) to flow under a given torque (M):

$$\theta_c = \left(\frac{\tau_c\Omega}{2|M|}\right)^{1/2} \tag{5.33}$$

5.2.5 Squeeze or Stretch Flow of a Long Band between Two Parallel Planes

We consider a flow similar to that described in Section 5.2.1, except that now the shape of the fluid volume is a band of width $2L$ much smaller than its length D (see Figure 5.6), but its thickness $2b$ remains much smaller than both these lengths. By reason of symmetry there is no motion in direction z and, under the lubrication assumption, the pressure drop in the direction x of the length governs the flow and balances the fluid shear in direction y of its thickness. The momentum equation over a fluid portion of thickness dx and situated above the horizontal midplane yields:

$$\tau_p = b\frac{\partial p}{\partial x} \tag{5.34}$$

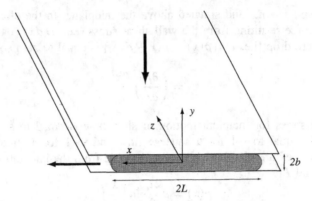

Figure 5.6 Schematic representation of a squeeze flow of a band between two parallel plates.

This equation is valid only when there is a flow; otherwise the stress and pressure distribution are a priori unknown. By integrating (5.34) between 0 and $x = L$ for an incipient motion or a very slow flow ($\tau_p \rightarrow \varepsilon\tau_c$), we obtain the pressure distribution:

$$p(x) = \varepsilon \frac{\tau_c}{b}(L - |x|) \tag{5.35}$$

which, after integration over the whole surface of fluid in contact with one plane, gives the total force exerted on the planes:

$$F = \varepsilon \frac{\tau_c D L^2}{b} \tag{5.36}$$

5.2.6 Flow Instabilities

Fractures

Fractures are observed at the periphery of pasty materials with high yield stress as they start to flow outside the gap when squeezed between two disks [29,30] (Figure 5.7) such as at the periphery of an excessively dry dough squeezed by a roll. Since this effect occurs or is visible only at the periphery of the sample, it may be considered at first sight to have no impact on the determination of rheological characteristics on the basis of the preceding calculations applied to the flow occurring within the gap. However, such phenomena might reflect flow characteristics of the bulk differing from those assumed in the calculations, but at this stage, it is difficult to explain these effects from a complete physical or mechanical point of view.

Fingering

A phenomenon often observed when stretching a paste between two plates is the formation of fingers at the periphery of the samples evolving in treelike structures

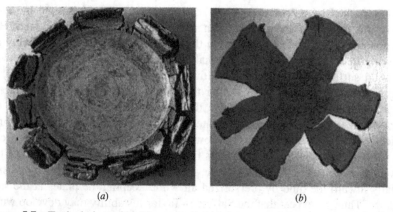

(a)   (b)

Figure 5.7 Typical view of highly concentrated clay pastes [cement paste (a) and kaolin paste (b)] after squeezing between two striated disks at a velocity of approximately 1 mm/s: as it flows out of the gap, the material separates into several bands. (From Ref. 30, Figure 10.)

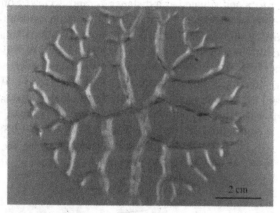

Figure 5.8 Final view of a thin layer of (commercial) hair gel sample initially squeezed between two plates and then stretched by slowly separating the plates.

after complete separation (Figure 5.8) [31–33]. Such structures are commonly observed in various practical situations when one separates two solid surfaces initially in contact via a thin layer of mud, glue, paint, puree, or another substance. The experimenter in rheometry can also frequently observe this effect when separating a cone and a plate or parallel disks (for a small gap) after a test with a thin layer of pasty material. This phenomenon occurs with simpler liquids, for example, Newtonian, but in that case only at a sufficiently high rate of separation of the solid surfaces.

These different characteristics are typical of the Saffman–Taylor instability [34] for simple [35] or yield stress fluids [36], which occurs when a viscous fluid is pushed by a less viscous one, for example, a liquid pushed by a gas, and results from a competition between surface tension and viscous effects. Here an apparent critical difference between the usual Saffman–Taylor instability and fingering in stretch flows is the fact that in the latter case the distance between the solid surfaces changes in time. However, as shown by equation (5.12), as long as the gap remains much smaller than the typical length of the surface of contact between the fluid and the solid, the velocity of the air–fluid interface (U) is much greater than the separation rate (V). As a consequence, the separation basically induces a flow of the fluid toward the central axis (or the line of contact for two tilted plates), which results from the depression inside the fluid. This flow is at each instant very close to the flow that would be obtained by pushing radially inward the fluid by pressurized air while keeping the plates at the same distance. Thus we expect that the Saffman–Taylor instability may develop within a squeeze or stretch flow under conditions analogous to those for constant gap.

Let us examine in more detail the physical origin of this phenomenon with the help of a rough description analogous to that proposed by Homsy [35]. We consider the flow of a material in a given direction x between two parallel plates separated by a small distance $2b$. When the flow is stable, the air–fluid interface is perpendicular to x and situated at a distance L, and is curved only in the plane (x, y) with a radius of curvature R. From the lubrication assumption it is easily shown that equation (5.34) is still valid. Let us now consider a small perturbation of the air–fluid interface of the form $\eta = \varepsilon_0 \exp(ikz + \omega t)$, in which k and ω are two parameters ($2\pi/k = \lambda$ is the *wavelength* of the perturbation). The length of the fluid in the direction x now depends on z (see Figure 5.9). We assume that the curvature and the contact angle of the interface remain constant in the planes (x, y); the interface is deformed only along z. Thus the curvature of the interface in the plane (x, z) is approximately $1/\eta'' = -1/k^2\eta$, so that from (3.58) there is now an additional term ($\gamma_{LG}k^2\eta$) in the pressure difference along the interface. Let us consider a "band" of fluid along the direction x. The additional force per unit

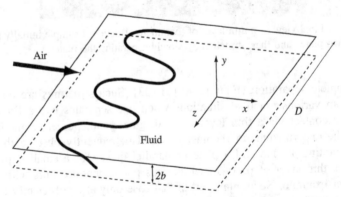

Figure 5.9 Schematic representation of the destabilized gas–fluid interface (dark line) for a fluid pushed by a gas between parallel plates separated by a short distance (Hele–Shaw cell).

length (dp) applied to this fluid band and resulting from the local displacement η may be roughly estimated. It is equal to the sum of the viscous force due to the additional displacement (which may be negative), which from (5.34) is approximately equal to $(\partial p/\partial x)\eta$, and to the force due to surface tension effects:

$$dp = \left[-\varepsilon \left(\frac{\tau_p}{b}\right) - \gamma_{LG}k^2\right]\eta \tag{5.37}$$

If this additional force is of the sign of η, that is, if $\varepsilon = -1$ and $\lambda > \lambda_c$ with $\lambda_c = 2\pi\sqrt{b\gamma_{LG}/\tau_p}$, the additional pressure will tend to push even more the regions that have already been curved toward positive values of x, and will tend to delay even more the advance of the regions that have already been curved toward negative values of x. Thus the perturbation will be amplified and the instability will develop. It appears from (5.37) that, for this to occur, the viscous forces must be sufficiently large compared to the forces resulting from surface tension. Note that when $\varepsilon = 1$, that is, when the liquid pushes the gas, the flow is always stable.

In reality, a perturbation is not expressed as simply as above but it may be decomposed in an infinite sum of such periodic perturbations of different wavelengths λ. Among them there is a wavelength of maximum growth rate (λ_m), which should tend to overwhelm the other waves because the corresponding instability process is stronger for it, and the instability will tend to develop with this wavelength. In fact, this is true only in the very first instances of the flow since the approach described above and more sophisticated theories are valid only within the frame of very small perturbation amplitudes. The preceding approach tells us that, as long as $\varepsilon = -1$, for a given fluid velocity there is always a range of wavelengths in which the flow is unstable, which means that the flow of a fluid pushed by a less viscous is a priori always unstable. However, in practice the instability will be visible and will take the form of fingering only if the perturbation has a wavelength smaller than the length of the interface, that is, if λ_m, which tends to rapidly overwhelm the other wavelength, is smaller than D for a straight flow and $2\pi R$ for a cylindrical layer of viscous fluid between two solid plates.

A more complete theoretical treatment for a Newtonian fluid pushed by a gas with negligible viscosity finally yields

$$\lambda_m = 2\pi b \left(\frac{\gamma_{LG}}{\mu U}\right)^{1/2} \tag{5.38}$$

for a straight flow [34] (where U is the average velocity of the front). Since the wall shear stress for a Newtonian fluid is of the order of $\mu U/b$, this expression is close to that of the critical wavelength deduced by the approximate approach given above. For a cylindrical layer of fluid such as that considered in Sections 5.2.1 and 5.2.2, we have [37]

$$\lambda_m = \frac{2\pi R}{\left(\dfrac{\mu U}{\gamma_{LG}}\dfrac{R^2}{b^2} + \dfrac{1}{3}\right)^{1/2}} \tag{5.39}$$

Complementary developments have been proposed for different types of (non-yielding) non-Newtonian fluids [38–40]. For all these fluids we emphasize that for sufficiently slow flows the instability does not appear in practice because the wavelength of maximum growth tends toward infinity as the velocity tends toward zero, and thus exceeds the characteristic length of the air–fluid interface. For yield stress fluids this is not the case; as the fluid velocity tends toward zero, the wall stress tends to a finite value, namely, the yield stress, so that the wavelength of maximum growth is larger than a finite value for very slow flows. For the straight flow of a yield stress fluid the wavelength of maximum growth rate for very slow flows was found to be [36]

$$\lambda_m = 2\pi \sqrt{\frac{3\gamma_{LG}b}{\tau_c}} \tag{5.40}$$

so that the criterion for the development of an apparent instability is

$$\tau_c > \frac{12\pi^2\gamma_{LG}b}{D^2} \tag{5.41}$$

In the case of a radial flow we have

$$\lambda_m = 2\pi R \left(\frac{3\gamma_{LG}b}{\gamma_{LG}b + \tau_c R^2}\right)^{1/2} \tag{5.42}$$

so that the criterion for the apparent onset of instability is

$$\tau_c > \frac{2\gamma_{LG}b}{R^2} \tag{5.43}$$

These criteria concern only the onset of the instability between fixed plates, that is, the possible development of instability at a given time under given boundary conditions, and thus can be applied to stretch flows [41], which, as demonstrated above, are instantaneously similar. The only difference concerns the variations of the separating distance in time, which implies that the stability criterion evolves in time. This means that under some circumstances the instability might first develop and then tend to damp as b increases. However, since we are dealing with yield stress fluids that stop flowing under insufficient stress, the fingers already formed behind the average interface never disappear. In practice, the criterion (5.41) or (5.43) is almost always fulfilled. For example, for typical values such as $R = 4$ cm, $\gamma_{LG} = 0.05$ N/m and $b = 1$ cm, the criterion (5.41) is expressed as $\tau_c > 0.4$ Pa. This means that most stretch flows of yield stress fluids, which correspond to flow with smaller fluid layer thicknesses and larger yield stresses, will be unstable and will tend to develop fingering.

5.3 SPREADING OR COATING FLOWS

Here we consider the situation in which the fluid spreads over a solid surface as a result of a volume or a surface force. Such situations are encountered in industrial or environmental processes where the fluids are poured on an inclined surface and left to flow under the action of gravity, applied onto a solid surface and then spread by another solid surface in relative tangential motion, or placed on a rotating disk and forced to flow over it as a result of the centrifugal force. The common characteristic of these flows is that the fluid motion is mainly of the shear type, that is, the dominant fluid velocity is parallel to the boundaries.

Such flows are a priori complex; the velocity components and the surface of contact between the fluid and the solid vary with both position and time. Here we will focus on sufficiently simple conditions under which an approximate, analytical description of some flow characteristics is possible for yield stress fluids. This is particularly the case when one considers only quasistatic or stoppage processes, for which the velocity can be assumed equal to zero and the maximum shear stress, generally along the wall, assumed to be equal to the yield stress. Under these conditions we consider three simple flow types associated with the three processes described above: gravity flow of a sheet of fluid over a planar surface (Section 5.3.1), spreading due to the relative motion of two solid surfaces (Section 5.3.2), and spreading over a rotating disk (Section 5.3.3). In the following text we will assume that the lubrication assumption is valid; that is, the ratio of the fluid thickness to its longitudinal extent (in the flow direction) remains much smaller than 1 and the gradient of the velocity in the y direction is much larger than in the other directions. Some instabilities may also occur within such flows; they are reviewed in Section 5.3.4. At last, since typically in spreading or coating flows the fluid increases its gas–fluid interface as it spreads, it is critical to examine the role of surface tension on flow characteristics (Section 5.3.5). Note that other, more complex, types of spreading or coating flows have been treated by empirical or complete analytical approaches, such as some steady flows of yield stress fluids in various confined [42] or free surface geometries [43], roll coating flows [44], journal bearing flows [45,46], and flows of a rivulet of yield stress fluid with surface tension effects [47].

5.3.1 Gravity Flow of a Yield Stress Fluid Over an Inclined Plane

Uniform Flow

Let us consider the uniform flow of an infinitely large and long sheet of a yield stress fluid of thickness h over an inclined plane of slope i. The basic characteristics of this flow type have been described in Section 1.4.1; fluid layers parallel to the plane glide over each other. The stress distribution is given by (1.55), which implies that the shear stress varies linearly from $\rho g h \sin i$ along the bed to 0 at the free surface. As a consequence, there is a critical height, $y_c = h - \tau_c / \rho g \sin i$, beyond which the shear stress is smaller than the yield stress, and the fluid remains rigid in the region contained between the planes $y = y_c$ and $y = h$. The thickness $(h - y_c)$ of this so-called *plug* is

$$h_c = \frac{\tau_c}{\rho g \sin i} \tag{5.44}$$

It follows that no uniform flow is possible as long as the total thickness of the sheet h is smaller than this critical thickness h_c. Moreover, if one is capable of progressively decreasing the thickness of the sheet while keeping the slope constant, the fluid completely stops flowing when reaching the critical thickness. Conversely, a sheet of constant thickness h lying over a horizontal plane starts to flow when the slope reaches the critical value i_c such that $\sin i_c = \tau_c / \rho g h$.

Two-Dimensional Spreading

Here we consider a layer of material of finite length along the direction x and lying over an inclined plane (Figure 5.10). The width (D) of the layer [i.e., in the lateral direction (z)] is so large that edge effects in this direction are negligible. As a consequence, the velocity component in the direction z is zero everywhere and the flow characteristics are identical in each plane (x,y); thus, the flow is "two-dimensional". The flow of a yield stress fluid under such conditions can be described with the help of numerical simulations [48,49]. In that case the flow characteristics are the single solution of momentum and mass conservation along with initial and boundary conditions. On the contrary, in the static regime it can take various arbitrary shapes, obviously still depending on the previous flow history, as long as the yield criterion is reached nowhere in the material (see Figure 5.10). For example, we can coat different, uniform sheets of thicknesses smaller than h_c that do not flow afterward. In fact, the shape of the deposited layer can be univocally related to the stress distribution, and thus may be predicted, only if it corresponds to the ultimate stage of a flowing fluid slowly evolving toward stoppage, since in that case, in the absence of inertia effects, the yield stress is asymptotically reached in the previously flowing regions.

Here we assume that the thin layer has been formed by material spreading under gravity until complete stoppage. We consider either a previous downward or upward flow (see Figure 5.11) of the whole material along the direction x, with the material front in the flow direction. Under these conditions, during stoppage, the shear stress in the regions of largest stress values (here along

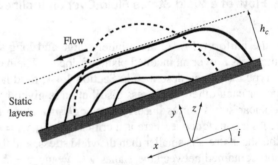

Figure 5.10 Different possible aspects (in two dimensions) of static layers (dark lines) of a yield stress fluid on an inclined plane. The critical height is represented by a dotted line parallel to the solid plane. The (virtual) heap represented by the dashed line cannot remain at rest because its thickness is somewhat larger than the critical thickness.

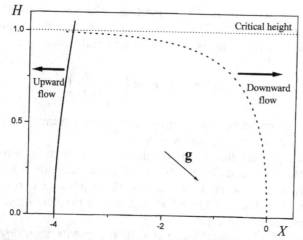

Figure 5.11   Fluid depth profile after stoppage for an upward or a downward yield stress fluid flow for a two- or three-dimensional layer in the direction of steepest slope (see the definition of dimensionless variables in the text). The present curves correspond to the dimensionless form of (5.48) with $\tan i = 1$, in a frame of reference parallel to the solid bottom surface (inclined relative to the horizontal).

the wall) reaches its minimum possible value: τ_c. Moreover, in contrast with lubricational flows considered in Section 5.2.1 the pressure cannot be considered as constant in each cross section; because of significant gravity effects, it now also depends on y. Within the frame of the lubrication assumption, the main velocity variations are in the y direction, and the momentum equation in the direction y is expressed as $0 = -\partial p/\partial y - \rho g \sin i$, from which we deduce the "hydrostatic" pressure distribution of general use in hydraulics

$$p(x, y) = \rho g(h(x) - y) \sin i \qquad (5.45)$$

in which we neglected the pressure at the fluid–gas interface.

Let us now apply the momentum equation in integral form to a portion of fluid of length dx and perpendicular to the bottom. In projection along the direction x we obtain

$$\rho g h D\, dx \sin i - \varepsilon \tau_c D\, dx - \rho g D \frac{h^2(x + dx)}{2} \cos i + \rho g D \frac{h^2(x)}{2} \cos i = 0 \qquad (5.46)$$

where ε is equal to 1 and -1, respectively for downward and upward flows. The first term of (5.46) corresponds to the gravity force, the second one to the resistant force due to shear along the wall, and the two last terms to the pressure force over each extremity of the fluid portion. After simplification this equation becomes

$$-h\frac{dh}{dx} = (\varepsilon h_c - h) \tan i \qquad (5.47)$$

Taking the origin ($x = 0$) as the intersection of the free surface profile with channel bottom, equation (5.47) may be integrated to give the profile of the free surface of the fluid layer, that is, $h(x)$ or, equivalently, $x(h)$:

$$x \tan i = h + \varepsilon h_c \ln \left(1 - \frac{h}{\varepsilon h_c} \right) \tag{5.48}$$

This equation may also be expressed in dimensionless form, $X = H + \varepsilon \ln(1 - \varepsilon H)$, in which $X = x \tan i / h_c$ and $H = h/h_c$.

This model predicts that the free surface at the front is perpendicular to the plane: $dh/dx \rightarrow \pm \infty$ for $x = 0$. For a downward flow over an inclined plane ($i \neq 0$), $h \rightarrow h_c$ when $x \rightarrow -\infty$ and the depth $h_c/2$ is reached for $x = -0.19 h_c / \tan i$ (see Figure 5.11), which means that the fluid depth tends rather rapidly toward the critical depth.

In the case of a horizontal plane ($i = 0$) the gravity term disappears from equation (5.46), which may be integrated to give the free surface profile:

$$h(x) = \sqrt{\frac{2\tau_c}{\rho g} x} \tag{5.49}$$

In that case $h \rightarrow \infty$ when $x \rightarrow \infty$, which means that there is no asymptotic depth. With a yield stress fluid, it is possible to form a heap of arbitrary central thickness over a horizontal plane.

It is worth noting that this provides a very good approximation of the profile of the free surface as long as the fluid depth varies *gradually* as a function of x; otherwise it is a priori only a rough approximation. In particular, the validity of the lubrication assumption on approach of the front of the flow is questionable since there, as a result of the mass conservation, the velocity component in direction y is not negligible.

Spreading from a Source Point

In practice it is easier to pour a limited amount of fluid over an inclined plane and let it flow in all directions than to induce the flow of a wide sheet of material as described above. However, the flow is obviously also more complex; the velocity component in direction z is no longer equal to zero—this is a three-dimensional transient flow that must be coped with. Let us first consider the spreading of a yield stress fluid supplied from a given point; the fluid obviously flows in the direction of the steepest slope, but it also constantly spreads in the lateral direction, thus leading to a surface of contact of width increasing with the distance from the source [50–52]. This contrasts with the initial theory due to Hulme, [53] who considered that, because of its yielding character, such a fluid would form, after some time and distance, a mouth of constant width. In reality the width increase results from the fact that as long as the yielding criterion is overcome as a result of some flow in one specific direction (here x), the fluid will at the same time flow in any other direction along which some finite stress of any amplitude applies to it. This is the case along the lateral edges; because of the necessary decrease in fluid depth on approach of the line of contact between air, fluid, and solid, there is, as long as the fluid has not stopped flowing in the main direction, a pressure gradient that induces some finite stress and thus some flow in the lateral direction.

Here we consider the final shape of a given amount of fluid poured from a vessel over a plane above a given point (see Figure 5.12). Once again the flow is initially well developed and thus it is expected that the yield stress will ultimately be reached after stoppage in the region of maximum stress, again along the wall, but now the flow direction is not uniform. At least the front of the material moves in the direction of steepest slope of the plane, and the material heap is symmetric relative to its midplane parallel to the direction of steepest slope (x) and perpendicular to the solid plane (see Figure 5.12). By reason of symmetry, in this midplane the velocity component in direction z is equal to zero. Within the frame of the lubrication assumption it follows that the momentum equation as expected in the two-dimensional case (5.46) again applies in a fluid layer around this midplane [54,55]. Consequently the fluid depth profile in the midplane of the deposit exactly corresponds to that given by (5.48). The complete fluid depth profile, from the downward front to the upward front, is again given by two profiles of the type (5.48) intersecting at the source point such as that represented in Figure 5.11. This result also applies to the profile of the free surface in the midplane of a deposit formed by the slow spreading of a fluid amount over a conical surface [56].

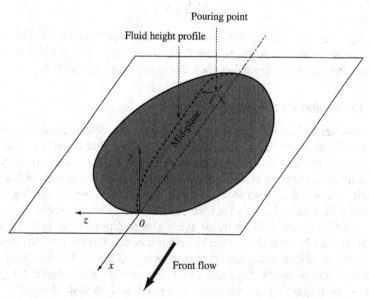

Figure 5.12 Schematic representation of the deposit of a yield stress fluid poured over an inclined plane.

It is remarkable that according to this approach, the critical fluid depth should not be reached anywhere in the final deposit. This might appear in contradiction with the result that no uniform flow can be obtained for a fluid depth below the critical one (see Section 5.3.1), and thus the final fluid depth after stoppage of a uniform flow cannot be smaller than h_c. In fact, gradually varied flows can take place at fluid depths smaller than h_c because of the additional pressure gradient [see equation (5.46)] due to fluid depth variations along x. Note that although some approximate analytical expressions for the shape of the deposit in the other directions may be found [54], the effective lateral extent depends on the initial and boundary conditions during spreading, in particular the fluid discharge. The complete characteristics of the final deposit shape can be deduced from numerical simulations involving a specific constitutive equation of the type (2.20) [55,57,58].

In the case of a horizontal surface ($i = 0$) the free surface profile is again given by (5.46), but this result is now valid in all directions because of the symmetry of the problem by rotation around the central axis. We can thus deduce the expression for the central height (h_0) of the deposit of radius R:

$$h_0 = \sqrt{\frac{2\tau_c R}{\rho g}} \tag{5.50}$$

It is also valuable to compute the fluid amount

$$\Omega = \frac{8\pi}{15} \sqrt{\frac{2\tau_c}{\rho g}} R^{5/2} \tag{5.51}$$

from which we can deduce the yield stress as a function of the fluid extent ($2R$) over the plane. Let us recall that these results are a priori valid only for a maximum thickness much smaller than the extent, say, $h_0 < R/10$.

5.3.2 Spreading Over a Rotating Surface

Let us consider a fluid amount lying on a disk at rest. We assume that the disk starts rotating at the initial instant. Intuitively we feel that a pasty material will remain at rest for low rotation velocities and start to flow and spread radially for sufficiently higher velocities. Indeed, this situation has a strong analogy with the gravity spreading over an inclined plane, except that now the driving force is the centrifugal force, which varies with the distance from the axis and increases with the rotation velocity. Since this is again a three-dimensional transient flow, it should obviously be treated in detail by numerical simulations [59,60]. However, it is possible to obtain some information concerning the basic characteristics of such flows from a rough approach analogous to that used above for gravity spreading, once again within the frame of the lubrication assumption.

We consider the incipient motion or stoppage of a layer of yield stress fluid lying on a rotating disk. We use the cylindrical frame of reference (O, r, θ, z)

fixed in the laboratory and in which Oz is the axis or rotation of the disk. As for the flow over an inclined plane (see Section 5.1.1), the fluid layer may have various shapes and remain at rest relatively to the disk as long as the driving force does not overcome some critical value related to the yield stress of the fluid.

Let us consider the shape of the fluid layer after flow stoppage. We may express the momentum equation in radial direction over a small vertical portion $(\Delta\Omega)$ of fluid $(h, dr, r\,d\theta)$ within a heap lying (at rest relative to the disk) on a disk rotating at a given rotation velocity (ω). Assuming a hydrostatic pressure distribution (5.45), the resulting momentum equation is similar to (5.46) except that the volume force term due to gravity $(\rho g \,\Delta\Omega \sin i)$ is now replaced by the inertia force due to the rotation $(\rho \,\Delta\Omega \omega^2 r)$. Two regions may then be distinguished at equilibrium:

- Close to the axis of rotation there exists a region in which the material has not moved; thus the shear stress will not necessarily have reached the yield stress and will be a priori unknown. In the particular case of a cylindrical layer of uniform thickness h, the pressure terms in (5.46) vanish throughout the undeformed region, and the outer boundary of this undeformed region is the cylindrical surface situated at the critical distance $(r < r_c)$ for which the yield stress balances the inertia stress:

$$\tau_c = \rho h \omega^2 r_c \qquad (5.52)$$

  This equation also provides an expression for the *critical velocity* at which flow should begin when a cylindrical heap of uniform thickness and radius r_0 has been set on a rotating disk: $\omega_c = \sqrt{\tau_c/\rho h r_0}$.
- In the remaining part of the fluid $(r > r_c)$ the yield stress has been overcome and then ultimately reached during stoppage and the momentum equation is

$$H\frac{dH}{dR} + 1 - \Theta H R = 0 \qquad (5.53)$$

in which we have used $H = h/h_c$, $R = r/r_c$, $\Theta = \omega^2 h_c/g$ with $h_c = \tau_c/\rho g$. Note that, as long as the lubrication assumption remains valid, equation (5.53) defines the ultimate shape of the layer under equilibrium whatever the initial shape of the heap. The fluid depth profile may be determined numerically from (5.53) for different values of Θ.

5.3.3 Coating by the Relative Displacement of Two Almost Parallel Plates

Here we consider the flow induced by the relative displacement of two plates separated by a thin layer of fluid and forming a small angle θ. The surface of contact of the fluid with the upper plate at a given time has a width D perpendicular to the flow direction, and a length L in the flow direction (see

Figure 5.13). Once again we use the lubrication assumption and consider the specific case of very slow flow or incipient motion for which the shear stress may be approximated by the yield stress. In addition, gravity effects will be neglected. For two precisely parallel plates the flow is a simple shear, which simply requires that one exerts the tangential force $T = \tau_c DL$ to maintain the motion. Otherwise, the relative motion of the slightly nonparallel plates requires application of both tangential and normal forces. The tangential force arises mainly from the shear flow while the normal force arises from the plate inclination, which induces some slight squeeze flow of the material, the thickness of which is reduced by the coating process. The tangential force approximately keeps the value obtained for parallel plates, and in the following text we focus on the estimation of the normal force.

Proceeding as in the case of squeeze flow (see Section 5.2) we may apply the momentum equation to a fluid portion between x and dx. Here the shear stress at the wall is assumed to be equal to the yield stress, and the net stress counteracting the pressure difference results from the difference of directions of the boundary surfaces along which the yield stress acts. Thus we obtain

$$\frac{\partial p}{\partial x} = -\frac{\tau_c}{h(x)}(1 - \cos\theta) \approx -\frac{\theta^2 \tau_c}{2h(x)} \tag{5.54}$$

Between the plates the fluid thickness decreases linearly to a small unknown value, and for the sake of simplicity, we can assume that the fluid thickness $h = \theta x$ varies between 0 and θL. Under these conditions we can compute the pressure distribution (neglecting the pressure in the fluid out of the gap)

$$p(x) = \frac{\theta}{2}\tau_c \ln\frac{L}{x} \tag{5.55}$$

from which we deduce, after integration over the plate surface, the net normal force:

$$N = \frac{\theta L D \tau_c}{2} \tag{5.56}$$

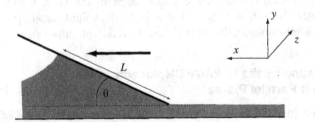

Figure 5.13 Schematic representation of a coating flow induced by the relative motion of two slightly nonparallel solid plates separated by a thin layer of fluid. Here, for the sake of clarity, the value of the relative angle (θ) has been increased.

5.3.4   Flow Instability

Apart from turbulence, which occurs in any situation for sufficiently large inertia effects, two types of instability, which do not seem to be related simply to inertia, may be specifically observed in spreading flows: fingering of the front and formation of roll waves in the flowing mass.

Fingering of a Spreading Layer

Thin layers of fluids flowing over a planar surface under gravity [61,62] or under a centrifugal force [63] were observed to evolve as fingers (see Figure 5.14). This phenomenon is typical of an instability, the origin of which remains somewhat unclear. First it was suggested that the rheological properties of the fluid do not seem to play a major role and that the instability results from some unbalance between interfacial and inertia effects [61]. Further interpretations assume that the free surface of the film forms a capillary ridge near the contact line (interface between the solid, liquid, and gas phases), a ridge that can be unstable to perturbations, leading to variations of the depth of the ridge [64,65]. Then the growth of fingers would be due to the fact that thicker regions of the fluid at the front flow more rapidly under the action of gravity [66]. Analysis of the instability in this frame have led to nontrivial criteria for the onset and fastest growing mode that could be approximated by the following simple result. For a Newtonian fluid the dominant wavelength of the finger pattern is [67]

$$\lambda_m \approx 14 \left(\frac{\gamma_{LG} h}{\rho g \sin i} \right)^{1/3} \tag{5.57}$$

in which h is the thickness of the fluid layer. It is expected that no instability appears if the width of the layer front is smaller than λ_m.

For yield stress fluids (described by a Herschel–Bulkley model) no complete theoretical treatment has been done, but an approximate expression, simply generalizing the result for the dominant wavelength for Newtonian fluids, was proposed for inclined plane flows [67]:

$$\lambda_m = 14 \left(\frac{\gamma_{LG} h}{\rho g \sin i} \right)^{1/3} \frac{1}{[\sin i_c (\sin i - \sin i_c) + (\sin i - \sin i_c)^2/(1 + (1/n))^n]^{1/3}} \tag{5.58}$$

Roll Waves

Roll waves frequently occur in sheet flows of water on steep slopes, for example, after a heavy rain on smooth roads or in steep riverbanks. In that case the flow evolves in almost periodic surges whose front velocity and depth exceed those of the corresponding uniform flow with the same mean flow discharge [68] (see Figure 5.15). This phenomenon may occur with more complex fluids, particularly yield stress fluids. Roll waves were frequently observed in natural mudflows [69,70] (see Figure 5.16). The conditions under which this instability occurs for a Newtonian fluid have been studied by various authors [71–74].

Figure 5.14   Typical aspects of the fingering instability for a hair gel (initially at rest around the center of the disk) spreading under the action of the centrifugal force due to the rotation of the disk. (Courtesy of H. Tabuteau.)

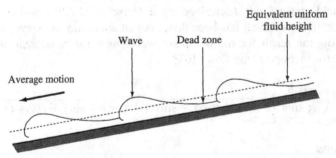

Figure 5.15   Formation of roll waves (in a longitudinal cross section) in free surface flows of viscous fluids over an inclined plane.

For example, for the laminar flow of a Newtonian fluid, the flow is unstable when the Froude number (Fr) is larger than a critical value:

$$\mathrm{Fr} = \frac{U}{\sqrt{gh\sin i}} > 0.527 \tag{5.59}$$

Figure 5.16   Roll waves in laboratory mudflows in a recirculating flume (for general experimental conditions, see Ref. 70).

in which U is the average fluid velocity in a cross section. However, an approximate approach, based on a linear stability analysis of momentum and continuity equations averaged over the fluid depth, provides [75], a general criterion for any rheological behavior type. The corresponding result appears capable of predicting rather well the results obtained with more complete approaches for simple fluids. According to this approach, a uniform flow over an inclined plane is unstable when

$$\left(\tau_p - h\frac{\partial \tau_p}{\partial h}\right) > \sqrt{gh\sin i}\,\frac{\partial \tau_p}{\partial U} \tag{5.60}$$

In order to explicit this criterion in the case of a yield stress fluid, we need to express the wall shear stress (τ_p) as a function of the fluid depth (h) and the mean velocity (U) through a fluid section. Let us consider the uniform flow of a Herschel–Bulkley fluid over an inclined plane. By integration of the momentum equation (1.55), in which one introduces the constitutive equation (4.15), we can deduce the velocity distribution:

$$y < y_c \Rightarrow v_x(y) = \frac{\alpha}{m+1}[y_c^{m+1} - (y_c - y)^{m+1}] \tag{5.61a}$$

$$h > y > y_c \Rightarrow v_x(y) = \frac{\alpha}{m+1}y_c^{m+1} \tag{5.61b}$$

with $m = 1/n$ and $\alpha = (\rho g \sin i / k)^m$. After integration of (5.61), we deduce the expression for the mean velocity:

$$U = \frac{\alpha}{m+1}\frac{y_c^{m+1}}{h}\left[h - \frac{1}{m+2}y_c\right] \tag{5.62}$$

Since we have $y_c = h(1 - \tau_c/\tau_p)$, equation (5.62) provides the relationship between τ_p, U, and h, from which we can determine the partial derivatives of (5.60) and show that this equation is equivalent to

$$\mathrm{Fr} > \frac{(m+1)G^2 - mG - 1}{(m+1)^2 G^2 + mG + 1} \tag{5.63}$$

in which $G = \rho g h \sin i / \tau_c$.

Since the right-hand side of (5.63) is always smaller than n, this equation in particular predicts that no stable flows can be obtained for a Froude number larger than n. As a consequence, in some cases the range of regimes for which uniform stable flows with yield stress fluids may be obtained can be rather narrow since they must satisfy two conditions: G must be larger than 1, as shown in Section 5.3.1 for flow to occur, Fr must be smaller than the critical value defined by (5.63). Note that a more sophisticated and more precise approach has been proposed [76].

5.3.5 Role of Surface Tension in (Stable) Spreading Flows

Here we examine under which conditions surface tension effects may play a role in the spreading flows of pasty materials. Basically surface tension may act in one of two ways: (1) for a constant gas–liquid interface the additional pressure induced by surface tension, which may vary with the position, can in turn affect the shape of this interface; or (2) if the gas–liquid interface changes, the additional surface energy can affect the flow characteristics.

Surface Tension Effect on a Constant Gas–Liquid Interface
The additional pressure effect (3.58) due to surface tension is typically of the order of γ_{LG}/R, in which R is the average radius of curvature of the interface. In general, the maximum variation of this additional pressure in a free surface flow corresponds to the difference between the values it takes at the front and along the uniform flow region. Assuming that the flow width is much larger than its thickness, the curvature of the front is of the order of the uniform fluid height h.

Since the curvature of the free surface is negligible in the uniform flow region, the maximum variation of the additional pressure due to surface tension effects is of the order of γ_{LG}/h. In parallel, the typical variation in pressure due to the gravity effect is $\rho g h$. As a consequence, the surface tension effect on a constant gas-liquid interface is negligible if

$$\frac{\rho g h^2}{\gamma_{LG}} \gg 1 \tag{5.64}$$

Surface Tension Effect Due to Interface Change

In order to establish a general but approximate criterion for negligible surface tension effect due to interface changes with a yield stress fluid, we will consider a specific, practical case of spreading flow. Assume that we have a cylinder of material of radius R and thickness b lying over a planar solid surface. We want to estimate the possible influence of surface tension on the deformation of this material when it spreads over the surface, that is, when the surface of contact between the solid and the fluid increases. Here we neglect gravity and surface tension effects resulting from the solid–fluid and solid–gas interfaces. Under these conditions we need to compare the energy (E), required for imposing some deformation to the material, to the work change associated with the corresponding fluid–gas interface change (W).

We may estimate the order of magnitude of the energy E for the rate of deformation tending to zero. Note that in view of estimating the dissipated energy, it is crucial to avoid the usual assumption of strictly rigid regions in some parts of the fluid, as was done in Sections 5.1–5.3 for simple descriptions of flow characteristics. In particular here, as the fluid heap spreads over the solid plane, we assume that it undergoes a transient, nonuniform flow for which all the fluid flows [4], even if some regions may be only slightly deformed. This means that the liquid regime is reached everywhere, so that at any given point within the material the stress tensor is expressed as $\mathbf{\Sigma} = \tau_c \mathbf{D}/\sqrt{-D_{II}} + f(\mathbf{D})$, in which $f(\mathbf{D}) = O(\mathbf{D}^n)$ when $\mathbf{D} \rightarrow 0$ with $n > 0$. The energy dissipated during the deformation over a time dt is

$$E = dt \int_\Omega \text{Tr}(\mathbf{\Sigma D})d\Omega = \tau_c dt \int_\Omega \text{Tr}(\mathbf{D}^2)/\sqrt{-D_{II}}d\Omega + dt \int_\Omega \text{Tr}(f(\mathbf{D})\mathbf{D})d\Omega \tag{5.65}$$

in which the second term becomes negligible compared to the first one when $\mathbf{D} \rightarrow 0$, so that to the first order one obtains:

$$E = 2\tau_c dt \int_\Omega \sqrt{-D_{II}}d\Omega \tag{5.66}$$

To estimate this energy, we need a sufficient knowledge of the velocity field within the fluid. We can consider two extreme cases, purely lubricational flow (see Section 5.2.1) or purely elongational flow (see Section 5.2.2), for which we could obtain good approximations of the corresponding velocity field. For a purely elongational flow (assuming perfect slip along the solid surface) the velocity field in the material corresponds to that found in Section 5.2.2, for which the local value of $\sqrt{-D_{II}}$ is $\sqrt{3}|d|/2$, in which $d = V/b = db/b\,dt$, where V is the velocity of the top free surface ($z = b$). It follows that the viscous energy dissipated during an elementary deformation db is

$$E_E = \sqrt{3}\tau_c\pi R^2 db \tag{5.67}$$

Let us now compute the surface energy change as a result of the deformation of the material cylinder. The change in the surface of the liquid–gas interface is equal to $dS = d(2\pi Rb + \pi R^2) \approx 2\pi d(Rb)$ while the changes in the solid–gas and solid–liquid interfaces are negligible because they result from the radius variations. Taking into account the volume conservation ($\pi R^2 b =$ constant) which implies $2dR/R + db/b = 0$, we finally obtain $dS \approx \pi R db$ and the surface energy change is $W = \pi R\gamma_{LG}db$. By comparing this expression with (5.67) we deduce that in the case $b \gg R$ surface tension effects are negligible when

$$\tau_c \gg \frac{\gamma_{LG}}{\sqrt{3}R} \tag{5.68}$$

For a pure lubricational flow $\sqrt{-D_{II}}$ is heterogeneous in the material, that is, it is a function of z and r, the distances of the current point respectively from the solid plane and from the central axis. Within the frame of the lubrication approximation, the main displacement is due to the relative gliding of fluid layers parallel to the solid surface, so that we can assume as a first approximation that $\sqrt{-D_{II}} \approx |\partial v_r(z, r)/\partial z|/2$. The integration over the entire fluid volume then yields $E_L = 2\tau_c dt \int_0^R \pi r|v_r(b, r)|dr$. Along the top free surface the fluid has no vertical velocity component, so the relation established for a fluid confined between two solid surfaces [see equation (5.12)] is valid for the local fluid velocity: $v_r(b, r) = -(r/2)(\partial b/b\partial t)$. Under these conditions we finally find, after integration

$$E_L = \frac{2\tau_c\pi R^3}{3b}\,db \tag{5.69}$$

Here the change in the surface of the solid–gas and solid–liquid interfaces must be taken into account. Since the thickness of the layer is small we have $dS_{SG} \approx -dS_{LG} \approx -dS_{SL} \approx -2\pi RdR$, and using (3.59) we deduce $dW_s \approx 2\pi R\gamma_{LG}(1 - \cos\theta)dR$. Now comparing this expression with (5.69) we conclude that in the case of $b \ll R$ surface tension effects remain negligible when

$$\tau_c \gg \frac{3(1 - \cos\theta)\gamma_{LG}}{R} \tag{5.70}$$

REFERENCES

1. A. Einstein, *Ann. Physik* **19**, 286 (1906).
2. G. K. Batchelor, *An Introduction to Fluid Mechanics*, Cambridge Univ. Press, Cambridge, UK, 1967.

3. B. R. Munson, D. F. Young, and T. H. Okiishi, *Fundamentals of Fluid Mechanics*, Wiley, New York, 1994.

4. G. G. Lipscomb and M. M. Denn, *J. Non-Newtonian Fluid Mech.* **14**, 337 (1984).

5. M. P. du Plessis and R. W. Ansley, *J. Pipeline Division, ASCE (American Society of Civil Engineers)* **2**, 1 (1967).

6. R. Hill, *The Mathematical Theory of Plasticity*, Oxford Univ. Press, Oxford, UK, 1950.

7. R. W. Ansley and T. N. Smith, *Am. Inst. Chem. Eng. J.* **13**, 1193 (1967).

8. A. N. Beris, J. A. Tsamopoulos, R. C. Armstrong, and R. A. Brown, *J. Fluid Mech.* **158**, 219 (1985).

9. K. Adachi and N. Yoshioka, *Chem. Eng. Sci.* **28**, 215 (1973).

10. D. D. Atapattu, R. P. Chhabra, and P. H. T. Uhlherr, *J. Non-Newtonian Fluid Mech.* **59**, 245 (1995).

11. M. Beaulne and E. Mitsoulis, *J. Non-Newtonian Fluid Mech.* **72**, 55 (1997).

12. J. Blackery and E. Mitsoulis, *J. Non-Newtonian Fluid Mech.* **70**, 59 (1997).

13. R. P. Chhabra and P. H. T. Uhlherr, in *Encyclopedia of Fluid Mechanics*, N. P. Cherimisinoff, ed., Gulf Publishing, Houston, 1988, **7**, Chapter 21.

14. G. Boardman and R. L. Whitmore, *Lab. Pract.* **10**, 782 (1961).

15. G. F. Brookes and R. L. Whitmore, *Rheol. Acta* **7**, 188 (1968).

16. P. H. T. Uhlherr, *Proc. 4th Natl. Conf. Rheology*, Adelaide, Australia, 1986, pp. 231–235.

17. L. Valentik and R. L. Whitmore, *Br. J. Appl. Phys.* **16**, 1197 (1965).

18. M. Hariharaputhiran, R. S. Subramanian, G. A. Campbell, and R. P. Chhabra, *J. Non-Newtonian Fluid Mech.* **79**, 87 (1998).

19. I. Machac, I. Ulbrichova, T. P. Elson, and D. J. Cheesman, *Chem. Eng. Sci.* **50**, 3323 (1995).

20. B. J. Briscoe, M. Glaese, P. F. Luckam, and S. Ren, *Colloid Surf.* **65**, 69 (1992).

21. T. Ferroir, H. T. Huynh, X. Chateau, and P. Coussot, *Phys. Fluids* **16**, 594 (2004).

22. C. J. Lawrence and G. M. Cornfield, in *Dynamics of Complex Fluids,*, Imperial College Press—The Royal Society, London, 1998, Chapter 27, p. 379.

23. G. H. Covey and B. R. Stanmore, *J. Non-Newtonian Fluid Mech.* **8**, 249 (1981).

24. A. Doustens and M. Laquerbe, *J. Theor. Appl. Mech.* **6**, 315 (1987).

25. C. Lanos and A. Doustens, *Eur. J. Mech. Eng.* **39**, 77 (1994).

26. J. D. Sherwood and D. Durban, *J. Non-Newtonian Fluid Mech.* **62**, 35 (1996).

27. J. D. Sherwood and D. Durban, *J. Non-Newtonian Fluid Mech.* **77**, 115 (1998).

28. G. H. Meeten, *Rheol. Acta* **39**, 399 (2000).

29. M. J. Adams, B. J. Briscoe, D. Khotari, and C. J. Lawrence, in *Dynamics of Complex Fluids*, Imperial College Press—The Royal Society, London, 1998, Chapter 29, p. 399.

30. N. Roussel and C. Lanos, *Appl. Rheol.* **13**, 132 (2003).

31. E. Lemaire, Y. O. M. Abdelhaye, J. Larue, R. Benoit, P. Levitz, and H. Van Damme, *Fractals* **1**, 968 (1993).

32. S. Tarafdar and S. Roy, *Fractals* **1**, 99 (1995).

33. D. Derks, A. Lindner, C. Creton, and D. Bonn, *J. Appl. Phys.* **93**, 1557 (2003).

34. P. G. Saffman and G. Taylor, *Proc. Roy. Soc. Lond.* **A245**, 312 (1958).

35. G. M. Homsy, *Ann. Rev. Fluid Mech.* **19**, 271 (1999).

36. P. Coussot, *J. Fluid Mech.* **380**, 363 (1999).

37. L. Paterson, *J. Fluid Mech.* **113**, 513 (1981).

38. S. D. R. Wilson, *J. Fluid Mech.* **220**, 413 (1990).

39. J. E. Sader, D. Y. C. Chan, and B. D. Hughes, *Phys. Rev. E* **49**, 420 (1994).
40. L. Kondic, P. Palffy-Muhoray, and M. J. Shelley, *Phys. Rev. E* **54**, R4536 (1996).
41. S. Z. Zhang, E. Louis, O. Pla, and F. Guinea, *Eur. Phys. J. B* **1**, 123 (1998).
42. R. B. Bird, G. C. Dai, and B. J. Yarusso, *Rev. Chem. Eng.* **1**, 1 (1982).
43. P. Coussot, *Mudflow Rheology and Dynamics*, Balkema, Rotterdam, 1997.
44. A. B. Ross, S. K. Wilson, and B. R. Duffy, *J. Fluid Mech.* **430**, 309 (2001).
45. J. A. Tichy, *J. Rheol.* **35**, 477 (1991).
46. S. L. Burgess and S. D. R. Wilson, *J. Non-Newtonian Fluid Mech.* **72**, 87 (1997).
47. S. K. Wilson, B. R. Duffy, and A. B. Ross, *Phys. Fluids* **14**, 555 (2002).
48. F. K. Liu and C. C. Mei, *J. Fluid Mech.* **207**, 505 (1989).
49. D. Laigle and P. Coussot, *J. Hydraul. Eng.* **123**, 617 (1997).
50. P. Coussot, S. Proust, and C. Ancey, *J. Geophys. Res.* **101**, 25217 (1996).
51. S. D. R. Wilson and S. L. Burgess, *J. Non-Newtonian Fluid Mech.* **79**, 77 (1998).
52. X. Huang and M. Garcia, *J. Fluid Mech.* **374**, 305 (1998).
53. G. Hulme, *Geophys. J. Roy. Astrophys. Soc.* **39**, 361 (1974).
54. P. Coussot, S. Proust, and C. Ancey, *J. Non-Newtonian Fluid Mech.* **66**, 55 (1996).
55. N. J. Balmforth, R. V. Craster, and R. Sassi, *J. Fluid Mech.* **470**, 1 (2002).
56. M. Yuhi and C. C. Mei, *J. Fluid Mech.* **519**, 337 (2004).
57. C. C. Mei and M. Yuhi, *J. Fluid Mech.* **431**, 135 (2001).
58. D. I. Osmond and R. W. Griffiths, *J. Geophys. Res.* **106**, 16241 (2001).
59. J. A. Tsamopoulos, M. F. Chen, and A. V. Borkar, *Rheol. Acta* **35**, 597 (1996).
60. S. L. Burgess and S. D. R. Wilson, *Phys. Fluids* **8**, 2291 (1996).
61. H. E. Huppert, *Nature* **300**, 427 (1982).
62. N. Silvi and E. B. Dussan, *Phys. Fluids* **28**, 5 (1985).
63. H. Tabuteau, *Spreading of Sewage Sludges*, Ph.D. thesis, ENGREF, Paris, 2005.
64. S. M. Troian, E. Herbolzheimer, S. A. Safran, and J. F. Joanny, *Europhys. Lett.* **10**, 25 (1989).
65. L. W. Schwartz, *Phys. Fluids A* **1**, 443 (1989).
66. M. A. Spaid and G. M. Homsy, *Phys. Fluids* **8**, 460 (1996).
67. J. R. de Bruyn, P. Habdas, and S. Kim, *Phys. Rev. E* **66**, 031504 (2002).
68. R. F. Dressler, *Commun. Pure Appl. Math.* **2**, 149 (1949).
69. T. R. H. Davies, *J. Hydrology* (New Zealand) **29**, 18 (1990).
70. P. Coussot, *J. Hydraul. Res.* **32**, 535 (1994).
71. T. B. Benjamin, *J. Fluid Mech.* **2**, 554 (1957).
72. C. S. Yih, *Phys. Fluids* **6**, 321 (1963).
73. S. P. Lin, *Phys. Fluids* **10**, 308 (1967).
74. C. O. Ng and C. C. Mei, *J. Fluid Mech.* **263**, 151 (1994).
75. J. H. Trowbridge, *J. Geophys. Res.* **92**, 9523 (1987).
76. N. J. Balmforth and J. J. Liu, *J. Fluid Mech.* **519**, 33 (2004).

CHAPTER 6

GRANULAR FLOWS IN FRICTIONAL REGIME

As we have seen in Section 2.3, a granular material exhibits a complex constitutive equation that may take different forms depending on flow regime. In this context we review different simple flow situations from which flow characteristics in some range of flow regimes may be at least approximately predicted or interpreted in terms of a given apparent rheological behavior. We start by usual viscometric flows, then consider different flow geometries of practical interest: free surface flows, conduit flows, compressive flows, and displacement of an object through a granular material. We focus on stationary slow flows and assume that frictional interactions predominate. Moreover, in most cases we assume that the behavior of the material follows the Coulomb model [equation (2.28)] with a constant coefficient of friction. In the absence of complete sets of experimental data, the question nevertheless remains somewhat open as to whether the values of the friction coefficient determined under different specific flow conditions are quite consistent. Even more critical is the fact that the Coulomb model does not provide a straightforward relationship between the kinematics (or some features of this kinematics) and the components of the stress tensor. In particular, the normal stress components are a priori unknown and in the following text, to solve the problem, we often assume a proportionality between two normal stress components by a factor k and eventually suggest some ideas for estimating its value depending on the flow under consideration. Afterward the flow characteristics for a finite flow rate can be described independently of any further assumption

Rheometry of Pastes, Suspensions, and Granular Materials: Applications in Industry and Environment
By Philippe Coussot   Copyright © 2005 John Wiley & Sons, Inc.

concerning the rheological behavior. In some cases (e.g., inclined plane flows) this provides a description of flow properties of the material even if the continuum assumption is not valid, such as when the flow thickness : grain size ratio is less than 10. These approaches nevertheless remain consistent as long as the corresponding flow laws are considered independently of any intrisic rheological behavior of the material determined under the continuum assumption.

6.1 VISCOMETRIC FLOWS

Because it is impossible to describe the rheological behavior of a granular material with a single model in a wide range of flow regimes, the usual viscometric tools partly loose some of their fundamental interest. Nevertheless, the flow geometries involved in viscometric flows make it possible to induce a flow with some simple characteristics that are useful for interpreting some rheological characteristics of the material in the frictional regime.

6.1.1 Couette Flow

Let us consider the flow of a granular material sheared between two concentric cylinders. In contrast with the Couette flow of a simple fluid, the variables of the flow may depend on z because gravity impacts the normal stresses acting between particles in contact. However, the flow is symmetric through rotation around the axis; in other words, the variables do not depend on θ. Under these conditions, if the free surface remains horizontal, the simplest velocity field corresponds to radial and longitudinal velocity components equal to zero. If the radial or longitudinal component of the velocity differed from zero at some point, there would be a net outward, radial or vertical, flow associated with an instability. For a stable flow, it follows that only the tangential velocity component differs from zero and depends only on r and z : $v_\theta = r\omega(r, z)$. Such a velocity field shows that the flow consists in the relative motion of concentric layers of material around the central axis.

Let us now consider the stress distribution within the material in the gap separating the two cylinders. In a material at rest in a conduit the normal stress in direction z tends to saturate beyond a critical depth (see Section 6.3) because the finite tangential force along the wall may partly counteract gravity effects. However, it is likely that this effect plays a minor role when the material flows tangentially to the wall, as in a Couette flow. Indeed, if a continuous network of contacts forms and is supported by the vertical walls, it is broken at each instant by flow, so normal stresses should partly relax. If, effectively on average, normal stress relaxation predominates, it is reasonable to assume that the normal stress (σ_{zz}) in direction z and the radial normal stress (σ_{rr}) are simply proportional to those resulting from a hydrostatic pressure distribution in a simple liquid, so that we have $\sigma_{rr} = k\rho g(h - z)$ (the origin of the z axis is taken at the outer cylinder bottom), in which k is an unknown parameter that we expect to be close to 1.

Because of the torque conservation, the shear stress distribution is still given by (1.84). Thus, it is along the inner cylinder that the shear stress value $(\sigma_{r\theta})$ assumes its largest value, so that in frictional regime, according to the Coulomb model, yielding occurs when $\sigma_{r\theta} = \sigma_{rr} \tan \varphi$. Neglecting edge effects, we deduce by integration the critical total torque applied to the inner cylinder for incipient motion: $M_c = r_1 \int_0^h 2\pi r_1 \sigma_{r\theta}(r, z)dz = k\rho g\pi r_1^2 h^2 \tan\varphi$, from which we get the expression for the coefficient of friction of the material:

$$\tan \varphi = \frac{M_c}{k\rho g\pi r_1^2 h^2} \tag{6.1}$$

In practice, the consistency of this approach, and thus the validity of equation (6.1) and all assumptions behind it (in particular the assumption of frictional regime), may be checked by carrying out experiments with different material heights (h); the torque should vary with the square height, and the coefficient of friction can be deduced from coefficient of proportionality.

Let us now consider the flow characteristics at different rates of rotation of the inner cylinder. At first glance torque–rotation velocity curves under increasing stress or flow rate ramps appear to be roughly similar to those obtained with yield stress fluids; there is a plateau at low rotation velocities and the torque significantly increases only beyond a generally large, critical, rotation velocity Ω_c [1–4]. However, more precise observations show that, depending on the flow history, different states of the material may be reached, which reflect the variations of the particle configuration in time. For example, it was shown that there is a significant hysteresis between the level of the plateau recorded for an increasing and a decreasing ramp of stress with polystyrene beads, glass beads, and sand [5] (see Figure 6.1); the torque in the incipient flow regime exceeds that in the flow stoppage regime. Direct observations of the material during such tests show that it tends to dilate at high velocities, which means that the lower torque level during stoppage likely corresponds to a looser configuration than for incipient flow. This effect has some analogy with that observed from triaxial tests (see Section 6.4); the force needed to impose a given deformation rate on the material depends on the previous flow history, but in that case it has also been shown to tend toward a unique value after a sufficient deformation. Under these conditions it is expected that, as for triaxial tests, the state of a granular material in a Couette flow tends to a reproducible, so-called critical, state, so that the torque level tends to reach a single, asymptotic value (M_c) in steady-state flow. This value may then be used as a reference, mechanical characteristic of the material associated to that specific geometry. In practice, in order to approach this critical state, one may induce a rapid flow by imposing a high rotation velocity, then progressively decrease this velocity to a low level and maintain this level for a sufficient time for the torque to reach a steady value (M_c).

Beyond Ω_c the torque–rotation velocity curve is easily reproducible; that is, it does not depend much on flow history. However, in this regime the material appears to significantly dilate at high apparent shear rates, an effect that depends

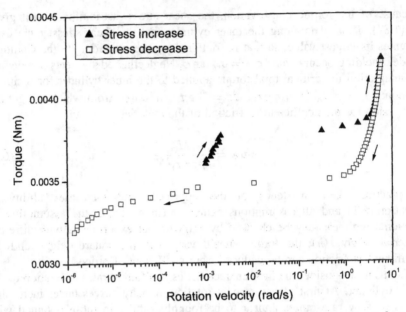

Figure 6.1 Torque versus velocity during increasing–decreasing torque ramps from 3.2 to 4.24 mN·m in 60 s (up) + 60 s (down) in a rheometer [vane geometry (six blades, diameter 3.6 cm; height 4.5 cm) with a roughened cylinder (diameter 4.7 cm)] filled with polystyrene beads (preparation: 50 s⁻¹ during 10 s, no rest). (From Da Cruz et al., *Phys. Rev. E* **66**, 051305 (2002).)

on boundary conditions. As a consequence, according to current knowledge, it is difficult to establish a correspondence between the torque–rotation velocity curve and the rheological behavior of the material in the frictional regime.

6.1.2 Annular Shear Cell

The cone–plate geometry is generally inappropriate for granular materials because the (small) gap at the truncation point can be of the order of the particle size, which may lead to a strong friction. With simple parallel disks the upper tool initially in contact with the granular material may rapidly loose contact with it under flow because of the slight tendency of noncohesive grains to move out of the gap. Thus it is necessary to confine the material in a given volume and submit it to a controlled normal force. This is the origin of another geometry inspired from viscometry and often used for studying the rheological behavior of granular materials: the annular shear cell. It consists in an annular channel of low depth filled with the material, which can then be sheared by an upper disk (lid) in rotation around the axis. The major advantage of this geometry is that either the normal stress or the material volume can be controlled from the upper disk. Its disadvantage is that, because of the material confinement within the closed channel by lateral walls, the stress distribution within the sample remains partly uncontrolled. Nevertheless, observations similar to those for Couette flows are generally made with this geometry [1]; the torque increases with the velocity essentially at high rotation rates and the plateau at low velocities depends on the flow history. Again, the steady-state plateau level (M_c) for the torque obtained after a rapid preshear may be used to deduce a basic rheological parameter of the material, namely, its coefficient of friction.

As a first approximation, we can assume that the flow develops in the form of material layers perpendicular to the central axis and gliding over each other. We can neglect edge effects and assume that the stress tensor does not vary significantly with the distance from the rotation axis. In that case the normal stress in the vertical direction is roughly equal to the sum of the stress due to the additional normal force applied to the lid and to the stress due to gravity. At low rotation velocities of the upper disk, the shear generally localizes in a thin region in contact with the disk, where the average normal stress is lowest; in this region the stress due to gravity acting in the material is zero and the normal stress is equal to the ratio of the normal force (N) to the surface of contact between the lid and the material ($\pi(r_2^2 - r_1^2)$). In addition, the torque applied to the material is $M = \int_{r_1}^{r_2} 2\pi r^2 \tau \, dr = 2\pi(r_2^3 - r_1^3)\tau/3$. From the Coulomb model, we finally obtain

$$\tan\varphi = \frac{\tau}{\sigma} = \frac{3M_c(r_2^2 - r_1^2)}{2N(r_2^3 - r_1^3)} \tag{6.2}$$

6.2   FREE SURFACE FLOWS

6.2.1   Granular Heaps over Horizontal Surfaces

When a granular material is progressively poured over a horizontal solid surface, it forms a heap, the shape and size of which depend on pouring history and geometric parameters. In particular, different shapes of heap with different angles of the edge slopes (relative to the plane) may be obtained, but it appears that these angles cannot be larger than a critical value, the *maximum angle of repose* φ_{max} of the material; when one tries to form a heap with a larger edge slope, the material flows and the lateral slope decreases until it reaches the maximum angle of repose. Since this result is valid for any slope of the free surface of the heap, the limiting shape of a heap is a cone of slope φ_{max}. For example, if the grains are poured over a fixed point, they rapidly form such cones of increasing size.

The main advantage of such experiments is that φ_{max} may be related to the angle of friction of the material for incipient motion $\tan\varphi_{start} = \mu_{start}$ (see Section 2.3.1). Let us consider a virtual heap of material made of two planar free surfaces of infinite width (the problem is thus two-dimensional) and inclined at angle θ to the horizontal (see Figure 6.2). When the conditions for which the heap is unstable are just reached, a failure is expected to occur along some surface along which the Coulomb criterion is satisfied. Let us assume that this is a planar surface (AB) forming an angle i to the horizontal. In the absence of inertia effects, the momentum balance over the volume of material situated

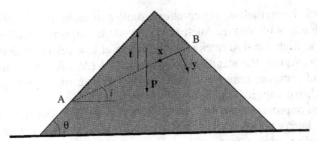

Figure 6.2   Two-dimensional granular heap lying on a horizontal surface.

above the plane (AB) is $0 = \mathbf{P} + \mathbf{t}$, in which $\mathbf{P}$ is the material weight and $\mathbf{t}$ the average stress vector along the surface (AB). We can decompose these vectors into their components along the surface (AB) and the perpendicular direction so as to write $\mathbf{P} = (P \sin i)\mathbf{x} + (P \cos i)\mathbf{y}$ and $\mathbf{t} = T\mathbf{x} + N\mathbf{y}$. The momentum balance in particular implies that $T/N = \tan i$. In parallel, the rheological behavior of the material in the frictional regime provides some information concerning the stress tensor; at the beginning of motion, the shear stress along the slip surface is related to the normal stress via (2.28), which here yields $T/N = \tan \varphi_{start}$, from which we obviously deduce $i = \varphi_{start}$. Since necessarily $i < \theta$, this means that there can exist a volume of material in the heap tending to slip along a surface of the type (AB) only if the angle of repose of the heap (θ) is larger than φ_{start}; otherwise the heap will remain at rest. Although the demonstration is not as easy, a similar result may be obtained with a conical heap [6]. The general result is that all conical heaps with an angle of repose smaller than φ_{start} are stable, which means that the maximum angle of repose (φ_{max}) of a granular heap is φ_{start}.

It is worth noting that the height of the heap does not play any role in this analysis, which implies that the angle of repose is independent of the heap size. This is in fact true only on a scale of observation much larger than the particle size. In practice, the shape of the heap is not perfectly conical; the slope may slightly change (in general, decrease) on approach of the bottom, as a result of the specific interaction between the particles and the plane. On such a scale the coefficient of friction between the grains and the solid plane may play a major role in the structure of the heap [7].

More critical is the fact that even at a much larger scale the final, mean angle of repose θ depends on the way the material has been set up. For example, if it forms sufficiently progressively, the final angle corresponds to the maximum angle or repose. If we now slightly perturb this heap or pour some more grains onto the top, one or several avalanches occur, which may leave the heap with a slope slightly smaller than the previous one. Depending on its conditions of preparation (heaps may be obtained according to various procedures), the slope angle of a granular heap may thus vary between two (generally close) limits, say, φ_{min} and φ_{max} [8], which are characteristics of the rheological behavior of the material, and the angle of friction (φ) of the material is in general situated somewhere between these two values, likely closer to φ_{min}. Finally, we note that segregation effects may also occur during this preparation; the particles tend to flow at different rates depending on their size and interactions with the other particles [9].

6.2.2 Granular Heaps over Inclined Surfaces

Let us now consider the case of a heap formed over an inclined solid surface. For sufficiently small inclination (i) we would expect to obtain a heap similar to that for a horizontal surface but truncated by the inclined, solid plane, with a maximum angle of repose still close to φ. Indeed, an analysis of the heap stability similar to that for a horizontal solid surface (see Section 6.2.1) leads to a similar result for the angle of repose relative to the horizontal. However, this result does not account for the interaction between the material and the solid surface. For a horizontal surface, this effect is negligible beyond a certain scale of the heap (see Section 6.2.1); it simply induces some difference between the average angle of repose and the slope close to the heap bottom. When this phenomenon develops over an inclined plane, the grains at the bottom can tend to flow downward, which may destabilize the rest of the heap. In particular, assuming that the grains cannot roll over the plane the heap is expected to be unstable if the coefficient of friction between the grains and the plane surface is smaller than some critical plane slope that depends on grains and plane properties. It follows that beyond a critical slope no stable heap can be obtained; a flow necessarily develops. This critical slope basically depends on the material characteristics and plane roughness, but it might also depend on the amount of material. This we will attempt to clarify in the following section by directly considering uniform or gradually varied flows.

6.2.3 Granular Flow

Free surface flows of granular materials have been the object of intense research from the beginning of the 1980s. Various experiments revealed the diversity of flow regimes or instabilities [10–13] as a function of boundary conditions. In parallel, theoretical works or simulations, generally assuming the predominance of collisional interactions, attempted to predict the properties of free surface flows of granular materials [14–18]. Subsequent theoretical approaches assumed the possibility of a continuous network of frictional contacts playing a significant role in the flow characteristics [10,19–21]. More recently a straightforward, yet incomplete, mechanical framework, in agreement with experimental observations, has been provided for the description of dense free surface flows of granular materials in the frictional regime.

Stoppage or Incipient Motion

Let us consider a granular layer of uniform thickness h over an inclined plane of slope i. The width and length of this layer are large, so edge effects can be neglected. Under such conditions, for a given thickness there exists a critical slope i_c beyond which the layer cannot remain at rest. This behavior type resembles that of yield stress fluids. However, for thickness values much larger than particle size, i_c is approximately independent of the layer thickness. This property, which contrasts with the behavior of yield stress fluids for which the critical slope decreases as the layer thickness increases, obviously finds its origin in the specificity of granular material behavior; the normal and tangential stresses at the wall increase proportionally with the layer thickness so that their ratio remains equal to $\tan i$. Incipient motion occurs when this ratio becomes larger than a fixed angle of friction, characteristics of the material and plane, but independent of material thickness. More puzzling is the fact that for small thicknesses the critical slope increases as the thickness decreases [22]. This, in fact, likely results from the interactions between the grains and the plane, which play an increasing role as the thickness : particle size ratio becomes smaller (or, equivalently, as the ratio of the normal force to the particle weight decreases).

Moreover, the conditions for incipient motion and stoppage do not coincide, and a more complete scheme of the interplay between h and i is as follows [23]. Following the addition of a large mass of material on a plane of initial slope of inclination i_0, flows occur and there eventually remains a layer of thickness $h_{stop}(i_0)$ at rest; now, if the plane is further tilted, the layer of material remaining on the plane starts to flow again only beyond a critical angle $i_c = h_{flow}^{-1}(h_{stop}(i_0))$, significantly larger than i_0, and the layer stops flowing when the thickness reaches $h_{stop}(i_c)$ (see Figure 6.3). There is thus a metastable region between the curves $h_{stop}(i)$ and $h_{flow}(i)$ in which one may observe either flow or rest depending on flow history. This hysteresis effect is similar to that observed in Couette geometry (Section 6.1) with granular materials and somewhat similar to that observed with pasty materials as a result of thixotropy (see

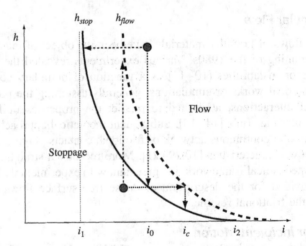

Figure 6.3 Critical depth of stoppage or incipient motion as a function of plane slope and flow history. Flow always occurs in the right region [$h > h_{flow}(i)$], whereas the material stops flowing or is at rest in the left region [$h < h_{stop}(i)$]. Between h_{stop} and h_{flow} the material may be at rest or in motion depending on flow history.

Section 3.3.1). It also bears some analogy with the effect of thixotropy for free surface flows of pastes over an inclined plane; after fluid stoppage over an inclined plane, the new (larger) slope at which the fluid again begins to flow is larger than the previous slope. Moreover, as for pasty materials, the new critical slope may increase with the time for reaching this critical slope as a result of aging effects (see Section 2.3.1).

The h_{stop} curve finally appears to be a robust characteristic of the granular material and plane. At a critical slope (i_1), h_{stop} tends toward infinity while at another critical slope (i_2), h_{stop} vanishes (see Figure 6.3). In this context Pouliquen proposed [24] representing this behavior by a decreasing exponential function making use of these two critical slopes characteristic of the material and the plane

$$h_{stop}(i) = h_a \ln \left(\frac{\tan i_2 - \tan i_1}{\tan i - \tan i_1} \right), \qquad i < i_2 \qquad (6.3)$$

in which h_a is a parameter depending on the material and plane characteristics.

Flow Rate

Systematic experiments of wide granular flows (large width : thickness ratio) with different materials and different plane roughnesses have shown [24,25] that the mean velocity (over the fluid depth) of uniform flows follows a simple law as a function of material thickness (Figure 6.4)

$$\frac{U}{\sqrt{h}} = \alpha + \beta \frac{h}{h_{stop}} \quad \text{when} \quad h > h_{stop}; \qquad U = 0 \quad \text{when} \quad h < h_{stop} \qquad (6.4)$$

in which α and β are independent of inclination and bead size but depend on the material and plane surface characteristics. Note that a similar variation of the velocity as a function of the fluid depth was found for channelized flows for glass spheres [26], and up to a critical flow rate beyond which $U \propto h$ [12]. This result [equation (6.4)] is fundamental since it shows that it is possible to predict the mean velocity of a granular flow as a function of the fluid depth as soon as one has determined $h_{stop}(i)$, α, and β from independent experiments. Note that as a first approximation it is often reasonable to take $\alpha = 0$. This curve is thus somewhat equivalent to a rheological characteristic of the material under free surface, frictional flows, even if for some of the corresponding data the flow thickness was only a few times the particle size.

Note that since flows can be obtained only for material thicknesses exceeding $h > h_{stop}$, it follows from (6.4) that a specificity of granular flows is that a priori one cannot obtain uniform flows at a velocity smaller than the critical value

$$U_c = (\alpha + \beta)\sqrt{h_{stop}} \qquad (6.5)$$

Note that this result is clearly analogous to that obtained for thixotropic pasty materials, which cannot flow steadily at an apparent shear rate lower than a critical value (see Section 3.3).

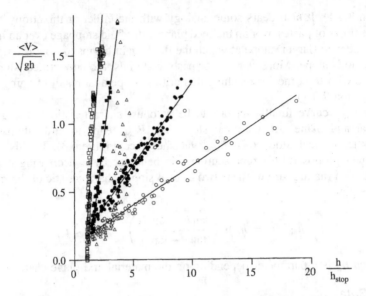

Figure 6.4 Froude number $(U/\sqrt{gh})$ as a function of the ratio h/h_{stop} for different systems of glass beads and different inclination angles of the plane (covered with the same beads). (From Ref. 25, Figure 5d.)

Finally, using (6.3) and (6.4), the uniform fluid depth h_0 associated with a given average velocity U and a given plane slope i may be expressed as follows:

$$h_0(U, i) = \left[\frac{h_a U}{\alpha} \ln \left(\frac{\tan i_2 - \tan i_1}{\tan i - \tan i_1} \right) \right]^{2/3} \tag{6.6}$$

Granular Fronts

For uniform free surface flows of fluids the wall shear stress (τ_p) may be determined from the momentum equation: $\tau_p = \rho g h \sin i$. This wall shear stress can also be considered as a function of the average velocity and fluid thickness: $\tau_p(U, h)$. For laminar flows this value reflects the viscous dissipation in the fluid depth and may thus be used as a first approximation for estimating the local wall shear stress due to viscous dissipation for a gradually varied or transient flow of *local* average velocity U (in a cross-section) and *local* fluid depth h. The usual numerical simulations of transient flows of yield stress fluids rely on depth-averaged flow equations that may be solved as soon as such an expression of wall shear stress is known [27]. This approach is, in fact, quite similar to that used for describing transient or nonuniform flows in hydraulics, except that in that case the wall shear stress now expresses energy dissipation by turbulence and should be determined by independent experiments (uniform flows) in the same geometry.

For granular flows in the frictional regime one cannot immediately use a similar approach. Indeed, for such a material type the wall shear stress cannot be considered solely as the result of viscous dissipations associated with a certain average velocity over a certain fluid thickness; the normal force also plays a role. Taking this into account, it has been suggested [28] that granular materials may behave as "generalized" Coulomb materials for which, in simple shear flows over inclined planes, the wall shear stress is simply proportional to the normal stress via a coefficient of friction μ. Since for an infinitely wide layer of material the normal stress at the wall results solely from gravity effects, we obtain

$$\tau_p(U, h) = \mu\sigma_{yy} = \mu\rho gh \cos i \tag{6.7}$$

Under these conditions we are now looking for the expression of a coefficient of friction $\mu(U, h)$, a function of the local average velocity and fluid depth.

In a method similar to that followed for laminar free surface fluid flows, this coefficient μ may be deduced from measurements in uniform flows, for example, at different slopes j, a case for which the momentum equation gives in particular $\tau_p = \rho gh \sin j$, so that we have $\mu = \tan j$. From (6.6) we deduce the expression for the friction coefficient as a function of local velocity and fluid depth:

$$\mu(U, h) = \tan i_1 + (\tan i_2 - \tan i_1)\exp\left(-\frac{\alpha h^{3/2}}{Uh_a}\right) \tag{6.8}$$

Then one may assume that the wall shear stress (6.7) expresses the local resistance to flow in a nonuniform or transient granular flow when the local average velocity and thickness are U and h. In particular, one can use it for predicting the shape of the front of a transient wave of granular material [28]. Using this expression in the momentum balance over a material portion of thickness dx in the direction x, we obtain

$$\rho gh \sin i - \mu(h, U)\rho gh \cos i - \frac{d \int_0^h \sigma_{xx} dy}{dx} = 0 \tag{6.9}$$

For pasty fluids following a constitutive equation of the type (2.20), although the normal stress components are all equal to the pressure within the frame of the lubrication assumption, this is not obvious for granular materials. A possibility consists in assuming that, as for conduit flows (see Section 6.3) there is a simple proportionality between the two main normal stress components: $\sigma_{xx} = k\sigma_{yy}$ [Pouliquen [28] suggested that k could take values between 1 and $(1 + \sin^2 i_1)/(1 - \sin^2 i_1)$]. In accordance with the momentum equation, since we have $\sigma_{yy} = \rho g(h - y)\cos i$, we deduce from (6.9) the equation for the profile of the free surface:

$$\tan i - \mu(h, U) - k\frac{dh}{dx} = 0 \tag{6.10}$$

The uniform fluid depth h_0 is obviously found as a function of the velocity by incorporating $dh/dx = 0$ in (6.10). Now, using the dimensionless fluid depth

$h^* = h/h_0$ and distance $x^* = x/h_0$, we get the dimensionless equation of the flow profile

$$k\frac{dh^*}{dx^*}\tan i - (\tan i_1 + (\tan i_2 - \tan i_1)\exp\chi h^{*3/2}) = 0 \qquad (6.11)$$

in which $\chi = \ln(\tan i - \tan i_1/\tan i_2 - \tan i_1)$. The profile of a granular flow that has just ceased because the slope or the thickness has progressively decreased is obtained from (6.11) by taking $h_0 = h_{stop}$.

It is worth noting that the slope of the free surface profile at the front ($h^* = 0$) is finite: $dh^*/dx^* = (\tan i_1 - \tan i_2)/2k$. This contrasts with the result with yield stress fluids, for which the slope of the free surface is in principle infinite at the front. The predictions of this approach have been successfully compared with experimental data under different flow conditions and with different systems [28]. Note that equation (6.11) predicts that the half-uniform depth is reached only after a distance of the order of 10–20 times h_0, which is much larger that the equivalent distance for a pasty material.

6.3 CONDUIT FLOWS

Let us consider the flow of a granular material through a vertical, cylindrical conduit. If the material is poured downward at a low rate, the grains fall quite freely through the conduit with some possible occasional interactions with the wall, but the influence of the conduit remains negligible. When the downward flow rate is increased, the density of grains in the conduit and the number of interactions with the wall are expected to increase. This might correspond to a dense, collisional regime, but there is as yet no clear theory for describing the corresponding flow characteristics. Beyond a critical rate of pouring, some jamming process may occur as a result of the formation of a continuous network of grains in contact between them and with the walls. In that case the flow rate is abruptly slowed down in some regions of the conduit. However, steady flows cannot realistically be expected in that case if the downstream edge remains free, that is, if the conduit is effectively an open cylinder. Indeed, since this is a cohesionless material, the grains situated at the bottom of a dense region will fall rapidly downward while the dense regions situated upstream will still flow slowly and instabilities can develop [29]. It seems that steady flows can be obtained only if the downstream boundary is partly closed, because in that case the smaller aperture at the bottom imposes a slow, dense, stable flow everywhere upstream. A typical situation of this type is a hopper with an aperture at its bottom.

Here we focus on slow, dense flows in conduit, which can naturally be considered to be in the frictional regime. Roughly speaking, two major effects govern the flow characteristics: (1) the interactions between the grain network and the wall, and in particular a kind of arching process that leads to a stress distribution strongly differing from that expected with simple fluids, with a saturation of the normal stress beyond a certain depth in the material; and (2) variation of the cross section before the aperture, which partly conditions the flow rate. In the following paragraphs we will first consider the stress distribution in a straight conduit and then examine the possibility for predicting the flow rate, assuming that it is governed entirely by inertia effects due to cross-sectional changes downwards.

6.3.1 Stress Distribution in a Straight Conduit

Here we consider the incipient motion or very slow flow of a granular material through a straight conduit and assume that the simple Coulomb model applies. Let us consider a cylindrical material portion of thickness dz (see Figure 6.5). For the sake of simplicity, we assume that the normal stress $\sigma_{zz} = \sigma$ is uniform in each cross section, depending only on z, and that it is proportional to the normal stress acting radially on the walls (σ_{rr}) by a factor k. These assumptions are certainly not strictly valid [6], but they make it possible to obtain a straightforward description of flow properties that in practice appears to be in good agreement with reality. The choice of the value of k is a critical aspect that has been discussed frequently in the literature [6]. Once again, in the absence of clear information, a value of the order of 1 may be used as a first approximation.

Under these conditions, the Coulomb model gives $\sigma_{rz} = \sigma_{rr} \tan \varphi^* = k\sigma \tan \varphi^*$, in which φ^* is the angle of friction of the grains with the wall, and the momentum balance applied to this material portion is

$$\pi R^2 [\sigma(z) - \sigma(z + dz)] - 2\pi R \, dz (\varepsilon k\sigma \tan \varphi^*) + \rho g \pi R^2 dz = 0 \qquad (6.12)$$

in which $\varepsilon = 1$ for a downward motion and $\varepsilon = -1$ for an upward motion. Taking into account the boundary condition at the free surface $\sigma(0) = 0$, the solution of

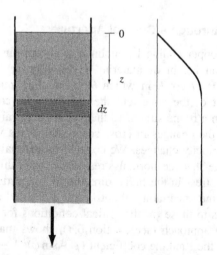

Figure 6.5 Normal stress distribution in a granular material slowly flowing through a straight, vertical conduit, according to the model of Janssen (see text).

equation (6.12) is

$$\sigma(z) = \frac{\rho g R}{2\varepsilon k \tan \varphi^*} \left[1 - \exp\left(-\frac{2\varepsilon k \tan \varphi^*}{R} z \right) \right] \tag{6.13}$$

For downward flows it follows from (6.13) that at small depths (i.e., $z \ll R/2k \tan \varphi^*$), the normal stress varies linearly with the depth

$$\sigma \approx \rho g z \tag{6.14}$$

while for larger depths (i.e., $z \gg R/2k \tan \varphi^*$), the stress becomes independent of the depth:

$$\sigma = \sigma_0 \approx \frac{\rho g R}{2k \tan \varphi^*} \tag{6.15}$$

For upward flows, such as those obtained by pushing the material by a piston, according to (6.13), the normal force increases exponentially with the depth. These characteristics strongly differ from those observed with a simple fluid under the same conditions; in any case, the normal stress (in fact, the pressure) increases linearly with the fluid depth.

This simple theoretical approach is initially due to Janssen [30], and more sophisticated approaches have been developed subsequently [6]. However, the validity of this initial simple theory was proved much more recently [31–33] (see Figure 6.6). In particular, it was shown that the data in slow downward and upward flows could be represented by the Janssen model using a parameter k changing by only 10% [34]. However, it was also shown that the Janssen model cannot be extrapolated to the case when an additional mass is set on top of the material without any contact with the wall [34,35].

6.3.2 Flow Rate through a Conical Aperture

Let us consider a hopper ending by inclined walls forming an open cone (see Figure 6.7). The total height of material in the cone is H, and the diameter of the aperture is $D = h \cot \theta$, in which θ is the cone angle and h the length between the summit of the cone and the aperture. We can describe the flow characteristics within a frame similar to that used for a straight conduit, but now taking into account the changes in cross-sectional surface with depth, which in particular induces velocity changes. We consider a material portion of thickness dz and again assume that the normal stress $\sigma_{zz} = \sigma$ in this portion is uniform and that the normal stress in the radial direction is proportional to σ by a factor k. There remains some uncertainty about k, but a detailed analysis, using some arguments analogous to those for the critical conditions for flow as described in the Mohr–Coulomb approach (see Section 6.4), shows that this parameter can be approximated by the Rankine coefficient $(1 + \sin \varphi)/(1 - \sin \varphi)$ [6]. A further discussion of this question may also be found in a paper by Eber [36]. Moreover, for the sake of simplicity, we will assume that the velocity component in the

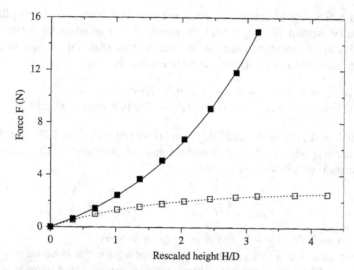

Figure 6.6 Resistance force to pushing (filled squares) upward and to pulling (open squares) downward a 62.5% bead packing in the steadily sliding regime at a velocity of 16 μm/s as a function of the height H scaled by the column diameter D (1.5-mm steel beads in a 36-mm duralumin column). The solid line and the dashed line represent the fit with the expression for the total force at the bottom of the column as deduced from equation (6.13). The parameters used were $k = 0.9$ for pushing and $k = 1$ for pulling forces. (Courtesy G. Ovarlez [33].)

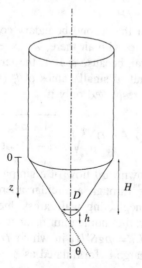

Figure 6.7 Hopper composed of a straight cylinder followed by a conical aperture.

direction z (i.e., v_z) is uniform in each section. The flow rate at a depth $H - z$ may thus be written $Q = v_z(z)S(z)$, in which Q is a constant for a steady flow.

Finally, the momentum equation applied to this material portion in the axis direction, now taking into account inertia effects, is

$$\rho[S(z+dz)v_z^2(z+dz) - S(z)v_z^2(z)] = \\ [S(z)\sigma(z) - S(z+dz)\sigma(z+dz)] - P(z)k'\sigma\, dz + \rho g S(z)dz \tag{6.16}$$

in which $k' = k \tan\varphi \sin\theta$, and $S(z) = \pi(H - z)^2(\cot\theta)^2$ and $P(z) = 2\pi(H - z)\cot\theta$ are respectively the radial surface and the perimeter of the material portion. Equation (6.16) transforms into

$$\frac{\partial\sigma}{\partial z} + \alpha\frac{\sigma}{H - z} = \rho g - \frac{\beta}{(H - z)^5} \tag{6.17}$$

in which $\alpha = 2(k'/\cot\theta - 1)$ and $\beta = 2\rho Q^2/\pi^2(\cot\theta)^4$.

We can now solve this equation by accounting for the boundary conditions at the top and bottom of the cone. First, we can consider that there is a straight conduit above the cone through which the granular material flows with a stress distribution accurately represented by the theoretical approach of Section 6.3.1. This gives a normal stress $\sigma(0) = \sigma_0$ along the limit between the cone and the straight conduit. Using this boundary condition, equation (6.17) may be solved by direct integration after multiplication of all the terms by $(H - z)^{-\alpha}$:

$$\sigma(\xi) = \frac{\rho g H}{1 - \alpha}[\xi^\alpha - \xi] + \frac{\beta H^{-4}}{4 + \alpha}[\xi^\alpha - \xi^{-4}] + \sigma_0\xi^\alpha \tag{6.18}$$

in which we used $\xi = 1 - z/H$.

Now we can account for the second boundary condition at the cone bottom; since the material is in contact with air there, we expect to have $\sigma(z = H - h) = 0 = \sigma(\xi = h/H)$. In general, because $\alpha > 1$, the terms in factor of ξ^α in (6.18) can be neglected in the limit of small values of ξ (i.e., $h/H \ll 1$). The second boundary condition can be respected only if

$$Q = \pi D^{5/2}\sqrt{\frac{(4 + \alpha)g}{2(\alpha - 1)\cot\theta}} \tag{6.19}$$

Let us now consider the downward flow of a simple liquid in a cylindrical container of section $S' = \pi R^2$ through a circular aperture of section $S = \pi D^2/4$. For a low-viscosity fluid, neglecting wall stress due to the (slow) fluid motion through the large container, the momentum balance over the complete fluid volume is expressed as $\rho Q^2/S = \rho g S'H$, in which H is the total fluid height. It follows that the fluid discharge is formulated as

$$Q = \frac{\pi R D}{2}\sqrt{gH} \tag{6.20}$$

which increases with the height of fluid.

Thus, in contrast with a simple, low-viscosity fluid, the discharge of a granular material from a hopper is independent of the material height. This basically results from the saturation of the normal stress with depth in the cylindrical part, which leads to a stress at the cone–cylinder interface (σ_0) independent of the material volume in the cylinder (in contrast with the pressure in a simple fluid). The dependence on $D^{5/2}$ in (6.19) should also be general since it basically results from the fact that the cross section of the flow in a cone reduces in proportion to $(H - z)^{-2}$. For example, leaving apart the influence of normal stress, that is, keeping only the inertia and gravity terms in (6.16), would immediately give the $D^{5/2}$ dependence for the average velocity.

These qualitative results appear to be in good agreement with experimental data; the flow rate through a given aperture is independent of the material height above as long as this height is not too small and the following empirical law was observed [37,38]

$$Q \propto (D - Kd)^{5/2} \tag{6.21}$$

in which K is a constant and d the diameter of the grains. For spherical particles $K \approx 1.5$ but is larger for angular particles. The simple $D^{5/2}$ dependence is recovered from equation (6.21) for $D \gg d$. For smaller values, the additional term in (6.21) may be seen as a natural correction to the abovementioned theoretical approach reflecting some jamming processes for an aperture diameter equal to only few grain diameters.

Let us now consider the case of a straight cylindrical hopper with an aperture in the middle of its bottom. At least in the ultimate stages of the flow, when most of the grains have moved out of the hopper, we expect that there remains a static conical region around the (circular) aperture and forming an angle to the horizontal equal to the maximum angle of repose. If we extrapolate this result to flow situations, we deduce that this dead region also exists during flow. As a consequence, the flow is equivalent to that in a hopper composed of a cylinder with an open cone at its bottom and forming an angle φ. If the cylinder : aperture diameter ratio is large, the discharge is again given by an equation of the type (6.19). This suggests that for intermediate cases an equation of this type also describes the flow discharge with a coefficient α depending on the characteristics of the system.

6.4 COMPRESSION FLOWS

Here we consider the case of a cylinder of granular material of length h, submitted to an isotropic, radial stress (σ_3) along its cylindrical surface and to a normal stress (σ_1) along its axis (in direction z) (see Figure 6.8). As long as the flow remains stable, the material undergoes a simple elongational flow as described in Section 5.2.2, and the stress tensor has only three nonzero components along the diagonal: σ_1, σ_3, and σ_3. In practice, we follow the deformation in the direction z (i.e., $\varepsilon_1 = \Delta h / h$) and the first normal stress difference (i.e., $N_1 = \sigma_1 - \sigma_3$). It appears that the response of the granular material in time depends on its initial

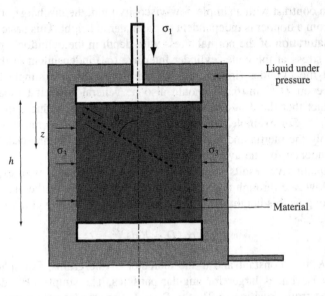

Figure 6.8 Schematic representation of the triaxial test. The dashed line illustrates the expected surface of shear banding, which develops when the Coumob criterion is reached.

state. Typically three types of $N_1 - \varepsilon_1$ curves may be observed (see Figure 6.9). For a dense material that has been, for example, vibrated and compacted before the test, the curve first reaches a maximum then progressively decreases toward an asymptotic value. For a loose material that has been, for example, formed by pluviation of grains, the curve progressively increases and tends to the asymptotic value mentioned above. The common asymptotic normal stress difference corresponds to the critical state described in Section 2.3.1; when this state has been reached, after a small deformation, the Coulomb criterion is satisfied somewhere in the material, which should consequently flow.

Let us try to determine along which surface the Coulomb criterion is reached within the material. We consider a surface element of normal vector $\mathbf{n} = \mathbf{x} \sin \theta + \mathbf{z} \cos \theta$ forming an angle θ relative to the cylinder axis. The stress vector at this surface is expressed as $\mathbf{t} = \mathbf{\Sigma} \cdot \mathbf{n} = \mathbf{x} \sigma_3 \sin \theta + \mathbf{z} \sigma_1 \cos \theta$. The tangential and normal components of this stress vector are thus respectively equal to

$$\tau = \mathbf{t} \times \mathbf{n} = \frac{\sigma_3 - \sigma_1}{2} \sin 2\theta \tag{6.22a}$$

$$\sigma = \mathbf{t} \cdot \mathbf{n} = \frac{\sigma_1 + \sigma_3}{2} + \frac{\sigma_1 - \sigma_3}{2} \cos 2\theta \tag{6.22b}$$

This means that the point of coordinates (σ, τ) belongs to the so-called Mohr–Coulomb circle of radius $(\sigma_1 - \sigma_3)/2$ centered in $((\sigma_1 + \sigma_3)/2; 0)$ (see Figure 6.10). If, for example, σ_3 is fixed, there exists a critical value of σ_1 for which this circle is tangential to the straight line defined by the Coulomb criterion

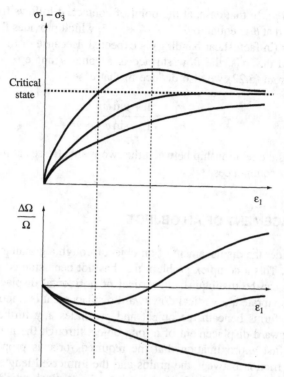

Figure 6.9 Stress (a) and relative volume change (b) as a function of deformation from triaxial tests for different initial states of a granular material. The horizontal dashed line corresponds to the critical state.

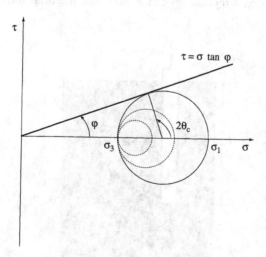

Figure 6.10 Mohr–Coulomb representation of shear and normal stresses in various directions.

(i.e., $\tau = \sigma \tan \varphi$). In that case, at the point of contact, it follows from geometric considerations that θ is equal to $\theta_c = \pi/4 + \varphi/2$, which provides the angle along which the flow (in fact, shear banding) is expected. Because of the symmetry by rotation around the axis, the flow surface is a cone of angle θ_c relative to the central axis. From (6.22) we also deduce in that case

$$\frac{\sigma_1}{\sigma_3} = \frac{1 + \sin \varphi}{1 - \sin \varphi} \tag{6.23}$$

which provides the relationship between the two main stresses at the critical state for which flow commences.

6.5 DISPLACEMENT OF AN OBJECT

Here we consider the displacement of an object through a granular material lying in a container. This a complex problem that has not been studied much, but two limiting cases can be distinguished: vertical or horizontal displacement. Unless the material is vibrated, a vertical downward displacement is impossible because some arching forms beneath the object and precludes any further motion. For the vertical upward displacement of a long object through the material, we may assume as a first approximation that the required force is proportional to the coefficient of friction between the grains and the immersed length of the object.

Let us now consider the horizontal motion of a vertical cylinder of diameter D through a granular medium in a container. This object is assumed to extend a distance H into the bucket of material (see Figure 6.11). The ratio of the distance between the object and the container walls to the cylinder diameter is large. In the frictional regime we can expect that the drag force element (dF)

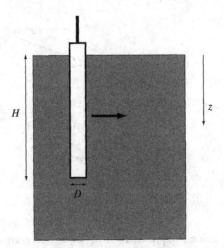

Figure 6.11 Scheme of a cylindrical object in motion through a granular medium.

exerted on a vertical portion (dz) of the cylinder surface at a given fluid depth (z) is simply proportional to the normal (radial) stress exerted on the cylinder and to the surface of the cylinder portion. When the object moves horizontally, the normal stress in the granular material, at least just around the object, likely relaxes, as in a Couette flow (see Section 6.1), so that the normal stress around the cylinder is equal to the average pressure in the material, which, neglecting the atmospheric pressure, is expressed as $\rho g z$, where ρ is the apparent density of the granular material. The local drag force thus is $dF \propto \rho g z D\, dz$. Note that we avoid considering the exact orientation and value of the local force, which are assumed to be independent of other variables considered here. After integration over the cylinder length, we obtain the total drag force

$$F \propto \rho D H^2 \qquad (6.24)$$

in which the coefficient of proportionality is independent of the velocity within the frictional regime only but depends on the frictional coefficient of the material. This basic approach was found to be consistent with precise experiments [39]. No theoretical approach has been developed for other object shapes, but we can expect that an approach similar to that described above, now using some coefficient depending on the specific shape of the object, could be developed. A series of empirical formulas for the drag force over objects of various cross-sectional shapes but still immersed vertically in the material have been established by Wieghardt [40].

6.6 FREE SURFACE FLOW OF A GRANULAR PASTE

6.6.1 Granular Paste Flowing over an Inclined Plane

We consider a uniform layer of granular paste lying at rest over an inclined plane. The stress distribution, namely, the tangential stress ($\sigma_{xy} = \tau$) in the direction x and the normal stress ($\sigma_{yy} = \sigma$) in the direction y, in this two-phase material, are given by equations (1.55) and (1.57). We can also consider separately the stress tensor within the interstitial fluid; it corresponds to an isotropic pressure p_f, which, assuming that one can follow a continuous path of fluid from the free surface to any point within the material, is given by a hydrostatic distribution

$$p_f = p_0 + \rho_f g (h - y) \cos i \qquad (6.25)$$

in which p_0 is the pressure of the ambient fluid. Because the complete stress tensor is the sum of the granular and the liquid terms [see equation (2.32)], the normal stress in the paste can also be expressed as

$$\sigma = p_f + \sigma' \qquad (6.26)$$

in which σ' is the granular normal stress term related solely to the interactions between the grains. From (1.57), (6.25), and (6.26), we deduce the expression for the granular normal stress term, also called the *effective normal stress*

$$\sigma' = \rho' g(h - y) \cos i \qquad (6.27)$$

in which $\rho' = \phi(\rho_P - \rho_0)$. Assuming now that the flow occurs when the Coulomb criterion specifically applies to the granular mass, which in such a simple shear is expressed as $\tau = \sigma' \tan \varphi$, we find from (1.55) and (6.27) the critical slope for which motion will occur:

$$\tan i = \frac{\phi(\rho_P - \rho_0)}{\phi\rho_P + (1 - \phi)\rho_0} \tan \varphi \qquad (6.28)$$

From this equation it appears that the critical slope for incipient flow of the granular paste is smaller than that of the equivalent, dry granular material.

6.6.2 Liquefaction of a Granular Paste at Rest over a Horizontal Plane

Let us now assume that the interstitial fluid moves relative to the granular mass that lies on a horizontal plane. For the sake of simplicity, we consider a flow at a mean rate U in the direction y. In that case Darcy's law provides the relationship between the pressure gradient and the fluid velocity

$$\frac{d(p_f + \rho_0 g(h - y))}{dy} = -\frac{\mu_0}{k_0} U \qquad (6.29)$$

in which k_0 is the permeability of the granular phase. We deduce the expression for the pressure distribution:

$$p_f = p_0 + \frac{\mu_0}{k_0} U(h - y) - \rho_0 g(h - y) \qquad (6.30)$$

From (6.26) and (6.30) with $i = 0$, we deduce the expression for the critical velocity for which the effective stress cancels:

$$U_c = \frac{\rho'}{\rho_0} \frac{k_0 \rho_0 g}{\mu_0} \qquad (6.31)$$

In that case the granular paste completely looses its resistance to flow since any tangential stress is capable of causing it to move.

REFERENCES

1. D. Prasad and H. K. Kytömaa, *Int. J. Multiphase Flow* **21**, 775 (1995).
2. G. I. Tardos, *Powder Technol.* **92**, 61 (1997).

3. G. I. Tardos, M. I. Khan, and D. G. Schaeffer, *Phys. Fluids* **10**, 335 (1998).
4. C. Ancey and P. Coussot, *C. R. Acad. Sci. Paris* **327**, 515 (1999).
5. F. Da Cruz, *Flow of Dry Grains—Friction and Jamming*, Ph.D. thesis, ENPC, Marne la Vallée, 2004 (in French)
6. R. M. Nedderman, *Statics and Kinematics of Granular Materials*, Cambridge Univ. Press, Cambridge, UK, 1992.
7. Y. Grasselli and H. J. Hermann, *Eur. Phys. J. B* **10**, 673 (1999).
8. R. L. Brown and J. C. Richards, *Principles of Powders Mechanics*, Pergamon Press, Oxford, UK, 1970.
9. R. Jullien and P. Meakin, *Nature* **344**, 425 (1990).
10. P. C. Johnson, P. Nott, and R. Jackson, *J. Fluid Mech.* **210**, 501 (1990).
11. C. E. Brennen, K. Sieck, and J. Paslaski, *Powder Technol.* **35**, 31 (1983).
12. C. Ancey, P. Coussot, and P. Evesque, *Mech. Cohesive-Frictional Mater.* **1**, 385 (1996).
13. E. Azanza, F. Chevoir, and P. Moucheront, *J. Fluid Mech.* **400**, 199 (1999).
14. P. C. Johnson, P. Nott, and R. Jackson, *J. Fluid Mech.* **210**, 501 (1990).
15. S. B. Savage, *J. Fluid Mech.* **92**, 53 (1979).
16. S. B. Savage and D. J. Jeffrey, *J. Fluid Mech.* **110**, 255 (1981).
17. J. T. Jenkins and S. B. Savage, *J. Fluid Mech.* **130**, 187 (1983).
18. C. K. K. Lun and A. A. Bent, *J. Fluid Mech.* **258**, 335 (1994).
19. H. M. Jaeger, C. H. Liu, S. R. Nagel, and T. A. Witten, *Europhys. Lett.* **11**, 619 (1990).
20. P. Mills, D. Loggia, and M. Tixier, *Europhys. Lett.* **45**, 733 (1999).
21. P. Mills, M. Tixier, and D. Loggia, *Eur. Phys. J. E* **1**, 5 (2000).
22. O. Pouliquen and N. Renaut, *J. Phys. II* (France) **6**, 923 (1996).
23. A. Daerr and S. Douady, *Nature* **399**, 241 (1999).
24. O. Pouliquen, *Phys. Fluids* **11**, 542 (1999).
25. GDR Midi, *Eur. Phys. J. E* **14**, 367 (2004).
26. J. W. Vallance, *Experimental and Field Studies Related to the Behavior of Granular Mass Flows and the Characteristics of Their Deposits*, Ph.D. thesis, Michigan Technological Univ., 1994.
27. D. Laigle and P. Coussot, *J. Hydraul. Eng.* **123**, 617 (1997).
28. O. Pouliquen, *Phys. Fluids* **11**, 1956 (1999).
29. Y. Bertho, F. Giorgiutti-Dauphiné, and J. P. Hulin, *Phys. Fluids* **15**, 3358 (2003).
30. H. A. Janssen, *Z. Ver. Dtsch. Ing.* **39**, 1045 (1895).
31. L. Vanel and E. Clément, *Eur. Phys. J. B* **11**, 525 (1999).
32. L. Vanel, P. Claudin, J. P. Bouchaud, M. E. Cates, E. Clément, and J. P. Wittmer, *Phys. Rev. Lett.* **84**, 1439 (2000).
33. G. Ovarlez, *Statics and Rheology of a Confined Granular Medium*, Ph.D. thesis, Paris XI Univ., Paris, 2002 (in French)
34. G. Ovarlez, C. Fond, and E. Clément, *Phys. Rev. E* **67**, 060302 (2003).
35. G. Ovarlez and E. Clément, *Phys. Rev. E* **68**, 031302 (2003).
36. W. Eber, *Phys. Rev. E* **69**, 021303 (2004).
37. R. M. Nedderman, U. Tüzün, S. B. Savage, and G. T. Houlsby, *Chem. Eng. Sci.* **37**, 1597 (1982).
38. W. A. Beverloo, H. A. Leniger, and J. Van de Velde, *Chem. Eng. Sci.* **15**, 260 (1961).
39. R. Albert, M. A. Pfeifer, A. L. Barabasi, and P. Schiffer, *Phys. Rev. Lett.* **82**, 205 (1999).
40. K. Wieghardt, *Ann. Rev. Fluid Mech.* **7**, 89 (1975).

CHAPTER 7

PRACTICAL RHEOMETRICAL TECHNIQUES

In practice several techniques are used in industry or geologic surveys for estimating the apparent viscosity of pasty or granular materials in terms of the resistance to flow they provide under complex or ill-defined flow conditions. The results of these tests are seldom subjected to a rheological interpretation based on a rigorous theoretical analysis of flow behavior as done with well-defined viscometric flows. In most cases, the data obtained with different materials are simply compared and, with some experience, it may be possible to distinguish critical values above or below which the material is thought to have the desired mechanical properties.

In this chapter, we show that the flows involved in many of the cases described are close to the simple nonviscometric flows described in Chapters 5 and 6. Hence, it appears possible to carry out a more detailed analysis of the data in rheological terms through the use of simplifying assumptions. For pastes, these are considered as simple, nonthixotropic, yield stress fluids for which the liquid regime is reached during most of the test, whereas for granular materials, these follow a Coulomb model (see Section 2.3.1). In doing so, we may extrapolate the basic rheological parameters (yield stress and friction coefficient) from these tests with pastes and granular materials.

In the following sections we review practical tests according to the classification of nonviscometric flows, that is, displacement of an object (Section 7.1), squeeze flows (Section 7.2) and spreading or coating flows (Section 7.3), along

Rheometry of Pastes, Suspensions, and Granular Materials: Applications in Industry and Environment
By Philippe Coussot   Copyright © 2005 John Wiley & Sons, Inc.

with conduit flows (Section 7.4) and granular flows (Section 7.5). In each of these cases, whenever possible, we propose an approximate analysis of the flow from which we may obtain a rheological parameter and discuss the conditions needed to ensure that the simplifying assumptions on which the measurements are based are valid. For tests involving squeeze flows, conduit flows, and displacement of an object, a small volume of test material is sufficient. However, for spreading or coating flows, it is necessary to use a much larger volume to ensure the lubrication condition (fluid depth : spreading distance ratio ≪ 1).

Note that in most of the following approaches for pastes we will be neglecting thixotropic effects eventhough they may significantly affect the data. In this context, it is likely that any practical test provides the value of an *apparent yield stress*, which is partly "procedure-dependent." To ensure its reproducibility, the test must be carried out according to a fixed experimental procedure so that, as a first approximation, fluids exhibiting different yield stresses would experience the same flow history.

Conversely one may focus on thixotropic effects and use some of the following practical tests for quantifying some specific characteristics of fluid thixotropy, namely, the apparent yield stress of the fluid after different times of rest (Section 7.6).

In Section 7.7, we review the main types of natural or industrial materials, which can be classified as pasty, slurry, or granular materials. We discuss their typical rheological behavior and suggest appropriate practical tests among those proposed in the previous sections for estimating some of their basic rheological characteristics.

7.1 TESTS INVOLVING DISPLACEMENT OF AN OBJECT THROUGH FLUID

When one considers the relative flow of an object and a fluid, two limiting cases can be distinguished: (1) the object is immersed in a volume of fluid much larger than the object volume and one observes how the object moves through it, and (2) the object is initially out of the fluid and one observes how it penetrates through it. Case 1 corresponds to the practical test called a *penetrometer*; case 2, to a test of displacement of an object.

7.1.1 Penetrometer

The principle of the penetrometer is to push a solid tool through a material under a given force F and measure its depth of penetration. Such tests are also used for solid materials; this is sometimes known as the *puncture test* [1]. For pasty materials, a typical penetrometer consists of a vertical cone that can plunge in a material lying in a container under the action of an applied force. Usually the cone is dropped under the action of its own weight. Its depth of penetration gives an indication of the "firmness" of the paste. The more rapidly (and less deeply) the object stops moving, the greater resistance to flow the fluid exhibits.

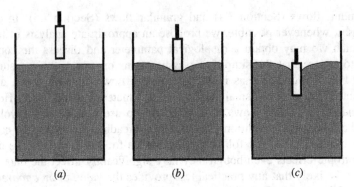

Figure 7.1 Principle of the penetrometer: (a) before the test; (b) before effective penetration (solid regime); (c) stoppage after some penetration (liquid regime).

When the object comes into contact with the fluid, the latter starts to deform in its solid regime. This continues up to a critical point of deformation at which the object then starts to penetrate the fluid (thus effectively separating it into two regions). As a first approximation, we can assume that the first stage (deformation) corresponds strictly to the solid regime while the second stage (penetration) corresponds strictly to the liquid regime (see Figure 7.1). When most of the fluid flow falls in the liquid regime, the ultimate depth of penetration reached after stoppage under a given force may be used to estimate the fluid yield stress (see text below). However, we first need to know beyond which depth the liquid regime is effectively reached almost everywhere along the object surface. Here we will consider only a very simple case—an object already partly immersed in the fluid at rest—and estimate the critical length of displacement of the object through the fluid beyond which the flow in the liquid regime is expected to be dominant.

Deformation in the Solid Regime

Let us consider the ideal case of a long cylinder of radius r_0 already immersed to a depth h in the fluid initially at rest. Then the cylinder is moved through the paste along its axis (see Figure 7.2) under the action of a force F, which induces a motion of a distance l_0. We assume that the container is a cylinder of radius R, much larger than r_0, so that the velocity components of the paste in the radial direction and the variations of the average paste depth in the container can be neglected. By neglecting edge effects as a first approximation, the shear stress does not depend on the depth; hence the stress distribution around the small cylinder is found from the force conservation along fictitious concentric cylinders at distance r within the material: $\tau = F/2\pi hr$. A fluid element situated at a distance r undergoes a deformation such that it moves to a distance $l(r)$ in the direction of the cylinder axis with, in the absence of slip between the paste and the penetrating cylinder, $l(r_0) = l_0$. As long as the material is in the solid regime, we have after stoppage $\tau = G\,dl/dr$ for a simple viscoelastic fluid.

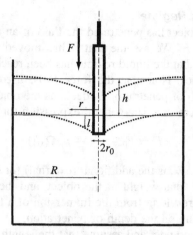

Figure 7.2 Deformation of fluid in the solid regime for a vertical cylinder initially immersed in the fluid, as a result of an applied force: initial (continuous horizontal line) and final (dotted line) position of the free surface of the fluid.

Integrating this equation, incorporating the abovementioned stress distribution, and considering the boundary condition in the absence of wall slip [i.e., $l(R) = 0$], we find the expression for the local displacement: $l = (F/2\pi Gh) \ln R/r$. At the transition between the solid and the liquid regime along the small cylinder, since we have $\gamma = dl/dr = \gamma_c$, we obtain the expression for the critical displacement $l_c = l(r_0)$ at the solid–liquid transition:

$$l_c = r_0 \gamma_c \ln\left(\frac{R}{r_0}\right) \qquad (7.1)$$

In order to estimate l_c, one may use the following value for the critical deformation $\gamma_c \approx 10\%$, which is the typical order of magnitude of the critical deformation observed for various pasty materials from rheometrical tests. The expression (7.1) thus provides the critical distance of displacement of the cylinder at which the material falls in its liquid regime along the cylinder surface, with the rest of the material remaining in its solid regime.

Let us now consider an object outside the material and deforming it at some point. Although more relevant and sophisticated theoretical approaches exist in the case of pure elastic solids [1], expression (7.1) may be used to obtain a rough estimate of the critical initial displacement of the central part of the material before penetration.

From a general perspective, the force applied to the object moving through the paste initially at rest can be expected to correspond mainly to the liquid regime if the resulting length of displacement is larger than l_c since in that case the fluid layers at most depths z and in contact with the object undergo a displacement significantly larger than l_c. Otherwise these are the viscoelastic properties of the paste in the solid regime that are observed.

Motion in the Liquid Regime

Now assume that an object has penetrated the fluid of an apparent depth h under the action of a force F. We assume that it has moved through the fluid of a distance $l(r_0) > l_c$, so that the liquid regime has been reached almost everywhere along the object surface. Note that the force to be used in the calculations is dependent on the depth of penetration. Indeed, as the object penetrates the fluid, the buoyancy force increases. The general expression for this force is

$$F = F_0 + mg - \rho g \Omega(h) \tag{7.2}$$

in which the first term, F_0, is the additional (external) force applied to the object, the second term is the total weight of the object, and the last term on the RHS is the buoyancy force resulting from the immersion of a fraction of solid tool of volume $\Omega(h)$, depending on the depth of penetration.

We first consider a cylinder and assume that the length of penetration is much greater than the cylinder radius ($h \gg r_0$) so that edge effects remain negligible. In that case the flow in the fluid around the cylinder is basically a simple shear, and the shear stress along the immersed part of the cylinder surface is equal to the yield stress at stoppage (see Section 5.1.2). The total force resulting from this stress balances the force applied to the cylinder (7.2) [here, with $\Omega(h) = \pi r_0^2 h$] so that

$$\tau_c = \frac{F}{2\pi r_0 h} \tag{7.3}$$

Let us now consider the case of a cone penetrating the paste (see Figure 7.3). The flow is more complex than in the case of the cylinder and it is difficult to estimate the apparent length of penetration beyond which the liquid regime is reached almost everywhere. We can nevertheless have a rough estimate of this value by using the formula (7.1), in which the mean radius of the immersed part of the cone (at the end of the test) is used.

Let us now consider the situation where the liquid regime has been reached before stoppage. The strain field is somewhat complex since the size of the object immersed in the fluid increases in time. This is a transient flow through a yield stress fluid, which cannot be described within the frame of the approximate theory of Section 5.1. A rough approach nevertheless consists in assuming that along the solid surface the flow resembles a simple shear and that the maximum shear stress is reached along this surface. At stoppage it is expected that the yield stress will be reached along an envelope of the type described in Section 5.1, which here should approximately coincide with the cone surface. The integration of this stress along the immersed surface gives the net vertical force exerted on the cone, from which we deduce the expression for the yield stress:

$$\tau_c = \frac{F}{\pi h^2 \tan \theta} \tag{7.4}$$

By neglecting buoyancy effects, the yield stress for a cylinder penetrating the fluid is essentially inversely proportional to the depth of penetration, whereas it

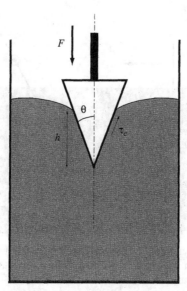

Figure 7.3 Schematic representation of the penetration of a cone penetromer through a paste.

is inversely proportional to the square depth in the case of a cone. In practice, this implies that the experimental results with the cone will appear more robust; some fluctuations of the yield stress value induce smaller variations of the measured depth with the cone than with the cylinder. This is likely the reason for the frequent use of this geometry in practice. However, conversely, for the cone, a small uncertainty in the depth measurement induces a larger uncertainty in the yield stress measurement, which implies that using a cylinder is in principle a more precise technique.

7.1.2 Displacement of an Object Immersed in Fluid

This test consists in measuring the minimum force necessary for displacing an object initially deeply immersed in a paste. Typical shapes used for the object are a sphere [2] or a plate [3]. Since the force is generally exerted by mechanical means, the object is handled by a stem that is also partly immersed and moves through the paste with the object (see Figure 7.4). A force in a given direction or a torque is exerted onto the stem.

For an object completely immersed and initially at rest, equation (7.1), in which one takes r_0 and R for the characteristic length of the object and the size of the container, respectively, also provides a rough estimate of the critical distance of displacement before reaching the liquid regime. When the object has significantly moved through the fluid, the liquid regime can be considered to have been reached and the total force will approximately correspond to the sum of the force exerted onto the object and that exerted on the stem by the fluid. The former term may

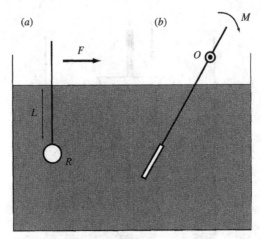

Figure 7.4 Different techniques of displacement of an object through a paste: (a) a sphere attached to a stem submitted to a force; (b) a plate attached to a stem submitted to a torque (relatively to an axis O).

be estimated from (5.11) and (5.12) for a sphere. For other shapes with aspect ratio still of the order of 1, one may get a rough estimate of the drag force by using the same equations, replacing the object by the minimum spherical envelope containing it (of radius R). Note that for a planar surface this approach fails when this surface is parallel to the flow direction. Indeed, in that case the critical surface approximately coincides with the surface of the plate and the yield stress is more precisely obtained from $F \approx S\tau_c$. The only constraint in that case is to avoid wall slip by using a sufficiently rough tool (see Section 3.2.8).

However, in the approaches described above, the drag force on the stem must also be taken into account. Let us, for example, consider the typical case of a cylindrical stem of radius r_0 and length L moving through a yield stress fluid. When the axis of the stem moves in its direction, neglecting edge effects, the drag force (under low velocities) exerted by the fluid is $2\pi r_0 L\tau_c$. When the axis of the stem moves in the plane perpendicular to it, the drag force can roughly be estimated from (5.10) using the value of k_c for the sphere of same radius, which leads to a drag force of $2\pi r_0 L k_c \tau_c$. Thus this drag force is a priori negligible compared to the drag force on the object only if the $r_0 L \ll 2R^2$. Otherwise this additional force must be accounted for to estimate the yield stress.

Finally, the yield stress of a paste can be roughly estimated by determining the critical diameter beyond which a sphere denser than the fluid falls through it under gravity. In that case the weight of the sphere minus the buoyancy force equals the drag force on the sphere. However, in practice, this test is difficult to carry out for two reasons: (1) a generally continuous set of spheres of increasing diameters is required, and (2) the critical velocity at the critical diameter is in theory zero so that at the approach of the critical radius an infinite amount of time is needed to determine whether the sphere moves or remains at rest (see Section 5.1.3).

7.1.3 Fall of an Object under Gravity through a Vibrating Fluid

Here we consider the possible displacement of an object deeply immersed in a fluid that is subjected to vertical vibrations at a frequency N and with an amplitude z_0. The exact solution of the flow problem is again rather complex, but we will assume that the object simply "sees" around it a fluid now undergoing a new pressure distribution resulting solely from vibrations. As long as the fluid remains static (motion "in mass") in the vibrated vessel, the solution of motion equations for the fluid in the absence of the object is a pressure field for the vibrated fluid simply equal to the hydrostatic distribution ($p_a + \rho g z$, where p_a is the atmospheric pressure) for the fluid at rest plus a term due to vibrations, namely, $\rho (2\pi N)^2 z_0 z \sin(2\pi N t)$, so that all proceeds as if the liquid density were replaced by an apparent, variable density ρ^*, defined as $\rho^* = \rho(1 + \Omega \sin(2\pi N t))$, in which $\Omega = (2\pi N)^2 x_0 / g$. Under these conditions the vibrations simply tend to periodically decrease and increase the buoyancy force given by equation (5.2) so that the apparent weight of the object oscillates around its average value. For a Newtonian fluid, since the velocity is proportional to the difference between the weight and the buoyancy force, the consequence of vibrations is that the fall velocity should oscillate around an average value corresponding to the velocity in the absence of vibration. Thus, if N^{-1} is much larger than the timescale on which the velocity is measured, one should record a constant velocity equal to that obtained in the fluid at rest.

For a simple yield stress fluid the resulting flow is more complex, but motion should occur when at any time the apparent weight of the object is sufficiently large, or equivalently when the apparent fluid density is sufficiently low [see equation (3.71)]. In practice, we expect that the object will begin to fall when one increases the frequency or the amplitude beyond some critical value. However, the object velocity just beyond these critical conditions is in principle infinitely small so that its detection is difficult in practice (see Section 5.1). In fact, because of thixotropic effects, one can again observe a bifurcation effect (see Section 5.1.3), which makes it easier to infer some apparent yield stress of the fluid by this technique (see Section 7.6) from the apparent critical frequency.

7.2 TESTS INVOLVING SQUEEZE FLOWS

Tests involving squeeze flows are frequently used in various industries for estimating the rheological characteristics of pastes because they are relatively easy to carry out; the sample simply needs to be set on the lower plate, then the force and the displacement in a given direction must be recorded or controlled, without any further control of the sample or the tool during the test. Most experiments use parallel plates that are moved closer to each other to squeeze the material. Three cases of squeeze flow are (1) simple squeeze flow, (2) imperfect squeeze flow, and (3) band squeezing. Another test is more appropriate for materials containing coarse particles; it consists in simply leaving a sample to "slump" over a horizontal surface under its own weight. This test in fact involves a "pure" squeeze flow only under some specific flow conditions while it involves spreading flow otherwise, and we will analyze it in detail here. Finally there is a test that involves stretching of a paste: the alveograph.

7.2.1 Simple Squeeze Tests

In order to relate in a straightforward way the yield stress to the critical force and fluid thickness, it is preferable to assume one of two extreme cases, elongational or lubricational regimes, respectively associated with a ratio of fluid thickness to sample diameter much larger or much smaller than one. However, elongational flows of this type are not easily obtained with pastes (except in the presence of significant slip at the wall), and, in practice, flows generally correspond to either the purely lubricational regime (Section 5.2.1) or the intermediate regime (Section 5.2.3). Experiments may be carried out under the simple conditions as described in Section 5.2 (see Figure 7.5) with either a constant volume or a constant diameter of fluid. The force–time curve for a constant velocity (or the velocity–time curve under a given force) may be interpreted in rheological terms, but as already stated (see Section 5.2), this is not an easy task. Here we again focus on the interpretation of data in terms of the yield stress of the material. It is worth noting that the following developments are valid for describing flow characteristics only if the resulting deformation (from the initial rest state) is greater than the critical deformation for the sample to reach the liquid regime. As a first approximation, this means that $\Delta b/b$ is larger than the critical deformation in simple shear γ_c.

In order to determine the yield stress of the material, one may impose the force F_0 and measure the corresponding critical thickness $2b_c$ between the plates at stoppage, or conversely impose a relative displacement of the plates and measure the corresponding force for ultimately reaching this distance under slow flow conditions. The volume of the paste involved in the calculations may remain constant ($\Omega_0 = 2\pi R^2 b$), if the plate diameter is larger than the final diameter

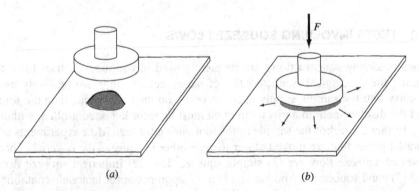

(a)　　　　　　　　　　　　　　*(b)*

Figure 7.5 Schematic representation of principle of the simple squeeze test; a solid tool approaches another solid surface (a) and eventually squeezes the paste lying on it (b).

of the fluid layer. Conversely, the layer diameter ($2R_0$) may be constant if it initially almost occupies the entire surface of the plates so that further squeezing pushes some fluid out of the gap. Using the generalized result for intermediate cases (from which appropriate results in limiting cases are recovered), namely, equation (5.31) with the form for the function f as suggested in Section 5.2.3, we obtain a general expression for the yield stress, under the constant volume test:

$$\tau_c = \frac{\sqrt{3}F_0}{\dfrac{\Omega_0}{b_c}\left(1 + \dfrac{1}{b_c}\sqrt{\dfrac{\Omega_0}{6\pi b_c}}\right)} \tag{7.5}$$

with $\tau_c = F_0\sqrt{18\pi b_c{}^5/\Omega_0{}^3}$ when $b/R \ll 1$, that is, in the lubricational regime. For a constant surface of contact we obtain

$$\tau_c = \frac{\sqrt{3}}{2\pi R_0{}^2}\frac{F_0}{\left(1 + \dfrac{R_0}{\sqrt{3}b}\right)} \tag{7.6}$$

with $\tau_c = 3b_c F_0/2\pi R_0{}^3$ when $b/R_0 \ll 1$.

7.2.2 Imperfect Squeeze Test

In some cases the fluid yield stress is not sufficient for the material to remain in a small heap at rest before squeezing. In order to avoid fluid spreading out of the disk envelope before starting the test, one may use a squeezing geometry with a vertical wall around the periphery of the disks (see Figure 7.6). This geometry has been referred to as the "imperfect" squeeze test [4]. As a first approximation, we can neglect the fluid flow above the disk, when it has exited from the gap between the peripherial wall and the upper disk, and assume that the energy dissipation is due essentially to a squeeze flow under constant diameter

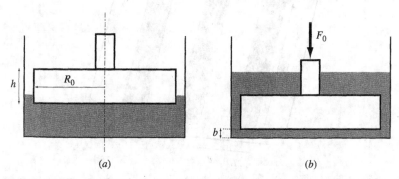

(a)   *(b)*

Figure 7.6 Schematic representation of the imperfect squeeze flow; a disk situated above a volume of paste in a cylindrical container (a) is moved toward the cylinder bottom (b) so as to squeeze the fluid and push it upward.

occurring in the gap between the disks, and to a simple shear flow through the gap between the tool and the outer cylinder. As a consequence, when the flow stops or starts, the required force is the sum of that determined in the previous paragraph under constant diameter and that equal to the yield stress applied along the two solid surfaces between the upper disk and the outer cylinder: $F_0 = (2\pi\,\tau_c R_0^3 f(R/b)/\sqrt{3}b) + (2\pi\,\tau_c R_0 h)$ (in which one may still use $f = 1 + R/b\sqrt{3}$). From this equation we can easily deduce the relationship between the critical distance for which motion stops and the fluid yield stress:

$$\tau_c = \frac{F_0}{(2\pi R_0^3 f(R/b)/\sqrt{3}b_c) + (2\pi R_0 h)} \qquad (7.7)$$

7.2.3 Band Squeezing

Certain squeeze tests mimic squeeze flows occurring during the use of civil engineering pastes. For example, for mortar glues, parallel layers of materials (of initial width $2L_0$ and separated by a distance d) are coated over a planar surface and then crushed by a transparent solid surface (see Figure 7.7) under a given force (generally a mass m lying on the plate). The final shape of the fluid layer, or more precisely the distance remaining between the bands, may be used to characterize the paste. Since the volume of each band is given, we have a relation between the geometric variables of the problem: $\Omega_0 = 4LDb$. Using the approach of Section 5.2.5 within the frame of the lubricational assumption, we can express the relation between the yield stress and the critical distance of lateral spreading of the material (L_c):

$$\tau_c = \frac{4\,mg\Omega_0}{D^2 L_c^3} \qquad (7.8)$$

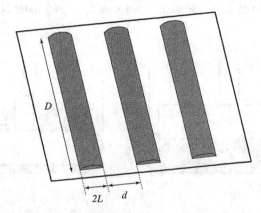

Figure 7.7   Schematic representation of principle of the band squeezing test; the paste bands have spread laterally as a result of their squeezing between the lower and upper (not shown here) plates.

Note that the bands come into contact as soon as the lateral spreading L_c is greater than $L_0 + d/2$, which provides a lower limit for the yield stress [from (7.8)].

7.2.4 Slump Test

This test is used in the food industry for controlling the behavior of corn-style cream, canned pumpkin pie filling, or canned applesauce, for example. In this field it is referred to as the *Adams consistometer* or the "USDA" consistometer. It is also used during fresh concrete preparation for controlling the strength of the material. Here, it is known as the *Abrams cone*. The first step of the test consists in filling with paste, through its upper aperture, an open cone or a cylinder lying on a horizontal solid surface. The cone or the cylinder, which will be referred to as the *container*, is then lifted upward so that the material can spread radially over the flat horizontal surface under the action of gravity (see Figure 7.8). The flow characteristics depend on the material behavior and on the dimensions of the container. Here we examine the conditions under which the resulting flow may be interpreted in rheological terms. The behavior of a paste sample of fixed volume (initially cylindrical) but with decreasing yield stress is then examined.

When the yield stress of the material is sufficient, the material remains solid; the sample is deformed only elastically. At sufficiently low yield stresses we expect that there is a region of the material, more or less close to the ground, for which

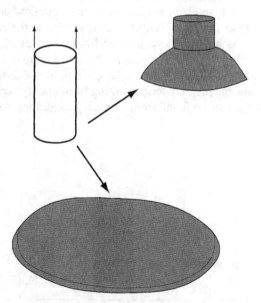

Figure 7.8 Schematic representation of principle of the slump test; the vertical cylinder (white) initially filled with paste is suddenly moved upward, leaving the paste to spread over the horizontal plane. Depending on the yield stress of the paste (see text), the fluid may yield either a shear slump (bottom) or an elongational slump (top).

the yield stress is overcome and the material flows. Two extreme, ideal cases can be distinguished (see Figure 7.8): (1) the sample height remains much larger than its diameter or there is perfect slip along the solid plane, so that we can expect that it mainly undergoes an elongational flow; and (2) the sample height finally becomes much smaller than its diameter at any given height and there is no wall slip, so we can expect that it, at least ultimately, undergoes mainly a shear flow.

"Elongational Slump"

Let us first consider case 1, the flow of a cylinder of paste (of initial height H) lying on a horizontal plane and submitted only to the action of gravity effects, with perfect slip along the bottom wall (see Figure 7.9). The detailed description of the flow is complex since, in contrast with the description in Section 5.2, edge effects are not negligible and the stress distribution depends on the height in the material. Here we will simply consider the equilibrium configuration of an initially cylindrical volume of material that has been submitted to the action of gravity. Qualitatively we can expect the material to flow at depths less than a critical value as a result of a sufficiently large force due to the weight of material located above. In contrast, the material should remain in its solid regime in regions situated above this critical height.

Under these conditions a simple, approximate approach consists in assuming that, in the first region, in the absence of inertia, due to the material situated above and acting on it with the weight $W(z) = \rho g (H - z) \pi R_0^2$, each horizontal layer of thickness dz, initially situated at a height z, is submitted to a simple elongational flow. Such an elementary layer is then squeezed until it reaches the critical radius $R_c(z)$ at stoppage. The second region is, on the contrary, assumed to be undeformed. As a consequence, in the limit of extremely slow flows and in particular immediately before stoppage, the stress tensor assumes the form described in Section 5.2.2, $\Sigma = -p\mathbf{I} + \tau_c \mathbf{D}/\sqrt{-D_{II}}$, in which $\mathbf{D}$ may be derived from (5.26), but now the pressure term resulting from gravity cannot be neglected. The expression for p may be found from boundary conditions, namely, $\Sigma \cdot \mathbf{e}_r = 0$

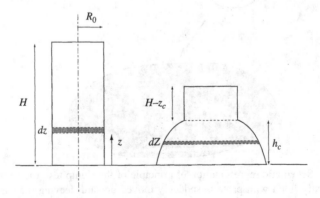

Figure 7.9 Cross section of the sample during an elongational slump test: initial state (left), final state (right).

(negligible pressure along the periphery of the sample), which gives $p = \tau_c/\sqrt{3}$. Using this detailed expression for the stress tensor, the momentum balance written on the volume of sample above the layer initially situated at the height z [i.e., $(-p + \tau_{zz})\pi R_c^2 = -W(z)$] provides

$$R_c = \left(\frac{\rho g(H - z)}{\sqrt{3}\tau_c} \right)^{1/2} R_0 \tag{7.9}$$

It appears that the material remains undeformed above the critical (initial) height z_c for which the weight of material above would lead to $R_c(z_c) = R_0$, so we find

$$z_c = H - \frac{\sqrt{3}\tau_c}{\rho g} \tag{7.10}$$

In order to determine the final fluid height, we now have to consider the deformation experienced by each of the fluid layers initially situated below z_c. From the mass conservation principle, the corresponding final thickness of each deformed layer is $dZ = (R_0/R_c(z))^2 \, dz$. It follows that the final height of deformed material, equal to the sum of these elementary final thicknesses, may be written as

$$h_c = \int_0^{h_c} dZ = \int_0^{z_c} \left(\frac{R_0}{R_c(z)} \right)^2 dz = \frac{\sqrt{3}\tau_c}{\rho g} \ln \left(\frac{H}{H - z_c} \right) \tag{7.11}$$

and the height difference, namely, the "slump," between the initial and final states ($\Delta H = z_c - h_c$) written in dimensionless form (i.e., using $\Delta H^* = \rho g \Delta H/\tau_c$ and $H^* = \rho g H/\tau_c$) is

$$\Delta H^* = H^* - \beta \left(1 + \ln \frac{H^*}{\beta} \right) \tag{7.12a}$$

in which $\beta = \sqrt{3}$. After having measured the initial height H and the slump ΔH, we may determine the yield stress of the material by estimating the value of H^* at the intersection between the curve ΔH^* versus H^* as given by (7.12) and the straight line $\Delta H^* = (\Delta H/H)H^*$. A similar approach may be followed for a truncated cone (top radius R_0, base radius R_H), but in that case the slump height is expressed as a function of the undeformed length, which is found from an implicit relationship as a function of the yield stress. More precisely, this leads to a modified version of equation (7.12a)

$$\Delta H^* = z_c{}^* - \beta \ln \left[\frac{(1 + 1/\alpha)^3 - 1}{(1 + (1 - z_c{}^*)/\alpha)^3 - 1} \right] \tag{7.12b}$$

in which $z_c{}^*$ is the dimensionless height above which the material remains undeformed, which may be found from $\tau_c/\rho g H = (\alpha/3\beta)\lfloor(1 + h^*/\alpha) - (1 + h^*/\alpha)^{-2}\rfloor$, with $h^* = 1 - z_c^*$, and $\alpha = R_0/(R_H - R_0)$.

For materials with such a large yield stress that they do not slump significantly under their own weight, it is possible to add a mass m at the top of the sample [5]. In this case a theoretical approach similar to that presented above but now with a fictitious additional thickness of material $z_0 = m/\rho R_0^2$ above H may be developed from which we find (in the case of a cylindrical sample)

$$\Delta H^* = H^+ - \beta \left(1 + \ln \frac{H^+}{\beta} \right) \qquad (7.13)$$

in which $H^+ = \rho g(H + z_0)/\tau_c$.

Slightly different results have been obtained by other workers for a cylinder [6–8] or for a cone [9,10], although they basically followed the approach outlined above; their expressions for the slump are identical to (7.12a) or (7.12b) but with a factor β equal to 2 instead of $\sqrt{3}$. In fact, they also assumed some kind of elongational flow but used a rough yielding criterion that consists in assuming that flow stops or starts in a layer when the maximum shear stress in the material, which may be demonstrated to be equal to $\sigma_{zz}/2$, reaches the yield stress value, instead of the relation that we could obtain from a proper three-dimensional approach using the Von Mises criterion.

"Shear Slump"

Let us now consider case 2, in which the slump takes the form of a radial spreading of a thin sheet of material over a horizontal surface (see Figure 7.8). Here, the material has in principle completely "forgotten" its initial shape. Under the assumption of perfect adherence on the plane, the flow characteristics are well described within the frame presented in Section 5.3.1. The shape of the final deposit obtained after complete stoppage is thus given by (5.50). From (5.51), we can deduce the yield stress as a function of the final fluid extent ($2R$) over the plane

$$\tau_c = \frac{225\rho g \Omega^2}{128\pi^2 R^5} \qquad (7.14)$$

an expression *a priori* valid when $R/H \gg 1$.

Discussion

In practice, the conditions of motion along the solid surface may be critical for determining the validity of one of the approaches described above. When the fluid widely spreads over the solid surface, wall slip is likely to remain insignificant because the material has not had enough time to develop a layer of different material along the boundary. The flow is transient, leading to partial renewal of the layer of material in contact with the solid surface with time. It is likely that a constant surface of fluid in contact with the wall, as in rheometers, is much more favorable situation for wall slip to develop. Consequently, the conditions of validity of the theoretical treatment of the slump test as a pure shear flow (case 2) are largely fulfilled, and it is reasonable to interpret the results in terms of effective yield stress of the material as soon as the material widely spreads over the horizontal surface. It is nevertheless important to ensure that the assumption of negligible surface tension effects is also valid, which, according to Section 5.3.5, is the case if (5.70) is valid.

When the fluid does not widely spread over the solid surface, the material along the fluid–solid interface does not change much during the flow so that wall slip is more likely to occur, but this effect can seldom be quantified a priori. Unfortunately, in practice, experiments often fall in intermediate cases, namely, R of the order of H, so that none of the preceding asymptotic cases can be considered to apply well. Actually existing data tend to confirm the capacity of equations (7.12) or (7.13) with $\beta = 2$ for roughly estimating the yield stress of the material from such an approach for any type of solid surface and in the limit of large α values [5,8,9]. In fact, numerical simulations [11] accounting for a three-dimensional expression of the constitutive equation show that the effective value of the slump progressively departs from the pure shear solution (associated with large deformation) as the yield stress increases, and finally corresponds closely to the pure elongational solution (with $\beta = \sqrt{3}$) for small slumps (small deformations). This means that these two solutions rather well cover the range of practical possibilities.

7.2.5 Chopin Alveograph

This test consists in inserting air under pressure at a given rate below a layer of paste and following the pressure–time curve up to the critical volume at which the bubble breaks. Different approaches of the corresponding flow have been proposed [12,13] for Newtonian or viscoelastic fluids. Here we propose a very simplified approach specifically devoted to the description of yield stress fluid flows. Bloksma [12] has also considered materials of this type within the frame of other particular hypotheses leading to more complex results.

We assume that the air "bubble" always keeps a semispherical shape (see Figure 7.10) with a radius R increasing in time, and that the paste volume (Ω_0) surrounding this bubble is constant. Moreover, we assume that the thickness of the paste around this bubble is uniform and equal to the difference between the radius R_2 of the outer, semispherical, free surface and the radius R_1 of the inner half-sphere: $2\pi(R_2^3 - R_1^3)/3 = \Omega_0$. In practice, the bubble volume increases as $\Omega = 2\pi R_1^3/3 = Kt$, in which K is the rate of increase.

Let us simply follow the growth of the virtual semispherical surface of radius R_m, which separates the paste into two equal volumes such that $R_m^3 - R_1^3 = 3\Omega_0/4\pi$. We assume that the paste layer undergoes a uniform simple elongation at a rate d associated with the growth of the virtual half-sphere of radius R_m: $d = dR_m/R_m dt$. Under these conditions we deduce from (5.27) the force needed to stretch the material

$$F = \frac{4\pi R_m^2 \tau_c}{\sqrt{3}} \tag{7.15}$$

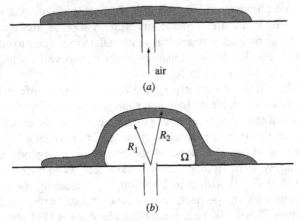

Figure 7.10 Schematic representation of principle of the alveograph; air is blown below the paste lying on a plane (a) so as to form a kind of paste bubble (b).

which is also equal to the force induced by the pressure on the inner semisphere: $F = 2\pi R_1^2 p$. We deduce the expression for the pressure as a function of time:

$$p = \frac{2}{\sqrt{3}} \tau_c \left(1 + \frac{\Omega_0}{2Kt} \right)^{2/3} \tag{7.16}$$

Two flow regimes thus appear: at short times ($t \ll \Omega_0/2K$) the pressure decreases as $t^{-2/3}$, while it tends toward an asymptotic value ($2\tau_c/\sqrt{3}$) over longer times ($t \gg \Omega_0/2K$). There is in fact an additional, initial regime when the paste is in its solid regime. We thus expect first an increase of the pressure due to the elasticity of the paste, then a rapid decrease of the pressure followed by a plateau, in agreement with the two flow regimes described above. Eventually the paste will crack, causing an abrupt, complete, relaxation of the pressure. Unfortunately, within the current frame of knowledge, the material characteristics associated to such a rupture cannot be related to its rheological behavior.

7.3 TESTS INVOLVING SPREADING OR COATING FLOWS

Such flows in particular include gravity spreading flows over planar surfaces, which constitute the simplest tests than can be used for characterizing a material. Indeed, in that case the only tool they require is a solid surface on which the material spreads, and the applied (controlled) force results from gravity effects. The basic test in this field is the natural shear slump test described in Section 7.2.4. A related test is the consistometer, which is simply a channelized slump. Slightly more sophisticated are the inclined plane test and the Ericksen gauge.

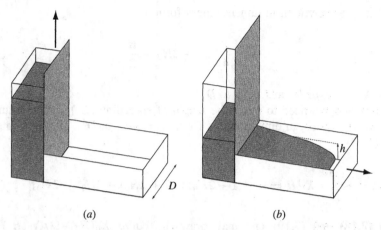

Figure 7.11 Schematic representation of principle of the consistometer; the box behind the gate is filled with the paste (a) and the gate is suddenly lifted so that the paste spreads in the channel (b).

7.3.1 Consistometer

The Bostwick consistometer, used in the food industry, consists of a stainless-steel level that is rectangular in cross section, containing two compartments (see Figure 7.11). The first compartment, which is initially filled with material, is separated from the second one by a gate, which is suddenly opened, leaving the material spreading along the channel under the action of gravity. In some cases the channel is slightly inclined. A similar consistometer is used for controlling the properties of fresh concrete: this is the "L box," which differs from the consistometer because grids may be added in the channel to simulate strands in real civil engineering structures.

Let us consider such a flow for a yield stress fluid without slip at the wall. We assume that the flow can be described within the frame of the lubrication approximation. The principles of the approach used in Section 5.3.1 are still valid; thus, the profile of the free surface may be found from the momentum balance, but now with an additional resistance to flow due to lateral wall effects. Here we assume that this lateral shear stress after flow stoppage is roughly equal to the yield stress, which was shown to be in agreement with various flow data with mud suspensions [14] for $h/D \leq 0.4$. Under these conditions the shape of the free surface after stoppage in an inclined channel is again described by an equation of the type (5.46) (in the same system of coordinates), where the second term is replaced by $\varepsilon \tau_c (D + 2h) dx$. For a horizontal channel, we obtain after integration

$$-\frac{\tau_c}{\rho g} x = \frac{Dh}{2} - \frac{D^2}{4} \ln \left(1 + \frac{2h}{D} \right) \qquad (7.17)$$

which may be rewritten in dimensionless form as

$$X = \frac{1}{4} \ln(1 + 2H) - \frac{H}{2} \qquad (7.18)$$

where $X = \tau_c x / \rho g D^2$ and $H = h/D$.

In practice, we need to relate the length of spreading (X_0) to the volume of material (ω), which, in dimensionless form ($\Omega = \omega \tau_c / \rho g D^4$), may be computed from (7.18):

$$\Omega = - \int_0^{H_0} X \, dH = -\frac{1}{8}[(1 + 2H_0) \ln(1 + 2H_0) - 2H_0^2 - 2H_0] \qquad (7.19)$$

From (7.18) and (7.19) one may draw the curve $X_0(H_0)-\Omega/X_0(H_0)$ (see Figure 7.12). In practice, when ω and D are known, measurement of the length of spreading l gives the value of $\Omega/X_0 = \omega/lD^2$, which, in the curve of Figure 7.12, is associated with a particular value X_0 from which we may obtain the yield stress value:

$$\tau_c = \frac{\rho g D^2 X_0}{l} \qquad (7.20)$$

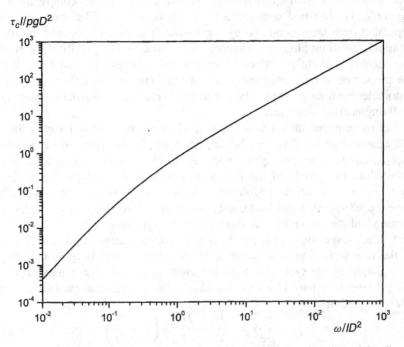

Figure 7.12 Theoretical dimensionless length of the sample as a function of the dimensionless volume at the end of a test with a consistometer.

7.3.2 Inclined Plane Test

The inclined plane test consists in leaving a certain amount of paste spread over an inclined plane under the action of gravity and analyzing the fluid depth profile in rheological terms. In contrast with the consistometer, edge effects resulting from side, downward, or upward walls are suppressed. Different procedures may be used depending on how the plane is coated and inclined [14–16]. An initial procedure consists in coating a uniform layer (of thickness h) of material over the horizontal plane and then inclining it until the critical slope (i_c) for which flow begins. In that case, under the assumption of negligible edge effects, the yield stress is equal to

$$\tau_c = \rho g h \sin i_c \tag{7.21}$$

However, this technique, which involves the observation of flow start, rarely provides reproducible results. Indeed, the detection of flow start on approach of the critical angle is difficult because it, in theory, requires a time of observation of a fixed displacement tending toward infinity (since the velocity tends toward zero when the slope angle tends toward i_c). Moreover, thixotropic effects, particularly the material restructuring, may play a significant role during the time required to reach this critical slope. Finally, this technique necessitates following exactly the same procedure before reaching each new slope angle, which implies an amount of testing very similar to that seen for systematic creep tests in conventional rheometry.

A more straightforward technique consists in gently pouring the fluid over a plane of fixed slope i and then either waiting for complete stoppage or further inclining the slope angle of the plane. The material finally forms a deposit (Figure 5.12) of approximately ellipsoidal shape, the lateral extent of which depends on the fluid behavior and on the pouring rate. The shape of the deposit in the longitudinal direction may nevertheless be fitted by a curve of the type (5.48), from which one may deduce the yield stress. More interestingly, if $h_0 \ll L$, in which L is the fluid extent in the direction of steepest slope the asymptotic fluid depth in the central region of the material should approximately correspond to the critical depth associated with the yield stress at this slope:

$$\tau_c = \rho g h_c \sin i \tag{7.22}$$

These theoretical interpretations are valid only if surface tension effects are negligible, that is, according to (5.70), if $\tau_c \gg 6(1 - \cos\theta)\gamma_{LG}/L$. In practice, since L is typically larger than 10 cm, this means that the yield stress must be much greater than 1 Pa.

Despite some advantages, the inclined plane test is not a test commonly used in industry. However, it is used for environmental fluids such as materials composing debris or mudflows. One may indeed analyze the shape of natural deposits of such fluids resulting from stoppage of free surface flows to estimate their apparent yield stress [17,18]. Here the deposits have been formed under uncontrolled conditions, but it is expected that the shape of the front of a deposit over a constant slope is given by (5.48). Moreover, the asymptotic fluid thickness far from the front should be equal to the critical thickness, which provides the yield stress value. This approach is in principle strictly valid only for a layer of yield stress fluid of wide extent in longitudinal and lateral directions relative to its thickness and lying over a uniform slope, in the absence of edge effects and inertia effects, which is particularly true in the ultimate stages of the flow. Since these conditions are rarely fulfilled in nature, determination of the yield stress via the theoretical approach of Section 5.3 is only approximate. However, this technique provides a simple means for roughly estimating, under the continuum assumption, the very large yield stress of these materials (typically of the order of 1000 Pa), which contain coarse particles, which would be virtually impossible with existing instruments appropriate for much smaller volumes.

7.3.3 Ericksen Gauge

For paints, a particular test of spreading over a solid surface is used. It consists in coating a layer of fluid over a planar surface and then brushing it with a crenelated plate in contact with the plane so as to form parallel bands of initial rectangular cross sections but with different initial thicknesses. Then the fluid spreads more or less under the action of gravity so that the bands may come into contact (see Figure 7.13). In practice, the band associated with the critical thickness for which this occurs is used as a characteristic of the paint. Another similar test consists in inclining the solid surface immediately following formation of the bands and then observing the vertical spreading under gravity. The band width is of the order of one centimeter, while the band thickness is of the order of several millimeters. It follows (see Section 5.3.5) that surface tension effects play a role as soon as the yield stress is of the order or smaller than $100\gamma_{LG}$, which typically assumes

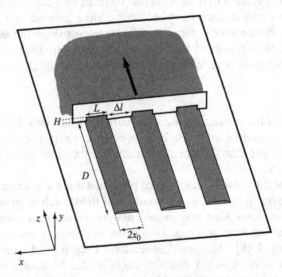

Figure 7.13 Schematic representation of principle of the Ericksen gauge during use.

values larger than a few pascals under the usual wetting conditions. Since this corresponds to the typical order of magnitude of yield stress values of paints, we can conclude that surface tension effects generally play a significant role in these tests. Thus, unless a value much greater than 1 Pa is assumed, it is not the yield stress of paint but the balance of gravity and wetting effects that mainly determine the final shape of the paint layers.

Here we consider the specific case for which wetting effects are negligible. We assume that the bands of fluid form as a whole just after the (instantaneous) motion of the crenelated plate. Thus one can consider the simple spreading of parallel bands over a horizontal plane under gravity. In this frame, due to the symmetry of the problem, no flow in the longitudinal direction (z) is expected. Thus the fluid spreads in the lateral direction under gravity, and the final shape of the cross section (in a plane (x,y)) of a given band is as given by (5.49). For a layer of half-width x_0, the volume of material is thus $\sqrt{8\tau_c/9\rho g}\,x_0^{3/2}$. Taking L and H respectively for the initial width and thickness of the band, we deduce $x_0 = (LH)^{2/3}(9\rho g/8\tau_c)^{1/3}$. Now, assuming that we have a set of parallel bands of fixed width but increasing initial thickness separated by an initial distance Δl, we find from these calculations the critical thickness beyond which the bands begin to join after spreading, $x_0(L_c) = (L_c + \Delta l)/2$, which gives an expression for the yield stress of the material:

$$\tau_c = \frac{9\rho g(L_c H)^2}{((L_c + \Delta l)/2)^3} \tag{7.23}$$

7.3.4 Atterberg Limits

The "Atterberg limits" are commonly used for characterizing the fine fraction of materials composing natural soils. They correspond to critical water contents for which some specific, mechanical criteria are fulfilled during standard empirical tests with pastes obtained by mixing the fine solid fraction (diameter <40 μm) of a soil with various amounts of water [19]. The *plasticity limit* is the critical water content below which a piece of pasty material fractures immediately when it is rolled with hands in the form of a thread with a diameter of about ~3 mm. The *liquidity limit* is the critical water content beyond which the two parts of a small amount of paste lying in a cup and separated into two regions tend to join after a given number of shocks.

These two tests are thus based on typical rheological characteristics of pasty materials, namely, fracture or flow beyond a critical stress. For the liquid limit test, the material, which is a yield stress fluid, forms a deposit in the cup that does not flow significantly after it has been separated into two parts. The lateral shapes of the two layers so formed can as a first approximation be described within the frame of the lubrication assumption, and we expect a shape close to that predicted by equation (5.49) if the cup bottom is not too curved. When it is submitted to a shock, due to the inertia resulting from the impact, the apparent gravity increases during a short time. Under this larger apparent gravity the predicted thickness of the layer is smaller [see (5.50)], which leads to some further spreading of the fluid. A certain number of such shocks can obviously lead to a sufficient spreading of the two parts, which eventually join. However, it is difficult to relate this spreading more precisely to the rheological properties of the fluid because we would need to assume some specific rheological behavior and know the duration of the shock and the change in apparent gravity during the shock, which depend on the mechanical system.

7.4   TESTS INVOLVING CONDUIT FLOWS

In civil engineering for paints, cement pastes, or drilling fluids (Marsh cone), or in the food industry for relatively liquid preparations, a test is used that consists in allowing a material initially filling a container to flow under gravity through a bottom aperture, possibly followed by a short capillary. The time required for the material to drain completely is measured and considered as a mechanical characteristic of the material ("flowability"), but in general without further rheological interpretation. In parallel, similar flows of more viscous pastes are induced during extrusion processes commonly used in ceramic and food industries; in that case the fluid is pushed from a large cylinder through a small aperture. These two flow types differ only because the driving force in the former case is gravity while it is the additional force that plays this role in the latter case. In both cases the flow rate results from the balance between the driving force and viscous effects. Again the detailed flow characteristics cannot be easily determined analytically from a complete knowledge of the rheological characteristics of the material, and, in view of an approximate rheometrical analysis, we will only consider slow flows where the viscous dissipation results mainly from the yielding term of the constitutive equation (2.20).

Let us consider a cylindrical container of radius R ending (at its bottom) by a cylinder of radius r_0 (see Figure 7.14). We consider either a situation for which at the height h above the cylinder bottom there is a free surface and the stress is equal to the atmospheric pressure ($p = p_0$), or a situation for which at the height h an additional pressure ($p = p_1$) much greater than the atmospheric pressure is applied. The flow field is complex; in particular, we expect both some simple shear flow far from the orifice (upward), as in a capillary flow without edge effects (see Section 1.4.5), and some kind of coupled shear and elongation flow around the orifice, because as it flows, the cross section of the paste rapidly increases from a large to a small value.

Let us compare the orders of magnitude of the energy dissipations associated with a simple shear and an elongation of a cylindrical fluid portion of length L for an analogous average displacement dx during a time interval dt. For a yield stress fluid in slow simple shear through a capillary, we can assume that the shear rate is approximately constant in a region of thickness ε along the cylinder while the rest of the material is unsheared, so that we have $\dot{\gamma} \approx dx/\varepsilon\,dt$. For such a slow flow the shear stress in the sheared region can be approximated by

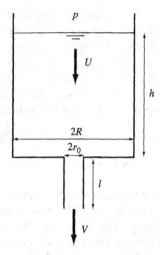

Figure 7.14 Schematic representation of principle of conduit flow.

the yield stress. It follows that the energy dissipation [as given by (1.19)], which is found by integrating the product $\tau\dot{\gamma}$ over the deformed volume, that is, the fluid layer of thickness ε, during this elementary motion is

$$W_S \approx \pi R L \tau_c dx \tag{7.24}$$

For the elongational flow the rate of deformation d is approximately equal to $dx/L\,dt$ and the normal stress along direction x is everywhere $2\tau_c/\sqrt{3}$ so that the energy dissipation in the deformed volume, which is here equal to the whole volume of material, is

$$W_E \approx \pi R^2 \tau_c dx \tag{7.25}$$

In fact, the effective flow that takes place at the exit of the cylindrical volume is more complex. As a first approximation, we can consider that this is a nonuniform elongational flow, where the radius of the elongated sample varies spatially. Let us compute the energy dissipated by a fluid portion still of length L but now of radius R', and advancing as a result of the same flow discharge as above, leading to an average displacement of the fluid of dx in the conduit of radius R. This implies that the displacement of this fluid portion is dx', such that $R^2 dx = R'^2 dx'$. From the approach described above, the energy dissipation is now equal to $\pi R'^2 \tau_c \, dx'$, which is identical to (7.25). This means that the energy dissipation in different fluid portions in the converging flow should be more or less similar and can be approximated by (7.25). Since, in addition, we can reasonably expect that L, the distance over which the flow is perturbed by the orifice, is of the order of R, the energy dissipations assuming either a converging elongational flow (W_E) or a shear flow (W_S) in a straight conduit are of the same order of magnitude.

On the basis of this result it now seems reasonable to assume, for the sake of simplicity, that the flow can be approximated by a uniform, simple shear flow in both conduits, which is equivalent to neglecting the perturbation induced by the presence of the reduction in cross-section on the upward and downward flow. Under these conditions let us apply the momentum balance over the fluid volume, assuming a hydrostatic pressure distribution, with negligible transient effects and a uniform velocity in the upstream (U) and downstream (V) conduits:

$$\rho V^2 \pi r_0^2 - \rho U^2 \pi R^2 = [p\pi R^2 - p\pi(R^2 - r_0^2) - \rho g h \pi (R^2 - r_0^2)]$$
$$+ \rho g \pi (h R^2 + l r_0^2) - 2\pi R h \tau_{p1} - 2\pi \tau_{p2} l r_0 \qquad (7.26)$$

in which τ_{p1} and τ_{p2} are wall shear stress respectively along the upper and outer cylinders. The first term on the RHS corresponds to the pressure difference; the second one, to the gravity force; and the last one, to the wall stress resisting motion. The LHS is the difference of inertia terms. According to the conservation of mass principle, we have $\pi R^2 U = \pi r_0^2 V$, which implies that $U^2 R^2 = (r_0/R)^2 V^2 r_0^2$, so that the second term on the LHS of (7.26) is negligible if $r_0^2 \ll R^2$. In the limit of slow flows τ_{p1} and τ_{p2} are close to τ_c and, since in general $Rh \gg lr_0$, the last term of (7.26) can be neglected. We finally deduce the expression for the average velocity:

$$V = \left(\frac{p}{\rho} + g(h + l) - \frac{2\tau_c Rh}{\rho r_0^2} \right)^{1/2} \qquad (7.27)$$

The condition for stoppage or incipient motion ($V \approx 0$) gives an expression for the yield stress of the paste as a function of the critical height of fluid in the cylinder (h_c) and the applied pressure:

$$\tau_c = \frac{(p + \rho g(h_c + l))r_0^2}{2Rh_c} \qquad (7.28)$$

This approach is valid as long as the wall shear stress in the small conduit is, despite flow, not much greater than the yield stress; otherwise the last term of equation (7.26) would no longer be negligible. In fact, it is often assumed that wall slip occurs in the downward conduit, with very concentrated suspensions generally extruded. In that case τ_{p2} depends on the flow rate (see Section 3.2) but is generally smaller than τ_c, so our approximation remains valid. Finally, we note that for sufficiently large flow rates $\tau_{p2}lr_0$ is no longer negligible and depends on V, and the solution of (7.26) depends on the form of the function F in (2.20).

7.5 PRACTICAL TESTS FOR GRANULAR MATERIALS

Currently, there is no conceptual framework for describing granular flows from a simple set of "rheological" parameters (see Section 2.3). However, most simple granular flows in the frictional regime can be described (see Chapter 6), albeit sometimes approximately, with the help of the Coulomb model. Consequently, practical tests with granular materials should attempt to determine the friction coefficient. Different practical tools exist that are used specifically to determine the friction coefficient of granular materials, but none of them provides the necessary ideal conditions for measurement. Indeed, the friction coefficient somewhat depends on the exact configuration of grains in the material, which in turn depends on the procedure and the geometry of the setup. The conditions under which the results provide the value of the friction coefficient of the material in its critical state (see Section 2.3) is strongly dependent on how each experiment is carried out.

7.5.1 Friction Coefficient from Quasistatic Flow Regime

The *Jenike cell* consists of two open containers in front of each other and filled with material (see Figure 7.15). A vertical load can be applied to them, and the force needed to induce a relative tangential motion is recorded. The ratio of the tangential to the normal force provides the coefficient of friction. This technique thus mimics the shear localization that occurs along the layer for which the Coulomb criterion is reached in a granular material. Its disadvantage is that the thickness of this layer is here imposed by the geometric characteristics of the setup.

The *shear box* consists of a rectangular cell with hinged sides and a freely moving lid (see Figure 7.16). A load is placed on the lid and the force required to induce shear is recorded, as is the vertical displacement of the lid, which gives a measure of the dilation of the material. Thus the shear is controlled, which has the advantage of avoiding shear localization occurring with other geometries at low apparent shear rates, but the trade off is a partly uncontrollable stress distribution due to the relative motion of the material and the lateral boundaries. Moreover, this technique, which artificially avoids shear localization, is not necessarily appropriate for representing the behavior of granular materials under other flow conditions.

The Jenike cell and the shear box can provide interesting and straightforward information concerning the coefficient of friction of the material, but the *triaxial*

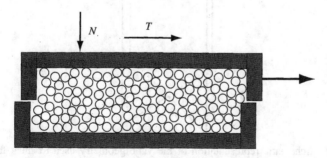

Figure 7.15 Schematic representation of principle of the Jenike cell.

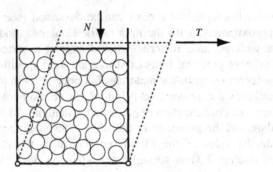

Figure 7.16 Schematic representation of principle of the shear cell.

test (see Section 6.4) is often preferred in practice as it enables one to obtain reproducible results under controlled conditions. Indeed, the density and normal stress components can be followed in time under different pressure conditions, which provide a complete set of information concerning the rheological behavior of the material. Its main disadvantage is that it specifically concerns compressive flows from which the rheological properties of the material under shear conditions have to be extrapolated.

The *scissometer* used in soil mechanics may be used for granular materials. It simply consists of a vane plunged into the material and to which an increasing torque is applied (see Figure 7.17). When a failure occurs in the material, the vane starts to rotate, generally as the torque reaches its maximum value, which is

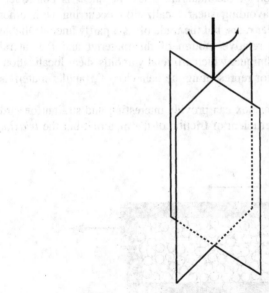

Figure 7.17 Schematic representation of the vane geometry used to shear the material in the scissometer.

considered as a characteristic of the material. In general, the failure corresponds to a shear localization along the virtual cylindrical envelope of the vane, which is the surface of cylindrical revolution along which the shear stress is maximum (due to momentum conservation). An approach similar to that for a Couette flow (see Section 6.1.1) could thus a priori be used since the shear distribution is analogous. However, if for a continuous flow we could invoke some normal stress relaxation leading to approximation of the radial normal stress by the hydrostatic pressure (see Section 6.1), at the initiation of flow the validity of this assumption would be doubtful because stress relaxation generally would not have time to occur. If one nevertheless assumes that stress relaxation precisely occurs just after the failure, the approach of Section 6.1.1 may be used to determine the friction coefficient.

Free surface flows may also be used to obtain some (less complete) information concerning the coefficient of friction of the material. For this purpose it is sufficient to pour a certain volume (much greater than the volume of one grain) over a horizontal surface. The slope of the conical heap thus obtained approximately corresponds to the angle of friction. Note that since this angle may vary (see Section 6.2.1) with the procedure used to obtain the heap, it is preferable in practice to systematically use the same procedure for different materials that are to be compared. The angle of friction may also be determined with the help of a particular instrument; the material is placed in a rotating drum (see Figure 7.18), and then, if the rotation rate is not too high, the average slope of the free surface of the material will correspond to the angle of friction [20].

To conclude, a setup used in practice for characterizing granular materials and especially sands is the "L box," which is similar in principle to the consistometer for pastes (Section 7.3). In that case, since this a granular flow over a horizontal solid surface, we expect the material to ultimately form a heap with a free surface inclined with an angle equal to the maximum angle of friction. However, the extent of material, which is often the typical data recorded, generally cannot be related to the coefficient of friction, because it depends on the properties of the (complex) flow through the bottom aperture under the influence of the material amount in the column.

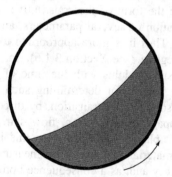

Figure 7.18 Schematic representation of principle of the rotating drum.

7.5.2 Flows in Frictional Regime

The characteristics of free surface flows, that is, the flow rates under different slopes and thicknesses, may be predicted (Section 6.2.3) from the preliminary determination of h_{stop} (i). Thus it is, for example, possible to induce a series of spreadings of large longitudinal extent and finishing with stoppage over different plane slopes. The final thickness far from the front (in the approximately uniform region) provides the value of h_{stop} (i). Let us recall that this value depends on both the material properties and the interaction between the wall and the material. The flow front after stoppage of the free surface flow of a granular material may also be used for determining the different parameters defining the curve h_{stop} (i) in (6.3).

Usual flow tests consist in leaving a given amount of material flowing through a hopper with an aperture at its bottom (as described in Section 6.3) and measuring the time for complete emptying, which is retained as a characteristic of the material. This corresponds to a complex flow whose characteristics were discussed in Section 6.3. Thus, using these results, the average discharge should in principle be related to material behavior. However, the interaction between the walls and the material may also play a significant role, so that such an interpretation is not straightforward.

Another test consists in vibrating the granular material and observing its density changes. The result depends on several material characteristics such as the difference between its initial density after preparation and its maximum packing fraction, the coefficient of friction between the grains, and the angularity of particles. There is no clear relationship between the results of this test and some rheological parameters of the materials.

7.6 STUDY OF THIXOTROPY USING PRACTICAL RHEOMETRICAL TESTS

We have seen in Section 2.2.4 that the effects of thixotropy can likely be observed in either the solid or liquid regime. The use of practical (nonviscometric) tests for characterizing the thixotropy in the liquid regime is nevertheless problematic. Indeed, in that case the form of the constitutive equation is unknown but should involve determination of several parameters, and the tests involve complex uncontrolled flows. Thus it is more appropriate to study the thixotropy of materials in the solid regime (see Section 3.1.5), for example, by measuring the evolution of the elastic modulus with the time of rest. However, practical tests are not sufficiently precise for determining such a parameter. In general, these tests focus on the solid–liquid transition, by directly determining the critical stress for flow stoppage or, less often, flow start (see previous sections). When thixotropic effects play a significant role this is more likely an apparent yield stress of the material which can be measured. However, this stress depends on the flow history and, as a consequence both on the rest time preceding the test and the procedure (and in particular the timing for imposing different stress values) used for determining it. In simple shear the basic approach consists in observing the evolution of the apparent yield stress [τ_c (λ) as defined by (2.25)] as a function of the time of rest between the preshear and the time at which a given stress is again applied [21] (see Section 2.2.4). In that case the flow history is well controlled so that the evolution of the apparent yield stress reflects a property closely correlated to a simple rheological model. For practical tests, the flow history may not be well controlled, so it is necessary to review the most appropriate experimental procedures for determining an apparent yield stress, the evolutions of which would reflect at best those observed under simple, controlled conditions. The practical tests described above generally involve determination of the yield stress either for initiation of flow under an approximately controlled stress ramp or for cessation of flow under an uncontrolled stress ramp. In the following sections we review the most appropriate procedures for the main practical rheometrical techniques, starting by examining in further details the response to simple shear creep tests of a paste after a sufficiently long time of rest.

Creep Tests in Simple Shear after Rest

Here we consider the ideal case of a material presheared and then left at rest during a certain time and submitted to a given stress level. The response of the material to different stress levels exhibits (see Figure 7.19) the viscosity bifurcation described in Section 3.3.1; thus, for a stress above a critical value, the material flows at a relatively high rate whereas for a stress below this value it flows much more slowly and then stops flowing altogether. However, it is worth noting that when the material has been left at rest for a sufficiently long time, it barely begins to flow if the applied stress is below the critical value, whereas initially it does not flow then rather abruptly starts to flow after a certain time when the applied stress exceeds a critical value (see Figure 7.19). This is a kind of collapse that occurs at a critical stress, which obviously increases with the time of rest. This abrupt increase in the apparent flow rate provides a clear practical criterion for estimating the apparent yield stress since one should observe a marked difference in the apparent flow for two different stress values (above and below the critical value). It is also worth noting that the time at which this collapse occurs increases as the applied stress approaches the apparent yield stress (see Figure 7.19), a trend that, in contrast, tends to make precise measurements difficult since in principle one should wait for an infinite time for being sure to have overcome the apparent yield stress of an infinitely small value. Note that this delayed collapse is accurately predicted by the basic thixotropy model described in Section 2.2.4 (see Figure 7.20), confirming its capacity to predict fundamental qualitative trends of the flow of thixotropic fluids.

Let us now examine the response of such a material under a stress ramp. We have already examined this situation in Section 3.1.5; after an initially viscoelastic response, there is a plateau in the apparent flow curve, the level of which increases with the time of preliminary rest (Figure 3.14). When the stress is then

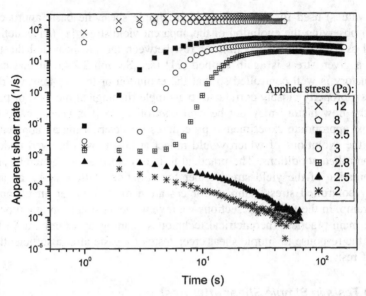

Figure 7.19   Creep tests with a bentonite suspension after a rest period of 150 s following preshear.

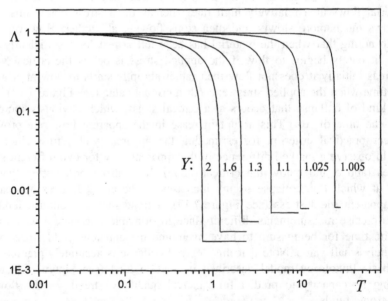

Figure 7.20   Dynamic characteristics of the destructuration of a thixotropic fluid following the model (2.24) with $\eta = \eta_0(1 + \lambda^n)$ (see Section 2.2.4) under creep tests for an initial structure state value λ_0: "dimensionless" structure parameter $\Lambda = \lambda/\lambda_0$ as a function of the dimensionless time $T = t/\theta\lambda_0$ for different values of $Y = \tau/\tau_c(\lambda_0)$ in the case $n = 1.2$ and $\lambda_0 = 100$. (From Ref. 22, Figure 2.)

decreased, the apparent flow curve falls below the curve in the increasing stage (see Figure 3.14). Considering the behavior for creep tests under fixed values (see text above), we expect the stress at the plateau level to roughly correspond to the critical stress value at which an abrupt increase in flow rate should occur. This measurement is more precise as the duration of the stress ramp (for the same stress range) increases; however, this interval must remain much shorter than the characteristic time for significant restructuring; otherwise the recorded plateau level will be affected by the time factor. For its part, the apparent yield stress observed in the flow curve for a decreasing ramp can be seriously affected by the flow history so as to nearly fail to reflect the initial state of structure associated with a preliminary rest.

From these considerations in simple shear, we deduce the basic rules to follow for estimating a relevant apparent yield stress from practical rheometry:

- Tests under decreasing stress, leading to a measurement of a critical force for flow stoppage, are not appropriate because this force is in general strongly affected by the flow until stoppage (the material restructuring reached during the rest time is partly lost during this flow).

- Tests under increasing stress are appropriate but their total duration must be sufficiently long for precise determination of the apparent yield stress and sufficiently short compared to the time of preliminary rest.

Test Involving Displacement of an Object

This type of test may be used for determining the apparent yield stress of a paste after some rest; one can increase progressively the force exerted on the object and thus determine the critical value at which the object seems to abruptly begin to move significantly. It is also possible to use the vibration test (see Section 7.1.3); the critical frequency for apparent motion should increase with the time of rest.

Tests Involving Squeeze Flow

The squeeze flow test under a given load is not appropriate; the material initially flows and then stops flowing for a critical stress distribution that depends on the flow that it has undergone during this stage and also depends, in an indeterminate manner, on the material's initial state. This is the same problem encountered with the slump test. For squeeze (or stretch) flows under a variable force applied to one of the plates, the absolute values of the stress in the fluid vary with the squeezed distance, so it is in general extremely difficult to relate the evolution of the velocity to some critical strength.

Tests Involving Spreading Flow

Spreading flow tests involving measurement of a critical slope or thickness for fluid stoppage over a solid surface are inappropriate since the material undergoes a uncontrolled flow history, which tends to destructure it in an indeterminate way. It is nevertheless possible to study the thixotropic character of the fluid by measuring the critical slope for flow start of a uniform fluid layer left after different times at rest over a plane [22].

Finally, it is also possible to measure the initial maximum pressure for incipient motion of the fluid left at rest in a conduit (see Section 7.4).

7.7 APPLICATIONS TO INDUSTRIAL AND ENVIRONMENTAL MATERIALS

Here we review the main types of natural or industrial materials falling in the range of pasty, slurry, or granular materials. We examine their components and how their mechanical behavior fits in one of the main classes discussed previously. Then, assuming that usual laboratory rheometry can always be carried out, we discuss the most appropriate practical rheometrical means for determining their basic rheological characteristics. We obviously cannot pretend to be exhaustive on either the material types and characteristics or the practical techniques. We simply wish to provide some markers for a scientific approach of these problems, relying on the developments of this book.

7.7.1 Foodstuffs

In order to predict their flow properties in industrial processes, it may be necessary to know the rheological behavior of foodstuffs from which various physical properties are inferred for control or prediction purposes. During mastication several characteristics may be appreciated by consumers via the mouth, tongue, or nose, when the food is mixed or dispersed and comes in contact with saliva. In this context, different characteristics such as "crispy," "juicy," "soft," "creamy," "crunchy," "chewy," and similar adjectives have been used to describe the way the food reacts when submitted to various constraints in the mouth. The generic term for qualifying these properties is *texture* [1]. This texture may be (only) partly appreciated from the viscosity of the material, but the great advantage is that the rheological parameters of a material in principle constitute objective characteristics.

Foodstuffs can be concentrated emulsions (butter, salad dressing, mayonnaise, etc.), suspensions (soups, marmalades, compotes, jams, etc.), foams (whipped cream, chocolate foams, etc.), or intermediate materials that are difficult to classify as either a gel, a suspension, or an emulsion because they, for example, contain small solid particles, large polymeric chains and a certain fraction of oil or water droplets embedded in another liquid phase (yogurt, ketchup, mustard, peanut butter, puree, jelly, etc.). In most of these materials, the solid elements are either sufficiently dispersed or soft or flexible so that hard interactions remain negligible. Thus, most foodstuffs to be consumed fall in the "paste" range. These fluids clearly exhibit a yield stress that render them capable of stopping flow under the usual conditions: in the spoon, in the container, and so on. The practical means for determining the yield stress and for studying thixotropy are all the nonviscometric flows described in this chapters except perhaps those involving spreading or coating flows, which require either sufficiently low yield stress or large fluid volumes. Note that many foodstuffs, which are mixtures of different liquid and solid phases, can easily develop interfacial problems such as wall slip.

There are also various pastes used during industrial processes, for various preparations such as biscuits or cakes, which then solidify or are cooked. These pastes can contain a significant fraction of coarse solid particles interacting via direct contacts. A critical example in this field is chocolate, which is made of fat, sugar, milk, cocoa, and some additives. Each type of particle may develop significant colloidal interactions below a critical size. When the volume fraction of the solid phase (sugar, cocoa, and milk) is sufficiently large, both colloidal interactions and direct contacts can become significant. Depending on the relative importance of these different interactions, the chocolate behavior can be either of the paste or granular paste type.

Finally, there are also various types of foodstuff powders, most of them eaten in small amounts mixed with other materials, such as salt, sugar, cereals, rice, and pepper. Unless they are moistened, these materials are of the granular type and may be characterized using techniques described in Section 7.5.

7.7.2 Cosmetic and Pharmaceutical Materials

Various cosmetic or pharmaceutical products such as gels, creams (emulsions), varnishes, lipsticks, and foams are applied onto the skin surface, through which they are expected to penetrate or to remain at rest with a certain thickness over a period of time. These materials must have a sufficiently high apparent viscosity at low flow rate to remain in place on the skin, but they must offer a moderate resistance to flow during the application or for penetrating the skin. For these applications, such materials are obviously basically fine soft jammed (pasty) materials, the yield stress of which may be characterized with the help of various techniques described in this chapter except for spreading or coating flows, which require large fluid volumes.

In the pharmaceutical industry many dry powders are also used, with different particle sizes. These granular materials may be characterized by employing the techniques described in Section 7.5.

7.7.3 Environmental Materials

Lavas

Lavas are made mostly of liquid magma, bubbles, and crystals. As the fluid flows out of the vent, thermal exchanges—mainly via radiations—occur so that both the liquid viscosity and the volume fraction of crystals increase [23]. Two limiting scenarios may thus be encountered. Generally for basaltic lavas, the crystallization temperature of which is relatively low, the material eventually stops flowing mainly because the liquid viscosity tends toward infinity, in particular along the interface with air. The aspect of the corresponding deposits (*pahoehoe*) is often smooth, ropelike, wrinkled, or scaly because during the final stage of the flow, the flow takes place in the form of successive slow bursts of lavas at different places. There remains some controversy as to whether these materials exhibit a yield stress, which would be due to the formation of a crystal framework of anisotropic crystals [24,25]. On the contrary, lavas with a larger silica fraction seem to tend to stop flowing as the volume fraction of crystals tends toward the maximum packing fraction. Here, it is likely that we are dealing with granular pastes. The aspect of the corresponding deposits (*aa*) is rough, and in some cases the free surface is composed mainly of large grains or boulders.

Since it is very difficult to take a field sample and maintains its characteristics and parameters (in particular crystal and gas bubble contents) constant until a rheometrical test can be performed in the laboratory, there have been some attempts to determine directly the rheological properties of lavas with using specific instruments inserted in the flowing fluid in situ [26] (per the penetrometer principle (see Section 7.1)). However, this remains extremely delicate, and an alternate way consists in directly interpreting the deposits in terms of yield stress (see Section 7.3.2) or coefficient of friction (see Section 7.5.1) for *aa* flows. For pahoehoe flows, although the effective yield stress existence is unclear, it is likely that determining an apparent yield stress in such a way and using it to model natural flows provides a reasonable way to roughly predict flow characteristics, because viscous dissipation is analogous. A last means consists in measuring the free surface velocity and fluid depth of channelized lava flows. Then an analysis of the type developed in Section 1.4.1 can be used [equation (1.65)].

Debris Flows

Various types of flows occur on mountain slopes as a result of soil destabilization and mixing with water. For example, *debris flows* or *mudflows* form in mountain streams after intense rainfalls and propagate downstream as successive surges [27]. *Lahars* are similar flows forming over volcanic slopes covered with easily removable ashes. Some landslides can also transform into debris flows if they move over sufficient distances for a process of liquefaction to occur. Roughly speaking, these different materials have similar components such as water, clay, silt, sand, pebbles, and boulders, although with different solids fractions [28]. The (colloidal) water–clay mixture forms a typical pasty material in which the coarser grains are embedded. When the volume fraction of these grains is not too large (i.e., not too close to the maximum packing fraction), the behavior of the mixture is governed by the interstitial clay–water suspension and thus is generally of the paste type [29]. When the volume fraction is close to the maximum packing fraction, the material is of the granular paste type. In that case the yield stress is equal to the sum of the yield stress of the equivalent suspension without direct contact and to the yielding term due to the granular network and proportional to the normal stress [see equation (2.30)]. The corresponding value remains reasonable only if the clay fraction is not too large, which explains why in nature one tends to observed either *muddy debris flows* of the paste type (with a significant fraction of clay), or *granular debris flows* of the granular paste type (with a very small clay fraction), so that the yield stress of the interstitial fluid is small and the granular term is likely dominant in the yielding term. In nature two main broad wide classes of debris flows from deposits may be observed: muddy debris flows form smooth deposits with a front shape resembling that predicted for simple yield stress fluids (see Section 5.3) or granular debris flows form chaotic deposits with sharp angle fronts. In this context it is possible to determine the yield stress of muddy debris flows from natural deposits on steep slopes by applying the methods described in Section 5.3.1 [30]. The coefficient of friction of the granular phase of granular debris flows may be determined from the front angle of deposits. Also, the apparent viscosity may be deduced from measurements of the free surface velocity and fluid depths for different flows (see Section 1.4.1).

Sewage Sludges

Sewage sludges are the residues of wastewater treatment, composed mainly of water, organic matter, and in general a small fraction of mineral matter (clay and silt). Since they basically consist of long, soft organic chains dispersed in water, the resulting materials are of the pasty type. Note, however, that their yield stress changes in time as a result of an aging process resulting from chemical and microbiological evolutions, which affects storage properties. The "liquid" sludges have a water content of about 98% (in weight) and a small yield stress (of the order of a few pascals), decreasing exponentially with water content [31]. Here, spreading flows such as those of the inclined plane test or a conduit flow test may constitute good practical solutions. On the contrary, "pasty" sludges have a water content of about 85% in weight and a large yield stress (say, >100 Pa) [32]. In that case the slump test (possibly the modified version involving an additional weight), the squeeze test, or the penetrometer are appropriate tests. The thixotropic character of these materials remains to be explored in detail but, as a first approximation, for storage, it should be of negligible impact as compared to that of aging processes.

Mining Slurries

Materials formed either during the transport or after remixing of water with mining products or deposits, such as bauxite and coal, are concentrated suspensions of solid particles in water. The colloidal fraction is in general small but may be sufficient to form an interstitial pasty fluid that lubricates the relative motion of coarser grains and governs the behavior of the complete material. As a consequence, the material can be of either the pasty or granular pasty type depending on the solid fraction and components.

Snow

Snow is made of water crystals of various shapes, possibly with some liquid. The solid density is generally small. The crystals develop complex interactions via direct contacts, sometimes by fusion and then solidification, and sometimes with liquid bridges. Under flow (avalanches) the crystals may melt, tangle, break, or aggregate together, forming much larger solid elements. Moreover, this granular structure of low density may dilate in a much stronger may than for a usual dry granular material. As a consequence, it is difficult to determine the predominant type of interactions within an avalanche. Two extreme types may be considered (the effective snow avalanches are in general intermediate): *aerosol avalanches*, in which the snow density is rather low and the flow is analogous to a gravity density current [33]; and *dense avalanches*, for which the snow density is much larger and appears to propagate like a viscous fluid [34]. In such a context a rheometrical approach of snow remains problematic.

7.7.4 Civil Engineering Materials

Civil engineering materials are often in the form of fluid systems when they are prepared for use in products such as paints, concrete, mortars, concrete, cement, ceramic slips, bitums, powders, and drilling fluids. Besides various specific constraints, the two very general and basic constraints of these materials are that they be both sufficiently easy to handle during their preparation and use and sufficiently resistant at rest or after drying or settling. These constraints are obviously somewhat contradictory, and it is a constant challenge in industry to improve the formulation of these materials in order to better resolve these issues while at the same time avoiding new perturbing problems. In particular, in view of the final resistance of the structure, it is common to seek the largest fraction of grains within a matrix of thin particles and liquid, which implies the need to fit the complete formulation in order to maintain a sufficiently low viscosity. Finally, we note that because of the abovementioned constraints and because these materials contain many particles larger than 1 μm, they are often around the frontier between pasty and granular materials.

Drilling Fluids

Drilling fluids have several functions. They are pumped down so as to lubricate the advance of the drill bit through the rocks. Because of their sufficient apparent viscosity, they are then able to transport debris upward, and because of their yield stress and thixotropy, they are even capable of supporting them at rest after flow stoppage. They are also used to consolidate the walls of the wellbore; when a large pressure is applied to the fluid, the water flows through the rock pores, leaving a much more viscous material (filtercake) along the walls. Drilling fluids are basically clay–water suspensions or water-in-oil emulsions with clay. Various additives are often added to adjust the rheological properties. These fluids are generally considered as thixotropic yield stress fluids (with a relatively low yield stress, typically of the order of 5 Pa). During their use rock debris and clay are incorporated, and it is likely that the yield stress of the material progressively increases with the added solid fraction, as in muddy debris flows. Only if the volume fraction of rocks reaches the maximum packing fraction could the material possibly become a granular paste, which would be virtually unable to move through the annulus between the drillpipe and the rock formation.

There exist standard techniques for in situ characterization of the rheological properties of original drilling fluids that involve simple rheometrical test (velocity ramp) and stress overshoot analysis. Other practical rheometrical tests as described earlier in this chapter might be used, such as spreading or coating flows.

Glues

Glues are generally formed of polymeric materials that solidify at ambient temperature and "adhere" onto the solid surfaces that they join. In their liquid state they are viscoelastic fluids, possibly with a yield stress. The most practical rheometrical test in this field is the separation of two solid surfaces with a glue layer between them (peeling), since this reproduces under controlled conditions the main criterion for the qualitative appreciation of glue adhesion in practice. Here we only consider the case of a liquid glue, a short time after gluing. In general, one of the solid surfaces is lifted up from one of its sides, which could lead to a flow as described in Section 5.2. For typical pasty materials, the material that may adhere to the solid surface following the motion of the remaining fluid plays a negligible role; however, this is not the case with such polymeric viscoelastic materials—as a result of the long polymer chains, the glue barely separates into two parts, so long filaments joining the two solid surfaces remain. As a consequence, the induced flow generally involves several effects, which precludes any straightforward interpretation in rheometrical terms: flow toward the smallest gap region, fingering (see Section 5.2.6) or cavitation, or stretching of fluid filaments linking the two surfaces. By using different tests, it is possible to study some of these effects individually: standard squeeze flow, standard elongational test under the condition $R \ll h$.

Fresh Concrete

Fresh concrete is a mixture of water, cement, sand, and gravel in various proportions, along with different additives (superplasticizers, nanoscale silica particles, flying ashes, etc.) [35]. The basic colloidal fraction is found in cement, to which specific additives must be added. The rheological behavior of fresh concrete should at first sight have strong similarities with debris flows; as long as the volume fraction of noncolloidal particles is not too close to the maximum packing fraction, the behavior is of the paste type, governed by the cement–water suspension; otherwise the material is of the granular paste type. This classification is of critical importance since the apparent yield stress [including the two terms of equation (2.30)] for a paste may be much smaller than that of the same material in the form of a granular paste (although with a slightly larger solid fraction). Thus fresh concrete of the granular paste type may be impossible to handle because of its very high apparent yield stress. The different additives appear to play a critical role; as additional colloidal particles increase the interstitial fluid fraction and the strength of the interstitial paste; polymers tend to disperse the colloidal particles and thus decrease the strength of the interstitial paste. Another problem is the extreme sensitivity of the behavior to the exact composition of the material. An additional problem with fresh concrete is sedimentation, which may be in part affected or counteracted by these additives. From a general point of view, fresh concrete has been considered mainly as a yield stress fluid, possibly thixotropic. Also, irreversible behavior evolutions (aging) begin almost as soon as the material has been prepared. This aging process may be slowed down or accelerated by additives. In practice, almost all practical rheometrical techniques described in Chapter 7 may be used in the laboratory to evaluate the material behavior (yield stress and thixotropy), but the slump test is still generally used in situ in view of qualitative comparison of materials.

Paints

Paints are composed mainly of solid particles (pigments), polymers in solution or in suspension (latex), in water or in an organic solvent. Various additives may also be added, such as surfactants or defoaming agents. The pigments give the paint its final color, while the polymers ensure the cohesion of the film after drying. The final material must have a yield stress that enables it to initially remain on the brush and then to suddenly stop flowing after it has been spread over a solid surface. However, in order to avoid handling problems and severe stoppage thicknesses, this yield stress must not be too great. In general, since the yield stress and thixotropy of paints are relatively low (say, of the order of 1 Pa), primarily spreading or coating flows (see Section 7.3) can be used in practice because other tests are sensitive to greater forces.

Extruded Pastes

Various materials are formed by extrusion of very viscous pasty or granular pasty materials (ceramics, bricks, porcelain, catalyst pellets, etc.). The internal structure of these materials corresponds closely to those considered in this book but at very high solid fractions. Their rheological behavior can no doubt be described using the methods presented in this book, but they exhibit a very high yield stress and a high critical shear rate (see Section 3.3), and they often fall in the class of granular pastes. This implies that in practical flows under moderate rates through straight conduits the flow easily localizes in a very thin layer or the material slips along the solid wall. This explains why both some bulk yield stress, associated with bulk deformation, and wall shear stress associated with certain types of wall slip, are considered as characteristic rheological parameters of the material [36]. However, bulk yield stress depends solely on material behavior, whereas wall shear stress reflects the interaction between the wall an the material. This wall interaction may include both wall slip (Section 3.2) effects or flow in the discrete regime (see Section 3.3.4). This phenomenon is of critical practical importance since it affects the shape of the extrudate [37]. An additional typical effect with such materials is phase separation; under the high pressures exerted onto the fluid, the water may migrate through the material [38,39].

REFERENCES

1. M. Bourne, *Food Texture and Viscosity—Concept and Measurement*, Academic Press, New York, 2002.
2. M. Schatzmann, P. Fischer, and G. R. Bezzola, *J. Hydraul. Eng.* **129**, 796 (2003).
3. J. Guo and P. H. T. Uhlherr, *Proc. XIIth Int. Congress on Rheology*, Quebec City, Canada, A. Aït-Kadi, J. M. Dealy, D. F. James, and M. C. Williams, eds., 1996, p. 731.

4. T. Suwonsichon and M. Peleg, *J. Food Eng.* **39**, 217 (1999).
5. J. C. Baudez, F. Chabot, and P. Coussot, *Appl. Rheol.* **12**, 133 (2002).
6. J. Murata, *Mater. Constr.* (Paris) **17**, 117 (1984).
7. B. Rajani and N. Morgenstern, *Can. Geotech. J.* **28**, 457 (1991).
8. N. Pashias, D. V. Boger, J. Summers, and D. J. Glenister, *J. Rheol.* **40**, 1179 (1996).
9. W. R. Schowalter and G. Christensen, *J. Rheol.* **42**, 865 (1998).
10. S. Clayton, T. G. Grice, and D. V. Boger, *Int. J. Miner. Process.* **1613**, 1 (2002).
11. N. Roussel and P. Coussot, to appear in *J. Rheol.* **49** (2005).
12. A. H. Bloksma, *Cereal. Chem.* **34**, 124 (1957).
13. B. Launay, *Cereal. Chem.* **67**, 25 (1990).
14. P. Coussot and S. Boyer, *Rheol. Acta* **34**, 534 (1995).
15. P. H. T. Uhlherr, K. H. Park, C. Tiu, and J. R. G Andrews, in *Advances in Rheology*, B. Mena, A. Garzia-Rejon, and C. Rangel-Nafaile, eds., Springer-Verlag, 1984, Vol. 2, p. 183.
16. Q. D. Nguyen and D. V. Boger, *Ann. Rev. Fluid Mech.* **24**, 47 (1992).
17. P. Coussot, D. Laigle, M. Arattano, A. Deganutti, and L. Marchi, *J. Hydraul. Eng.* **124**, 865 (1998).
18. K. Whipple and T. Dunne, *Geol. Soc. Am. Bull.* **104**, 887 (1992).
19. J. E. Bowles, *Engineering Properties of Soils and Their Measurements*, McGraw-Hill, New York, 1992.
20. P. Evesque and J. Rajchenbach, *C. R. Acad. Sci. Paris* **307**, 223 (1988).
21. N. J. Alderman, G. H. Meeten, and J. D. Sherwood, *J. Non-Newtonian Fluid Mech.* **39**, 291 (1991).
22. H. T. Huynh, N. Roussel, and P. Coussot, *Phys. Fluids* **17**, 033101 (2005).
23. A. R. McBirney and T. Murase, *Ann. Rev. Earth Planet. Sci.* **12**, 337 (1984).
24. M. O. Saar, M. Manga, K. V. Cashman, and S. Fremouw, *Earth Planet. Sci. Lett.* **187**, 367 (2001).
25. S. R. Hoover, K. V. Cashman, and M. Manga, *J. Volcanol. Geotherm. Res.* **107**, 1 (2001).
26. H. Pinkerton and R. S. J. Sparks, *Nature* **276**, 383 (1978).
27. T. R. H. Davies, *Acta Mechanica* **63**, 161 (1986).
28. P. Coussot and M. Meunier, *Earth-Sci. Rev.* **40**, 209 (1996).
29. P. Coussot, *Mudflow Rheology and Dynamics*, Balkema, Amsterdam, 1997.
30. A. M. Johnson and J. R. Rodine, in *Slope Instability*, D. Brunsden and D. B. Prior, eds., Wiley, New York, 1984, Chapter 8.
31. P. T. Slatter, *Water Sci. Technol.* **36**, 9 (1997).
32. J. C. Baudez and P. Coussot, *J. Rheol.* **45**, 1123 (2001).
33. E. J. Hopfinger, *Ann. Rev. Fluid Mech.* **15**, 45 (1983).
34. J. D. Dent and T. E. Lang, *Ann. Glaciol.* **4**, 42 (1983).
35. P. F. G. Banfill, *Rheology of Fresh Cement and Concrete*, Chapman & Hall, London, 1990.
36. J. Benbow and J. Bridgwater, *Paste Flow and Extrusion*, Clarendon Press, Oxford, UK, 1993.
37. A. T. J. Domanti, D. J. Horrobin, and J. Bridgwater, *Int. J. Mech. Sci.* **44**, 1381 (2002).
38. L. Chevalier, E. Hammond, and A. Poitou, *J. Mater. Process. Technol.* **72**, 243 (1997).
39. S. L. Rough, J. Bridgwater, and D. I. Wilson, *Int. J. Pharmaceutics* **204**, 117 (2000).

INDEX

Rheometry of Pastes, Suspensions, and Granular Materials: Applications in Industry and Environment
By Philippe Coussot   Copyright © 2005 John Wiley & Sons, Inc.

Printed in the USA/Agawam, MA
August 28, 2018

682090.001